I0042094

Zinc

Zinc: Early Development, Applications, and Emerging Trends is a comprehensive book covering various aspects of the metallic element zinc that has a significant role in the growth and survival of humans. The United Nations Organization has aptly declared it as a 'Life-Saving Material' since it helps in overcoming infections and strengthening the immune system. Zinc is an essential element next to iron, aluminum, and copper with abundant presence in nature, and is worth billions of dollars commercially. Besides the metal, its compounds and nanoparticles have also become attractive to researchers due to their enormous applications. The industrial and agricultural uses of the metal and its compounds are widespread.

An exclusive book covering all important aspects of zinc is yet to hit the market. This book, with contributions from experts in geology, chemistry, medicine (including dentistry and traditional systems), agriculture, veterinary science, biology, bioinformatics, and nanotechnology encompassing the latest developments in their fields related to zinc, fills this void. The uniqueness of the book is its interdisciplinary nature and potential use by researchers, students, and teachers of various specialties.

Features:

- Contributes to a better understanding of the complexities of zinc's role in various branches of science.
- Contains basic and practical information for a wider audience and researchers of different fields besides those interested in zinc.
- Provides detailed information on the usefulness of zinc in prophylaxis and treatment of diseases in modern medicine, dentistry, and traditional systems such as Ayurveda, Siddha, and Unani, as well as veterinary medicine.
- Highlights zinc's role in agriculture and food science with various crops and nutritive materials.
- Includes chapters on fast-growing areas – nanotechnology and bioinformatics.

Zinc
Early Development, Applications, and Emerging Trends

Edited by
Ethirajan Sukumar,
Krishnamurthy Vinoth Kumar,
and Annamalai Manickavasagan

CRC Press
Taylor & Francis Group
Boca Raton London New York

CRC Press is an imprint of the
Taylor & Francis Group, an **informa** business

First edition published 2024
by CRC Press
2385 NW Executive Center Drive, Suite 320, Boca Raton FL 33431

and by CRC Press
4 Park Square, Milton Park, Abingdon, Oxon, OX14 4RN

CRC Press is an imprint of Taylor & Francis Group, LLC

© 2024 selection and editorial matter, Ethirajan Sukumar, Krishnamurthy Vinoth Kumar and Annamalai Manickavasagan; individual chapters, the contributors

Reasonable efforts have been made to publish reliable data and information, but the author and publisher cannot assume responsibility for the validity of all materials or the consequences of their use. The authors and publishers have attempted to trace the copyright holders of all material reproduced in this publication and apologize to copyright holders if permission to publish in this form has not been obtained. If any copyright material has not been acknowledged please write and let us know so we may rectify in any future reprint.

Except as permitted under U.S. Copyright Law, no part of this book may be reprinted, reproduced, transmitted, or utilized in any form by any electronic, mechanical, or other means, now known or hereafter invented, including photocopying, microfilming, and recording, or in any information storage or retrieval system, without written permission from the publishers.

For permission to photocopy or use material electronically from this work, access www.copyright.com or contact the Copyright Clearance Center, Inc. (CCC), 222 Rosewood Drive, Danvers, MA 01923, 978-750-8400. For works that are not available on CCC please contact mpkbookspermissions@tandf.co.uk

Trademark notice: Product or corporate names may be trademarks or registered trademarks and are used only for identification and explanation without intent to infringe.

ISBN: 978-1-032-53521-0 (hbk)
ISBN: 978-1-032-53523-4 (pbk)
ISBN: 978-1-003-41247-2 (ebk)

DOI: 10.1201/9781003412472

Typeset in Times New Roman
by MPS Limited, Dehradun

Dedicated to

Dr. N. Venkatasubramanian
(1933–2022)
Former Principal & Professor of Chemistry
Ramakrishna Mission Vivekananda College, Chennai
who touched the lives of hundreds of his students and taught them
sincerity, honesty, and dedication besides science

Contents

Preface

Zinc is a unique metallic element with vast applications in many fields. The metal as well as its compounds including a plethora of nano derivatives finds extensive use all over the world. Its rich natural abundance contributes to the economic development of many countries. It may be interesting to note that the metal was found along with a few others, in the sediments of deep waters in the Red Sea area at the depth of 3534 meters. Several publications on various aspects of zinc and its compounds have appeared over the years; but a comprehensive collection right from the early development down to the latest applications and emerging areas is the need of the hour. Hence we have attempted to bring out this compilation with maximum information on this unique metal.

The details presented in the book cover Early Development, Applications and Emerging trends. The first part traces the exploration and use of zinc in the early years while in the second section, the use of zinc in various fields such as chemistry, biology, agriculture, medicine, etc., are discussed. The last section deals with emerging areas where zinc is an essential part with specific emphasis on nanotechnology.

We are confident that this book shall be highly beneficial to students, research scholars and scientists of various disciplines who wish to pursue further work on zinc. We are grateful to the experts for their chapter contributions which were screened for quality by the editorial board as well as external specialists. We thank the senior professors and Scientists of leading institutions for their impartial review of the chapters. Our special thanks go to Dr. B. Ilango, Scientist, University of Colorado, USA, for his interest in the project and Dr. S. Brindha, Research Scholar, Department of Historical Studies, Bharathi Women's College, Chennai for her help in the compilation work. We are grateful to the staff of editorial and production departments of CRC Press of Taylor & Francis Group especially Ms. Renu Upadhyay and Ms. Jyotsna Jangra for their overall support and efforts to bring out this book on time. Last but not least, we thank our well-wishers and family members for their understanding and unstinted support.

Ethirajan Sukumar
Krishnamurthy Vinoth Kumar
Annamalai Manickavasagan

Editors

Dr. Ethirajan Sukumar

Former Dean of Research, Saveetha Deemed University, Chennai (India)

He holds two Master's Degrees and PhD in Phytochemistry and Phytopharmacology from the University of Madras (India) and had a Post-Doctoral stint at the University of Virginia (USA). He worked as Assistant Professor at the Higher College of Technology (University of Technology and Applied Sciences), Muscat, Sultanate of Oman for eight years. Prior to full-time teaching, he was a Scientist/Research Officer in the CSM Drug Research Institute for Ayurveda, Tamil Nadu Medicinal Plants and Herbal Medicine Corporation and Central Research Institute of Siddha Medicine in Chennai (India) for more than 26 years. During his 40-plus years career in research, teaching, writing and administration. he has published two books, four book chapters and 85 research papers in peer-reviewed journals, translated three books (from a regional language to English) and presented many papers in various international and national forums and bagged the Best Paper award twice. He supervised 30 students for Doctoral, MPhil, and Master's programs in the multidisciplinary areas involving Chemistry, Biochemistry, Pharmacognosy, Pharmacology and Environmental Science. He delivered 55 invited lectures at universities, colleges, and conferences and also served as a Resource Person in many workshops and training programs. He chaired scientific sessions and delivered plenary lectures at international conferences and is associated with leading international/national journals as Associate Editor, Editorial Board Member and Peer Reviewer. His name is included in the *International Directory of Experts in Spices, Herbs and Medicinal Plants Research* published by the University of Massachusetts and in the *Who's Who in Science and Engineering* by Marquis Inc., USA.

Dr. Krishnamurthy Vinoth Kumar

Head, Research & Development Division, VPro Biotech, Puducherry (India)

He received his Masters and Doctoral degrees in Biotechnology from the University of Madras and has about 20 years of teaching and research experience. He had earlier training in research at the National Institute for Physiological Sciences, Okazaki, Japan, on Japanese Society of Promotion Science Fellowship. He has published 23 research papers and presented ten papers in Conferences and Seminars. He also trained students in their Masters and research courses. He is a Government of India nominated Member of the Institutional Animal Ethics Committee of a Medical College in the Union Territory of Puducherry. Presently, he is working in the development of vegetable-based healthcare products as Head of Research & Development Division.

Dr. Annamalai Manickavasagan
Associate Professor, University of Guelph, Canada
He obtained a Masters in Agricultural Food Processing from
Tamil Nadu Agriculture University, Coimbatore (India) and a
PhD in Biosystems Engineering from the University of
Manitoba, Canada. He is a licensed Professional Engineer
(PEng) in the province of Ontario, Canada. He worked with
Mc Cain Foods Limited (Canada) as a Scientist and Associate
Professor at Sultan Qaboos University, Sultanate of Oman.
At present, he is an Associate Professor (Food Processing) at the School of
Engineering, University of Guelph. He has diversified research and management
experience with academic institutions and industries. He has edited 9 books and
published 101 peer-reviewed journal papers and 20 book chapters. He has
supervised/is supervising 30 scholars for their Masters and Doctoral degrees.

Contributors

Zoha Abdullah
Department of Public Health Dentistry
Madha Dental College
Chennai, India

P. Abhinash
ICAR-National Dairy Research Institute
Karnal, India

K. Altaff
Department of Marine Biotechnology
AMET University
Chennai, India

Judie Arulappan
College of Nursing
Sultan Qaboos University
Muscat, Sultanate of Oman

Kanisht Batra
Lala Lajpat Rai University of
    Veterinary and Animal Sciences
Hisar, India

Sakshi Bhardwaj
Department of Life Science
Altem Technologies (P) Ltd
Bengaluru, India

B. Deivasigamani
Center for Advanced Studies in Marine
    Biology
Annamalai University
Parangipettai, India

Gautam R. Desiraju
Structural Chemistry Unit
Indian Institute of Science
Bengaluru, India

G. Deviga
Department of Chemistry
SRM Institute of Science and
    Technology (Deemed University)
Chennai, India

Ravikumar Dhanalakshmi
Department of Chemistry
Pachaiyappa's College
Chennai, India

Dasarathan Dharani
Department of Historical Studies
Bharathi Women's College
Chennai, India

A. Queen Elizabeth
Center for Advanced Studies in Marine
    Biology
Annamalai University
Parangipettai, India

F. Magdaline Eljeeva Emerald
ICAR-National Dairy Research Institute
Bengaluru, India

Sangita Ghosh
Department of Chemistry
Jadavpur University
Kolkata, India

Gopal Jeya
Department of Chemistry
Pachaiyappa's College
Chennai, India

K. Susan John
ICAR-Central Tuber Crops Research
    Institute
Thiruvananthapuram, India

Safreena Kabeer
Department of Food Process
  Engineering
SRM Institute of Science and
  Technology
 Kattankulathur, Chennai, India
and
Department of Food Technology
Faculty of Engineering
Karpagam Academy of Higher
  Education
Coimbatore, Tamil Nadu, India

Deep J. Kalita
Renuvix LLC
Fargo, North Dakota, USA

Kannabiran Krishnan
Department of Biomedical Sciences
VIT Deemed University
Vellore, India

S. Mathan Kumar
Qasr Al Shatti
Abu Dhabi, United Arab Emirates

Sushila Maan
Department of Animal Biotechnology
Lala Lajpat Rai University of
  Veterinary and Animal Sciences
Hisar, India

Venkataramanan Mahalingam
Indian Institute of Science Education
  and Research-Kolkata
Mohanpur, India

Annamalai Manickavasagan
School of Engineering
University of Guelph
Canada

S. Manivannan
ICAR-Indian Agricultural Research
  Institute
Dimaji District, Assam, India

M. Mariappan
Department of Chemistry
SRM Institute of Science and
  Technology (Deemed University)
Chennai, India

Rahul Mukherjee
Department of Geology
Central University of Punjab
Bathinda, India

Nagamaniammai Govindarajan
Department of Food Process
  Engineering
SRM Institute of Science and
  Technology
Kattankulathur, Chennai, India

M. G. Nithin
School of Dentistry
University of Queensland
Australia

Jitendra Kumar Pattanaik
Department of Geology
Central University of Punjab
Bathinda, India

Sukanya Paul
Department of Chemistry
Jadavpur University
Kolkata, India

Chintha Pradeepika
ICAR-Central Tuber Crops Research
  Institute
Thiruvananthapuram, India

P. Purushothaman
Department of Civil Engineering
SRM Institute of Science and
  Technology (Deemed University)
Chennai, India

Heartwin A. Pushpadass
ICAR-National Dairy Research Institute
Bengaluru, India

E. Rajkumar
Department of Chemistry
Madras Christian College
Chennai, India

Liju Raju
Department of Chemistry
Madras Christian College
Chennai, India

V. Ramanathan
Department of Chemistry
Indian Institute of Technology-BHU
Varanasi, India

Soundharya Ravindran
Department of Preventive Oncology
    (Research)
Cancer Institute (WIA)
Chennai, India

B. G. Seethu
ICAR-National Dairy Research Institute
Bengaluru, India

Dhivya Shanmugarajan
Department of Life Science
Altem Technologies (P) Ltd
Bengaluru, India

B. Shyamaladevi
Research, Chettinad Hospital and
    Research Institute, Chettinad
    Academy of Research and Education
Kelambakkam, Tamil Nadu, India

Sana Siddiqui
Deccan Dental Clinic
Hyderabad, India

Suvra Sil
Indian Institute of Science Education
    and Research-Kolkata
Mohanpur, India

Digvijay Singh
Department of Animal Nutrition
Guru Angad Dev Veterinary and
    Animal Sciences University
Ludhiana, India

Tao Ming Sim
Yong Loo Lin School of Medicine
National University of Singapore
Singapore

Neha Singh
Lala Lajpat Rai University of
    Veterinary and Animal Sciences
Hisar, India

Chittaranjan Sinha
Department of Chemistry
Jadavpur University
Kolkata, India

Vajiravelu Sivamurugan
Department of Chemistry
Pachaiyappa's College
Chennai, India

D. Srinivasan
Research and Development Division
VPro Biotech
Puducherry, India

Dinesh Kumar Srinivasan
Department of Anatomy
Yong Loo Lin School of Medicine,
    National University of Singapore
Singapore

R. Srinivasan
ICAR-National Bureau of Soil Survey
    and Land Use Planning
Bengaluru, India

Ethirajan Sukumar
Research and Development Division
VPro Biotech
Puducherry, India

P. Aditya Sukumar
ICAR-National Dairy Research Institute
Bengaluru, India

V. Kasthuri Thilagam
ICAR-Sugarcane Breeding Institute
Coimbatore, India

Gouri Tudu
Indian Institute of Science Education
   and Research-Kolkata
Mohanpur, India

Navin Venketeish
Department of Food Process
   Engineering
SRM Institute of Science and
   Technology
Kattankulathur, Chennai, India

R. Vijayalakshmi
Department of Preventive Oncology
   (Research)
Cancer Institute (WIA)
Chennai, India

R. Vijayaraj
Department of Marine Biotechnology
AMET University
Chennai, India

Krishnamurthy Vinoth Kumar
Research and Development Division
VPro Biotech
Puducherry, India

Sophia Cyril Vincent
College of Nursing
Sultan Qaboos University
Muscat, Sultanate of Oman

1 Understanding Zinc

Its Genesis, Distribution and Uses

Rahul Mukherjee, Jitendra Kumar Pattanaik, and P. Purushothaman

1.1 INTRODUCTION

Metals are the dominant groups of elements present in the periodic table. Among the trace elements, few metals are categorized as important based on their availability, utility and importance for both humankind and flora and fauna. Zinc, 24th dominant metal, is an essential micronutrient for organic growth and raw material for various product manufacturing. Zinc with an atomic number of 30, atomic weight of 65.38 g/mol and oxidation number of +2, occurs in five stable isotopic forms with masses 64, 66, 67, 68 and 70 and crustal abundance 50.9, 27.3, 3.9, 17.4 and 0.5 respectively (McDowell, 1992; Bradl, 2005; Goodwin, 2017). Zinc with bluish-white lustre has a density of 7.14 gcm^{-3} and melting point of 692.68 K (McDowell, 1992). Though zinc shows brittleness at the ambient temperature, it is malleable at 100–150°C temperature and shows ductility. Zinc being chalcophile in nature occurs mostly along with sulphide affinity. Zn is highly mobile in liquid environments with mass increment of 9% and is a good conductor of electricity (Porter, 1991).

Zinc is not a combustible metal and its thermal conductivity decreases with increase in temperature (Porter, 1991). It has diamagnetic properties and shows thermoelectric effects; due to its volatile nature in thermal processes. Zn is suitable for studying the naturally occurring isotopic variations (Porter, 1991) also found to occur in various forms of sulphates, nitrates, chlorates, oxides, carbonates and rarely as silicates (Barak and Helmke, 1993).

At the industrial plants, the first zinc oxide (ZnO) is produced from sulphide ore (Equation 1) and later it is carbon-reduced to form zinc metal (Ophardt, 2003).

$$2ZnS + 3O_2 \rightarrow \Delta 2ZnO + 2SO_2 \qquad (1.1)$$

The stability of zinc in dry air gets disturbed with increase in temperature and moisture content along with CO_2, which leads to the initiation of corrosion. However it is widely used to protect other metals from oxidation as it produces a thin light grey film during the initial stage of corrosion; which adheres to the surface preventing further corrosion (Dorton et al., 2010). Zinc availability in soils and water environments depends mainly on sorption characteristics of the element and its chelation capacity (Bradl, 2005; Sparks 2005). The element also forms complexes with Cl^-, PO_4^-, NO_3^- and SO_4^{2-}. Zinc's presence in an environment, on the other hand, depends on the concentration of the metal in geochemical reservoirs and rocks (Huston et al., 2015).

The abundance of zinc is about 0.00003% (~300 ppb) in the Universe, 0.0078% (~78 ppm) in the Crust (~52 ppm in the upper continental crust), 30 ppb in seawater, 5–770 ppm in soils and 0.0033% (~33 ppm) in human body (Yaroshevsky, 2006). In major rock types, the concentration of zinc varies from ~30 to 130 ppm (Stony meteorites: ~50 ppm; Acid rocks: ~40–60 ppm; intermediate rocks: ~73 ppm; basic rocks: ~100–130 ppm; ultrabasic rocks: ~30 ppm) (Yaroshevsky, 2006). Minimum ore grade for profitable extraction of zinc is 4 wt% while the concentration factor is 570 (Porter, 1991).

Intermittent distribution of zinc deposits with time involves relatively high zinc productivity for short time periods and low productivity for long periods due to the geodynamic evolution of the Earth. The total reserve of zinc in the world is estimated about 250 million metric tonnes (USGS, 2020). Australia (33.75%), China (22.5%), Russia (11.25%) and Mexico (11.25%) are the nations that have the largest zinc reserves of the world. India is the 7th ranked country in terms of zinc reserve with 10 million metric tonnes approximately (Figure 1.1).

At present, with the total mine production of zinc worldwide, China is the leader with 4.3 million tonnes (Figure 1.2).

Mine production of zinc in India is limited to 0.7 million tonnes (USGS, 2020). Global production of zinc metal has increased by around one million tonnes in the last ten years and reached up to 13 million tonnes in the year 2020. In an Indian

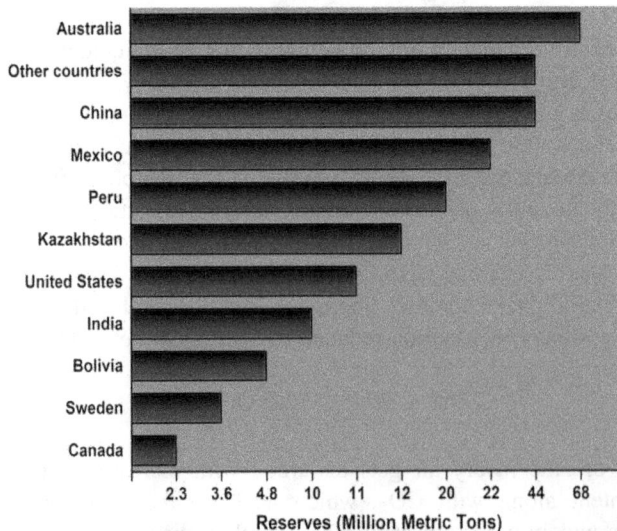

FIGURE 1.1 Global reserves of zinc.

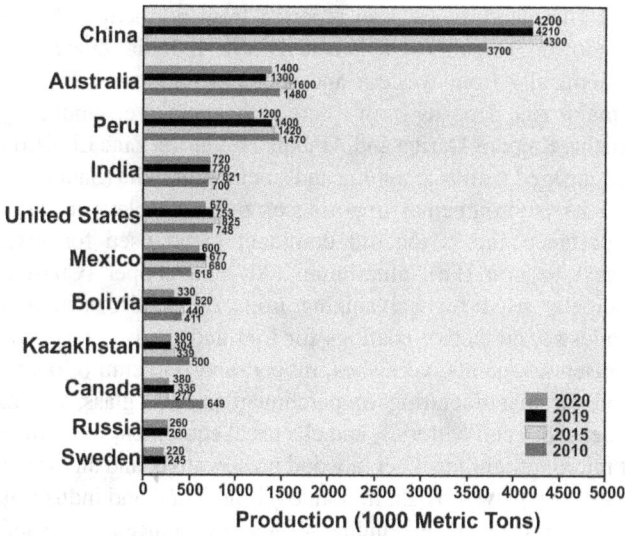

FIGURE 1.2 Country-wise mine production of zinc.

context, the production of primary zinc has been almost steady in the last ten years (Indian Minerals Yearbook, 2019) (Figure 1.3).

Along with the production, as per the demand, the price of zinc is also increased and reached US$ 2700 per metric ton in 2021 (USGS, 2020).

Zinc deposits in India are mostly localized in the Precambrian rocks of Peninsula with some minor occurrences in the lesser Himalayas though some sporadic

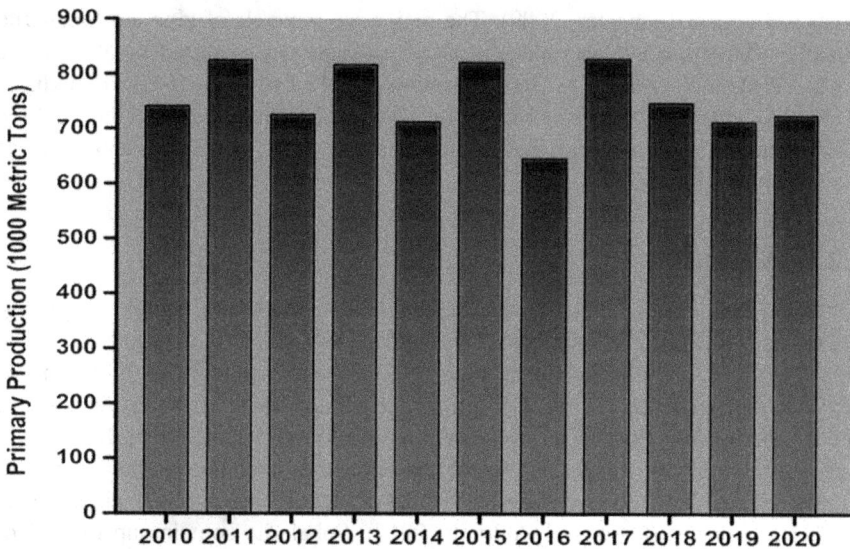

FIGURE 1.3 Primary zinc production in India.

occurrences of zinc deposits are also reported from northern and eastern regions. Rajasthan is endowed with almost 90% zinc reserves in India. Zinc along with lead is extracted economically from western and southern regions of India. Some of the examples of major zinc deposits from where zinc is mined economically are Zawar, Rampura-Agucha, Rajpura-Dariba and Ambaji. Hindustan Zinc Limited (HZL) is the only company engaged with zinc mining and smelting in India (Sabnavis et al., 2018).

Apart from its environmental importance, zinc also plays a major role in its industrial importance and is the 4th dominant metal used for various product generations next to iron (Fe), aluminium (Al) and copper (Cu) (Bradl, 2005). Though principally used for galvanizing iron, Zn is also widely used in the automobile industry, protective coatings for iron and steel, in cosmetics, powders, ointments, antiseptics, paints, varnishes, rubber, and linoleum (Goodwin, 2018). It is also needed for manufacturing of parchment papers, glass, automobile tyres, television screens, dry cell batteries, and electrical equipment. In agriculture, zinc is an important micronutrient fertilizer, a wood preservative, and an insecticide (Bradl, 2005). Since the role played by Zn in both environmental and industrial activities is of a wide range, proper understanding of its occurrence is a mandate. In this chapter, the association and occurrence of zinc in various geological environments are highlighted.

1.2   ORE MINERALS OF ZINC

Zinc ores normally contain 5–15% zinc. Approximately 80% of zinc mines are underground while 8% are mined from open pits, and the remainder are a combination of both types (Ray et al., 2014). Zinc with its chalcophilic nature, also shows lithophilic affinity and mostly occurs along with metallic lead and explored together. Zinc can also be associated with valuable metals such as gold, silver and copper (Sangster, 2009). The major ore minerals of zinc are Sphalerite (ZnS) and Smithsonite/Calamine ($ZnCO_3$) with some rare occurrences of Wurtzite (ZnS), Willemite ($ZnSiO_4$), Hemimorphite [$Zn_4(Si_2O_7)(OH)_2 \cdot H_2O$], Franklinite ($ZnFe_2O_4$), Zincite (ZnO) and Marmatite [(Zn, Fe)S] (Barak and Helmke, 1993). Zinc blende is a compound that comes in two forms: sphalerite and wurtzite (polymorph of Sphalerite).

1.2.1   SPHALERITE

Sphalerite is the major and most common zinc sulphide ore found in igneous, sedimentary and metamorphic rocks throughout the world. It is mainly found in veins and irregular displacement beds in limestone and contact metamorphic deposits. Approximately 95% of all primary zinc is extracted from sphalerite ores and it contains 67% zinc metal in its natural pure form (Haldar, 2020). It occurs in variable colours with submetallic lustre and often shows white to yellowish-brown streaks (Nesse, 2016). It possesses perfect dodecahedral cleavage with six numbers of cleavage planes (Klein and Philpotts, 2017). Sphalerite contains up to 25% of iron (Fe), which substitutes Zn in the crystal lattice. Increase in temperature favours high content of iron in sphalerite. The structure of sphalerite shows diamond-type network and may contain a considerable amount of manganese (Mn) and cadmium (Cd) (up to 5%) in solid solution (Labrenz et al., 2000; Schwartz, 2000). In addition to cadmium, sphalerite occasionally contains rare elements such as indium, germanium and gallium; which can be commercially recovered as by-products (Cook et al., 2009; Pohl, 2011).

Generally, Sphalerite is associated with galena, pyrite, marcasite, pyrrhotite, arsenopyrite, chalcopyrite, smithsonite, calcite, fluorite, rhodochrosite and dolomite. Contact metamorphism involving a combination of hot, acidic, zinc-bearing hydrothermal solutions and carbonates leads to the deposition of sphalerite. Based on the presence of various elements in the crystal structure of sphalerite, it shows distinct colours. In response to these different appearances, different names have been given for sphalerite (Nesse, 2016; Klein and Philpotts, 2017). An almost pure variety of sphalerite with less or no iron content is colourless to pale green in colour and fluorescence in nature is known as 'cleiophane' (Henn and Hofmann, 1985). Two polymorphs of sphalerite are wurtzite and marmatite, which share the same chemical compositions with different structures. Marmatite is an opaque, black variety of sphalerite with high iron content. Some of the sphalerite possesses red iridescence within the greyish-black crystals, known as 'ruby sphalerite' (Rager et al., 1996). When viewed under the microscope, sphalerite is isotropic and has very high relief with no pleochroism. Usually, sphalerite exhibits replacement texture with other sulphide ore minerals viz. pyrite, pyrrhotite, chalcopyrite, and galena. Iron-rich sphalerites often show a special texture named 'chalcopyrite disease', a texture where variable sizes of chalcopyrite blebs are unevenly distributed throughout a sphalerite grain. The main mechanism behind this kind of unusual texture is thought to be the co-precipitation of sphalerite and chalcopyrite in nature (Govindarao et al., 2018).

1.2.2 SMITHSONITE

Smithsonite is also known as zinc spar and often occurs as a secondary mineral in the oxidation zone of zinc ore deposits. The oxidation of sphalerite produces smithsonite in many sedimentary deposits worldwide. The common association of smithsonite is hemimorphite, hydrozincite, willemite, cerussite, azurite, malachite and anglesite (Driscoll et al., 2014). Among these, hemimorphite (zinc silicate) displays similarity with smithsonite making it difficult to distinguish between each other. Based on the following properties, one can differentiate these two similar-looking minerals by two methods: (i) In cold conditions, with dilute hydrochloric acid, smithsonite gives effervescence, while hemimorphite will not show this kind of effects and (ii) Smithsonite (~4.5) has slightly higher specific gravity than hemimorphite (~3.5) and possesses two sets of perfect cleavages at rhombic angles, while hemimorphite possesses only one set of perfect cleavage (Driscoll et al., 2014). The colour of smithsonite is variable and it generally formed at or near the surface as botryoidal masses (Choulet et al., 2014). The occurrences of smithsonite at the surface may be an indication of a major zinc sulphide deposit present below. The semi-transparent gem variety of smithsonite with bluish green colour and pearly lustre is known as 'bonamite' and its impure variety is known as 'dry bone ore'. The latter is massive, porous and friable in nature and often exhibits a honeycomb structure (Barnes, 1997). In solid solution series of smithsonite, substitution with manganese and iron leads to forming of rhodochrosite and siderite respectively.

1.3   ZINC GENESIS

Zinc is one of the important members along with lead and copper in the 'base metals' family, which includes high-volume and low-value metallic elements. In most of the base metal deposits, elements occur in association such as Cu-Zn, Pb-Zn and Cu-Pb-Zn (Sangster, 2009). These metals tend to form compounds with sulphur which makes these deposits polymineralic, besides polymetallic. Most of the base metal deposits are formed by hydrothermal processes, where metals are precipitated from hot brine solutions due to the change in physico-chemical conditions. As a part of base metal, zinc deposits are also formed as sphalerite from hot hydrothermal solution flow along the subsurface interconnected fractures (Haldar, 2020). In limestone terrane, these hot fluids flow through the cavities and form rich zinc deposits with patchy appearance (Sangster, 2009). Volcanogenic zinc deposits are formed in the areas of the ocean floor where underwater volcanic activity takes place. The major zinc ore found in the Cu-Pb-Zn deposits is sphalerite, while the other associated sulphides are pyrite, pyrrhotite, chalcopyrite and galena. Zinc deposits occur in diverse environments and include the following genetic types:

1.3.1   Volcanogenic Massive Zn - Sulphide (VMS) Deposits

VMS deposits have a higher degree of sulphide content (~50% to 80%) and are one of the major contributors of zinc, copper, gold and silver to the world's supply (~7%). These deposits are associated with mafic or felsic volcanic rocks including brecciated and pyroclastic varieties (Sangster, 2009). Massive sulphides of concordant nature and bulbous lens shape are present on or below the seafloor; while discordant, vein type, disseminated stockwork ore enclosed by highly altered host rocks are just below the massive sulphide ores (Shanks, 2014). The deposits are mainly strata-bound in nature, though often show stratiform characteristics (Haldar, 2020). Since their discovery, VMS deposits have become the exploration targets worldwide due to their polymetallic nature.

Metal sulphides are precipitated in the hydrothermal vents that present on the ocean floor known as 'black smokers', where hot hydrothermal fluids mix with cool ocean water (Haldar, 2020). Though a wide range of tectonic settings is related to VMS deposits, island arcs and spreading ridges are favourable sites. Zinc-sulphides (Sphalerites) are mainly limited to the upper part of VMS deposits. Pyrite is the chief ore mineral found in the VMS deposits (~50% to 90%) along with the mixtures of sphalerite, chalcopyrite and galena (~10%) (Shanks, 2014). The best example of VMS deposit is Kidd Creek deposit of Ontario, Canada, one of the largest and deepest base metal mines in the world (Sangster, 2009). In addition to this, some other important VMS deposits are Windy Craggy of British Columbia, Flin Flon deposit of Manitoba, Bathurst deposit of New Brunswick, Iberian Pyrite Belt of Spain. Southern Urals of Russia, Bisha deposit of Eritrea, Mt Read deposit of Tasmania, Wolverine deposit of Canada, Khnaiguiyah deposit of Saudi Arabia and Hokuroku deposit of Japan (Haldar, 2020).

1.3.2 SEDIMENTARY - EXHALATIVE (SEDEX) ZINC DEPOSITS

SedEx deposits are encountered in more than half of the world's zinc and lead resources, forming almost one-fourth of the world's production of the above metals (Goodfellow and Lydon, 2007; Wilkinson, 2014). Even though significant amounts of ore are produced, most of the SedEx deposits are not economically mineable. SedEx deposits are closely related to VMS deposits and are distinct in terms of their fine-grained clastic (shale and mudstone), calcareous clastic and organic-rich clastic host rocks and lack of intimate association with volcanic sequences (Turner and Einaudi, 1986; Misra, 1999). SedEx deposits are large and isolated or widely scattered ones in their host basins (Ridley, 2013). The genesis of SedEx deposits is explained by the simultaneous discharge of ore-bearing hydrothermal fluids from submarine vents in the ocean basins (Haldar, 2020). Sulphide deposits formed in a sedimentary basin by submarine venting of hydrothermal fluids. Though primarily deposition of sulphide minerals is stratiform in nature, later it gets hidden due to the remobilization of ore and post-deformational activities (Leach et al., 2010). These deposits are intimately associated with sedimentary rift zones and occur as rift-fill or rift-cover sequences (Sangster, 2009). These types of sequences are directly linked with syn-sedimentary faults. Generally, they are bimetallic in nature (Zn+Pb) and the grade of the mixed ore is almost 10%.

Most of the SedEx deposits are Zn-dominated relative to Pb and the total resources vary from 5–10 million tonnes (Sangster, 2009). Sphalerite and galena are the key economic minerals besides pyrite and some pyrrhotite. Other than these, sulfosalt minerals such as silver, arsenic, bismuth and antimony are also present in hydrothermal vents. Compositional layering present in the stratiform SedEx deposits are asymmetrically distributed and the thickness of the layers gradually decreases towards the edges of the deposits. Carbonate-hosted SedEx Zn-Pb deposits are stratiform in nature and formed at the temperature range of 100°–250°C. Some of the world's important SedEx deposits are Red Dog Mine of Northwest Alaska, McArthur River and Mt Isa of Australia, Rammelsberg deposit of Germany, Sullivan deposit of British Columbia, Selwyn Basin in Yukon, Qinling belt of central China, Rampura-Agucha and Rajpura-Dariba of India and Zambian copper belt (Kelley et al., 2009; Sangster, 2009; Haldar, 2020).

1.3.3 MISSISSIPPI VALLEY TYPE PB - ZN DEPOSITS (MVT)

The "Mississippi Valley Type" (MVT) deposits are low-temperature, carbonate-hosted (limestone, marl and dolostone) sulphide deposits with epigenetic strata-bound sphalerite and galena (Sangster, 1990; Paradis et al., 2007; Wilkinson, 2014). These deposits hold around a quarter of the world's zinc and lead reserves. MVT deposits are relatively smaller than SedEx and the grade of Zn+Pb mineralization is also <10% (Sangster, 1996). Their occurrence is mainly restricted to foreland thrust belts with some rare instances of rift zone association. The mineralization is mostly cavity-filling or open-space-filling, collapse breccias, faults and hydrothermal cavities (Sangster, 2009; Haldar, 2020). Hot to warm, saline aqueous solutions move through recently deposited carbonate platforms; react with limestone and change it to dolostone by replacing Ca with Mg (Sangster, 1996; Leach et al., 2010). Due to this reaction, fossils get destroyed and voids are generated in rocks making them preferable sites for Zn and Pb precipitation. The hot brine flowing through other sediments such as shale may pick up lead and zinc from them. The water then flows through dolostones which are now at a depth where oil migrates through the rock. The oil migrates upwards bringing sulphur with it. The sulphur from the oil interacts with the brines containing lead and zinc sulphide in the holes in the dolostone depositing galena and sphalerite. Bacterial action assists this process. Primary ore minerals in MVT deposits are generally galena (PbS) and sphalerite (ZnS). Weathered or altered MVT ores may contain anglesite ($PbSO_4$), cerussite ($PbCO_3$), smithsonite ($ZnCO_3$), hydrozincite (also a type of zinc carbonate), and secondary galena or sphalerite (McClenaghan and Paulen, 2018). There are numerous Zn-Pb-Ag sulphide deposits of these types along the Mississippi River in the US, Navisivik on Baffin Island, Pine Point in Canada, San Vicente in Central Peru, Silesia in Southern Poland, Polaris in British Columbia, Leonard Shelf and Admiral Bay in Western Australia (Leach et al., 2010; Haldar, 2020).

1.3.4 SKARN-TYPE ZN DEPOSITS

Skarn deposits are mined economically worldwide for their polymetallic association in terms of zinc, lead, copper, iron, nickel, gold, silver, tin tungsten and molybdenum (Sangster, 2009; Haldar, 2020). Their formation is generally restricted to the contacts of intrusive plutons and carbonate country rocks viz. limestone, dolomite or impure shales. These carbonate rocks are replaced, altered and metasomatized by the ore-bearing hydrothermal fluids that are released due to the cooling of adjacent ultramafic/mafic/felsic intrusives, leading to the precipitation of sulphides. Skarn deposits can be found along the faults/shear zones, at the seafloor and deeply buried metamorphic terrains (Zierenberg and Shanks, 1983). The preferred locations of skarn deposits are platformal continental margin carbonate rocks related to subduction or rifting. The grade of the Zn+Pb skarns ranges from 10–20% with some important by-products such as copper, gold and silver (Megaw et al., 1988). Generally, zinc in skarn deposits is present in between copper and lead, which forms at the proximal and distal positions respectively from the intrusives. Most of the zinc skarn deposits have higher concentrations of manganese and iron. Examples of some of the world-class economic skarn deposits are Pine Creek tungsten, California; Twin Buttes copper, Arizona and Bingham Canyon copper, Utah, USA; OK Tedi gold-copper, Papua New Guinea; Avebury nickel, Tasmania; and Tosam tin-copper, India.

1.3.5 IRISH-TYPE ZN DEPOSITS

Similar to MVT, Irish-type Zn-Pb deposits are also categorized under carbonate-hosted deposits having geology and genesis hybrid between SedEx and MVT deposits (Haldar, 2020). Though significantly smaller than SedEx deposits, Irish-type deposits are significant sources of zinc and lead worldwide. These types of deposits are stratiform to strata-bound in nature and often associated with normal faults, which act as pathways for the movement of hydrothermal fluids. Sphalerite, Galena, Iron-sulphides and barite are the major ore minerals hosted by thick, non-argillaceous carbonate rocks such as dolomitized platformal limestones (Hitzman and Beaty, 1996). Minor amounts of chalcopyrite along with sulfosalts minerals are not uncommon. These types of deposits are the products of mixing between the saline, acidic and relatively sulphur-poor, ore-bearing hydrothermal fluids and sulphur-rich fluids derived from seawater that is carboniferous in age (Hitzman and Beaty, 1996). The grade of zinc usually varies from 5%–6%, though it can reach up to 8% for some bigger deposits. Navan is the example of the largest Irish-type zinc deposits with 58 metric tons of reserves and 8.33% Zn (Fallick et al., 2001; Ashton et al., 2015).

1.4   USES OF ZINC

The major application of zinc is in the industrial sector. Zinc is resistant to corrosion in the presence of any reactive substances, salt water, hard fresh water and atmosphere. Therefore it is mixed with different metals to form alloys such as brass (copper and zinc) and bronze (copper, tin and zinc) that are of great utility. Zinc is a crucial element for growth and positive health besides playing an important role in wound healing and protecting from infections. Zinc deficiency in the body can cause constant fatigue, hormonal irregularities, poor growth and immunity and inhibit the ability to reproduce. Zinc supplements via food and medicines inhibit respiratory infections, improve eyesight, cure aches, common colds and old-age-related problems.

Zinc dust is utilized in the synthesis of zinc and sodium hydro-sulphites, which are vital bleaching agents in the wood pulp and textile industries. Zinc dust is used as a reducing agent for dye stuffs and other organic chemicals. It is also an important agent in gold refining and electrolytic refining of zinc from its ores.

Zinc is coated on steel, by the process of hot dip galvanizing, as galvanized steel prevents corrosion and becomes resistant to rough handling. Galvanizing is carried out in many ways such as electroplating, mechanical coating, sherardizing (also known as dry galvanizing) and zinc-dust painting. Zinc can be galvanized as a sheet, wire or rod on the basis of the requirements of the final product. Galvanized steel is utilized in buildings, bridges, automobiles, steel furniture, automobiles, telecommunication, household appliances and many other applications. Zinc coatings are applied as a base on the body panels of cars for paint. Sherardizing is suitable for bolts, nuts, nails, hinges and related fittings and hardware. Scaffolding fittings, door hinges in cars, spring steel, small chains, tent fasteners and canal tubes are commonly sherardized. Zinc dust paints are prepared using fine zinc dust, which is suspended in a machine that allows high pigmentation. Zinc dust paints are applied to any steel surface that is free from rust by spraying, dipping or brushing to protect ships' hulls, factory steelwork and car parts and to mend damaged zinc coatings.

Zinc alloys play a distinctive role in the industry of die-casting. Zinc die-castings show good strength, stiffness, ductility and resistance to impact. These are prepared with huge areas having a wall thickness of < 1 mm and tolerances are 0.25–25 mm. The main alloying component in these castings is aluminium. Zinc die-castings are broadly used in electrical equipment, hardware, electronic and automotive components.

Many chemical compounds of zinc are widely used in industries. The high melting point and small linear coefficient of expansion make it a useful component in the ceramic industry for the production of artistic glasses, enamels and grits. Zinc oxide is used in photocopying due to its photoconductivity and semiconducting properties. It is used for vulcanization in the rubber industry for the manufacture of tyres. To improve the hardness, light fastness of coating and flow characteristics, zinc oxide is used as a white pigment in paints. Zinc oxide is also used in fertilizers, cosmetic powders, lipsticks, ointments, and formulations for enamels and glazes. It is added in lubricants and a component of drying agents, adhesives and matte surface finishes. It absorbs sulphur dioxide from waste gases in stations employed for generating electricity. Zinc sulphate is used in fertilizer, disinfectants, glue clarifications, conservation and floating of timber. Zinc chloride has applications in batteries and textile industry. Naphthenates, octoates and stearates of zinc are used as lubricants and stabilizers in rubber and plastic industries. Zinc borate, zinc carbonate and nitrate are used in textile industries. Zinc phosphide is employed in pest control, zinc silicate in radio engineering and zinc metaphosphate in the glass industry, zinc phosphate in the paint industry as a pigment for protecting steel and iron are the other applications of zinc compounds.

Zinc is the most commonly used metal for a node in primary, secondary and rechargeable batteries as it has good electrochemical equivalence and electro-chemical behaviour and fairly good shelf life, low cost and is compatible with aqueous electrolytes. Zinc-carbon, zinc-chloride and zinc-bromide batteries are used in flashlights, radios, electronics, toys, electric vehicles and stationary storage for energy.

1.5 CONCLUSION

Zinc is the fourth dominant metal after iron, aluminium and copper, which possesses great applications in the life of human beings. Despite its low abundance in the universe as well as in the earth, zinc has both environmental and industrial importance. The occurrence of zinc in the rocks of the earth's crust is distributed over a vast geological time period due to various factors. Zinc generally occurs as sulphide ores in nature and is associated with different precious metals such as lead, gold, silver, tin and copper. The metallogenesis of zinc is complex and it can precipitate in diverse geological conditions. The important deposits of zinc include VMS, SEDEX, MVT and Skarn-types. Due to its multiple uses in context of human life, the demand for zinc is increasing along with its price day by day. To fulfil this high demand, many countries have increased production but still there is a very significant gap between the demand and production. Hence, there is an urgent need for the finding of suitable zinc prospects on the earth's crust and attempts should be made to understand its genesis that provides the basis of exploration. There is a need for the sustainable management of these newly identified zinc resources with the help of advanced exploration technologies, which will increase the production as well as serve mankind for long time.

ACKNOWLEDGEMENTS

RM and JKP are thankful to the Central University of Punjab, Bathinda, for providing infrastructural and administrative support to complete this chapter. PP is thankful to SRM Institute of Science and Technology for providing infrastructural and administrative support to complete this chapter.

REFERENCES

Ashton JH, Blakeman RJ, Geraghty JF, Beach A, Coller D, Philcox ME, Boyce AJ Wilkinson JJ. 2015. The giant Navan carbonate-hosted Zn-Pb deposit: A review. Current Perspectives on Zinc Deposits. Irish Association for Economic Geology, Dublin.

Barak P, Helmke PA. 1993. The chemistry of zinc, in Robson AD (Ed.), Zinc in Soils and Plants. Springer, Dordrecht, pp. 1–13.

Barnes HB. 1997. Geochemistry of hydrothermal ore deposits. 3rd Edition. John Wiley & Sons, New York.

Bradl HE. 2005. Sources and origins of heavy metals, in Bradl HE (Ed.), Heavy Metals in the Environment: Origin, Interaction and Remediation. Elsevier, Amsterdam, ISBN 978-0-12-088381-3.

Choulet F, Charles N, Barbanson L, Branquet Y, Sizaret S, Ennaciri A, Badra L, Chen Y. 2014. Non-sulfide zinc deposits of the Moroccan High Atlas: Multi-scale characterization and origin. Ore Geol. Rev. 56, 115–140.

Cook NJ, Ciobanu CL, Pring A, Skinner W, Shimizu M, Danyushevsky L, Saini-Eidukat B, Melcher F. 2009. Trace and minor elements in sphalerite: A LA-ICP-MS study. Geochim.Cosmochim. Acta. 73, 4761–4791.

Dorton KL, Wagner JJ, Larson CK, Enns RM, Engle TE. 2010. Effects of Trace Mineral Source and Growth Implants on Trace Mineral Status of Growing and Finishing Feedlot Steers. Asian-Australasian Journal of Animal Sciences, 23, 907–915.

Driscoll RL, Hageman PL, Benzel WM, Diehl SF, Morman S, Choate LM, Heather L. 2014. Assessment of the geoavailability of trace elements from selected zinc minerals. U.S. Geological Survey Open-File Report 2013–1309, Virginia, 78.

Fallick AE, Ashton JH, Boyce AJ, Ellam RM, Russell MJ. 2001. Bacteria were responsible for the magnitude of the world-class hydrothermal base metal sulfide ore body at Navan, Ireland. Econ. Geol. 96, 885–890.

Goodfellow WD, Lydon JW. 2007. Sedimentary-exhalative (SEDEX) deposits: Geological Association of Canada, Mineral Deposits Division, Special Publication 5, 163–183.

Goodwin, FE. 2017. Zinc Compounds. Kirk-Othmer Encyclopedia of Chemical Technology, John Wiley Online & Sons Inc. New York. John Wiley Online & Sons Inc. New York.

Goodwin, FE. 2018. Zinc and Zinc Alloys, in Warlimont, H , Martienssen, W (Ed.) Springer Handbook of Materials Data. Springer Handbooks. Springer, Cham.

Govindarao, B, Pruseth, KL, Mishra, B. 2018. Sulfide partial melting and chalcopyrite disease: An experimental study. American Mineralogist 103, 1200–1207.

Haldar SK. 2020. Introduction to Mineralogy and Petrology. Elsevier, Amsterdam, p. 419.

Henn U, Hofmann C. 1985. Green Sphalerite from Zaire. Journal of Gemmology 19, 416–418.

Hitzman MW, Beaty DW. 1996. The Irish Zn-Pb-(Ba-Ag) ore field: Society of Economic Geologists. Special Publication 4, 112–143.

Huston DL, Eglington BM, Pehrsson S. 2015. The metallogeny of zinc through time: links to secular changes in the atmosphere, hydrosphere, and the supercontinent cycle, in Archibald, SM, Piercey, SJ (Ed.) Current Perspectives on Zinc Deposits. Irish Association for Economic Geology, Dublin, pp. 1–16.

Indian Minerals Yearbook. 2019. Metals and Alloys: Part II: Lead & Zinc. Ministry of Mines, Government of India, Dublin, 58th Edition.

Kelley KD, Wilkinson JJ, Chapman JB, Crowther HL, Weiss DJ. 2009. Zinc isotopes in sphalerite from base metal deposits on the Red Dog district, Northern Alaska. Econ. Geol. 104, 767–773.

Klein C, Philpotts A. 2017. Earth Materials: Introduction to Mineralogy and Petrology, Second ed. Cambridge University Press, Cambridge, London, ISBN: 9781108110730.

Labrenz M, Druschel GK, Thomsen-Ebert T, Gilbert B, Welch SA, Kemner KM, Logan, GA, Summons RE, De Stasio G, Bond PL, Lai B, Kelly SD, Banfield JF. 2000. Formation of sphalerite (ZnS) deposits in natural biofilms of sulfate-reducing bacteria. Science 290, 1744–1747.

Leach DL, Bradley DC, Huston D, Pisarevsky SA, Taylor RD, Gardoll J. 2010. Sediment-hosted lead-zinc deposits in Earth history. Econ. Geol. 105, 593–625.

McClenaghan MB, Paulen RC. 2018. Mineral exploration in glaciated terrain, in: Menzies J, van der Meer JJM (Eds.), Past Glacial Environments (Sediments, Forms and Techniques). Elsevier, pp. 689–751.

Mcdowell, LR. 1992.Minerals in Animal and Human Nutrition. Academic Press, Inc, San Diego, California.

Megaw PKM, Ruiz J, Titley R 1988. High-temperature, carbonate-hosted Ag-Pb-Z n(Cu) deposits of northern Mexico. Economic Geology, 83, 1856–1885.

Misra KC 1999. Introduction, in: Misra KC (Eds.), Understanding Mineral Deposits. Kluwer Academic Publishers, Netherlands. 1–4.

Moezzi A, McDonagh AM, Cortie MB. 2012. Zinc oxide particles: Synthesis, properties and applications. Chemical Engineering Journal, 185–186, 1–22.

Nesse WD. 2016. Introduction to mineralogy, Third Edition. Oxford University Press, London, p. 512, ISBN: 9780190618353.

Ophardt, CE. 2003. Iron Blast furnace, Virtual Chembook. lelmhurst.edu/~chm/vch embook/326steel.htm

Paradis S, Hannigan P, Dewing K. 2007. Mississippi Valley-Type lead-zinc deposits, in Goodfellow WD (Ed.), Mineral deposits of Canada: A synthesis of major deposit-types, district metallogeny, the evolution of geological provinces, and exploration methods. Geological Association of Canada, Mineral Deposits Division, Special Publication no. 5, pp. 185–203.

Pohl WL. 2011. Economic Geology: Principles and Practice. Blackwell's, Oxford.

Porter FC. 1991. Zinc Handbook: Properties, Processing and Use in Design. CRC Press, New York.

Rager H, Amthauer G, Bernroider M, Schurmann K. 1996. Colour, crystal chemistry, and mineral association of a green sphalerite from Steinperf, Dill Syncline, FRG. Euro. Jour. Mineral., 8, 1191–1198.

Ray HS, Sridhar R, Abraham KP 2014. Extraction of nonferrous metals. East-west Press Pvt Ltd: New Delhi, India.

Ridley J 2013. Ore deposit geology. Cambridge University Press, Cambridge, England, pp. 398. ISBN: 9781139135528.

Sabnavis M, Jagasheth UH, Avachat H, Mishra M. 2018. Zinc Industry: the unsung metal of the economy, In: CARE Ratings: Professional Risk Opinion, Report, 31st October, 2018.

Sangster DF 1990. Mississippi Valley-type and SEDEX lead-zinc deposits: a -comparative examination, Institution of Mining and Metallurgy, Transactions Section B: 99, 21–42.

Sangster DF. 1996. Carbonate-hosted lead-zinc deposits. Society of Economic Geologists, Special Publication, Littleton, CO, 4, pp. 664.

Sangster DF. 2009. Geology of base metal deposits. Encyclopaedia of Life Support Systems: 6, Eolss Publishers, Paris, France, 91–116.

Schwartz MO. 2000. Cadmium in zinc deposits: economic geology of a polluting element. Int. Geol. Rev. 42, 445–469.

Shanks III WP. 2014. Stable isotope geochemistry of mineral deposits. Earth Systems and Environmental Sciences 13, 59–85.

Sparks DL. 2005. Toxic metals in the environment: the role of surfaces. Elements 1, 193–197.

Turner RJW, Einaudi MT. 1986. The Genesis of Stratiform Sediment-hosted Lead and Zinc Deposits: Conference Proceedings, Vol. XX, pp. 1–2, Stanford University Publications.

United States Geological Survey 2020. Mineral Industry Surveys: Zinc.

Wilkinson JJ. 2014. Sediment-hosted zinc-lead mineralization: processes and perspectives. In: Scott SD (Ed.), Treatise on Geochemistry 2nd ed. Elsevier 13: Geochemistry of Mineral Deposits, Amsterdam, Netherlands, 219–249.

Yaroshevsky AA. 2006. Abundances of chemical elements in the Earth's crust. Geochem. Int. 44, 48–55.

Zierenberg RA, Shanks WC. 1983. Mineralogy and geochemistry of epigenetic features in metalliferous sediment, Atlantis II Deep, Red Sea. Econ. Geol. 78, 57–72.

2 Zinc Metallurgical Heritage in India

V. Ramanathan and Gautam R. Desiraju

2.1 INTRODUCTION

Zinc (Zn) is a metallic element of hexagonal close-packed (hcp) crystal structure with a density of 7.13 grams per cubic centimeter. It has only moderate hardness and can be made ductile and easily worked at slightly above the ambient temperatures. In solid form, it is greyish-white, owing to the formation of an oxide film on its surface, but when freshly cast or cut it has a bright, silvery appearance. Its most important use, as a protective coating for iron known as galvanizing, derives from two of its outstanding characteristics: it is highly resistant to corrosion, and, in contact with iron, it provides sacrificial protection by corroding in place of the iron.

With its low melting point of 420°C (788°F), unalloyed zinc has poor engineering properties, but in alloyed form, the metal is used extensively. The addition of up to 45% zinc to copper forms the series of brass alloys, while, with additions of aluminum, zinc forms commercially significant pressure die-casting and gravity-casting alloys. In sheet form, zinc is used to make the cans of dry-cell batteries and alloyed with small amounts of copper and titanium to get an improved-strength sheet that has applications in the roofing and cladding of many buildings.

Zinc has wide utility and its world consumption falls into five areas. The most important one, approaching 50%, is the corrosion protection of iron and steel known as galvanization that involves application of zinc coatings to metal surfaces. About 15–20% is consumed both in brass and zinc-cast alloys and 8–12% is used both in wrought alloys and in miscellaneous uses such as chemicals and zinc dust. The chemical compounds of zinc, particularly zinc oxide, have important industrial and pharmaceutical uses.

2.2 METALS FROM ANTIQUITY

Eight metals have been known to humankind since antiquity. They are gold, silver, copper, tin, lead, mercury, iron and latterly zinc. It is interesting to note that the list featuring the weight percent of the top 25 most abundant elements in Earth's crust, which by the way accounts for 99.70%, does not contain these metals except for iron. Hence, it is very intriguing to know that ancient civilizations' endeavored to discover their surroundings that led them to the discovery of processing the metals. Among the 25 most abundant metals, Zinc gets placed in number 23 with 0.013% abundance (Mason, 1952).

DOI: 10.1201/9781003412472-2

Although some of the eight antique metals were found in their native metallic state, the others were processed from their ores. The two lowest in the electro-chemical series, namely gold and silver are mostly found in the native state and indeed obtaining them from their ores is trivially easy. Gold is usually recovered in its native state. As for copper, lead, tin, and mercury, they are also below hydrogen in the electrochemical series and their extraction is uncomplicated as it involves a reducing agent (mostly coke), and some type of heating that converts sulfide ores into oxides and drawing off the molten metal following reduction.

Copper is found in small quantities in nature (0.007% by weight in Earth's crust) and its extraction is mainly from its sulfide and oxide ores (Haynes, 2017). Lead and tin are found only as ores and the reduction is easy. More often these metals are not found in isolation but in combination with other metals. Hence even though the metals may be found in their native states, separating them from other metals is an arduous task. This is one of the reasons why in history we get to see the alloys of metals being in use even before the pure metals got in vogue. Mercury is a liquid at ambient temperatures and there is no question of obtaining it as such in nature but the general principles are the same as in the preceding examples.

Besides the use of these metals in tools and various other paraphernalia, their express use in preparing medical formulations in combination with various herbal ingredients has been one of the hallmarks of India's traditional medicinal systems. Interestingly, despite mercury being classified as one of the most toxic metals by the World Health Organization (Anonymous, 2003), this metal finds profound use in Indian Medicine (Charaka, Agniveśa, & Dridhabala, 2017). In fact, before the discovery of antibiotics, even in Europe it was a common practice amongst doctors to advise direct intake of mercury (Wainwright, 1910; Raisman, 1998). It is well documented that mercury was the only succor to treat the then-deadly disease of syphilis in 16th and 17th centuries in Europe. Thus, these six relatively unreactive metals have been known for thousands of years and have found major use in the growth of ancient civilizations in India, China and West Asia, where the ores are found in abundance.

Let us turn to the two metals in the list of those known to our ancients, that are above hydrogen in the electrochemical series and in this sense are more reactive, namely iron and zinc.

In Rig Veda, the work *Aayasi* (made of iron) which comes from the word *Ayas* connotes iron. In later periods, this word has also been interpreted to be a generic metal. It is also worth noting that the names of metals kept changing at different points in time. *Sisa* and *Trapu* are unambiguously identified as lead and tin. There is some degree of controversy about *hiranya* (gold and silver), *ayas* (silver, bronze), *shyaama* (iron, copper), and *loha* (copper, iron). In any case, it is clear that iron occurs here as either *shyaama* or *loha*. There is no identification of either mercury or zinc in the verse. One of the early names that gets inevitably associated with mercury is *Nagarjuna*.

Although mercury features in earlier Ayurvedic texts such as *Charaka Samhita* (where mercury is referred to as *paarada*), mainstream historians of science have largely identified it with the *taantrik* period which roughly starts from the 8th century CE. A large portion of various medical texts are dedicated to mercury and its various formulations used as internal and external medicine. In the Western alchemical literature, mercury is often anthropomorphized in the form of the King's son ably assisted by the servants, lead, copper, tin. and silver.

It is indeed safe to say that mercury was not known during the early Vedic periods and it is still not clear when exactly in the past this metal gained prominence in our country although there are quite a few articles that show that this metal was in vogue even in 500 BCE by alluding the Ayurvedic texts. Among others, the *Charaka Samhita* that date back to the 2nd century BCE, mentions the use of *rasa* in various medicinal preparations. For example, Chapter 5 of the text describes the preparation of 'Rasakarpoora', which is a mercury compound used to treat various diseases.

Another Ayurvedic compendium, the *Susruta Samhita* that dates back to at least the 6th century BCE mentions the use of *rasa* in medicinal preparations. For example, Chapter 40 of the book describes the preparation of *Rasakarpoora*, which is a mercury compound used to treat eye diseases.

Interestingly the *Atharvaveda* mentions the use of *rasa* in various medicinal and magical practices. For example, hymn 3.23 of the compendium describes the use of *rasa* in a potion to cure diseases. Today this is profusely used in several Siddha medicines.

In the second chapter of *Arthashastra* while describing the various forms of gold and their processing methods a verse appears as follows:

आकरोद्गतं सीसान्वयेन भिद्यमानं पाकपत्राणि कृत्वा गण्डिकासु कुट्टयेत्, कदलीवज्रकन्दकल्के वा निषेचयेत् । ९ ।

Here the term *Akarodgatam* literally means gold that is obtained from mines which is usually in combination with other metals. There are quite a few contemporary commentators who have interpreted the word *akarodgatam* as an amalgam, and in this way we get an impression that one of the early mention of this metal appears for the first time in *Arthashastra*.

2.3 ZINC METALLURGY IN INDIA

In any case, it is clear that zincin its pure form was not known in India as far back as when the other metals in the list above were understood. As mentioned earlier, the reactive iron in its pure form was known during the Rig Vedic era because the method of extraction (heating with a reducing agent) was in principle, similar to that employed for the other metals of antiquity. Hematite, its principal ore, is widely dispersed in great abundance and is very commonly found. The great strength of this metal and the relative ease of extraction possibly impelled its wide application so that it replaced copper and bronze for a variety of uses.

An indication that zinc posed special problems in its extraction is conveyed by the fact that there was a good knowledge of brass, its alloy with copper, in ancient India. Excavations at *Atranjikhera* revealed brass artifacts which date anywhere from 1200 to 600 BCE (Gaur, 1983). This was in any case before the pure metal was distilled. How and why would a metal be known in the form of an alloy before it was known in pure form? This fact alone provides a hint as to why and how our ancients realized two important facts about this purportedly new metal. The first was that there were widely distributed ores (most probably sphalerite and calamine) that resembled ores of other substances that were clearly identified as metals, and yet when later ores were treated in the usual manner (reducing agent, heating), did not result in one obtaining a pure metal. Secondly, they had realized fairly early that a mixture of copper ore and these new, unknown ores when smelted with a reluctant in the normal manner produced a new substance that we today call brass. They must have grasped that there was a new metal in these latter ores that was unusually difficult to extract in pure form.

All these become more curious and controversial in the sense that Mainz (2015) in his recent article, does not mention Indian metallurgical heritage. Except for the Delhi iron pillar, nowhere else in the article does the author mention the antiquity of metals in India. This article is in a way an exhibition of a century-long continuity in apathy towards Indian scientific heritage because in his acclaimed comprehensive review of the archaeometallurgy, titled 'Metals in Antiquity' Gowland (1912) is conspicuously silent on zinc and the Indian heritage of zinc. Given that we live in an internet age where access to information is relatively easier, the conspicuous absence of Indian metallurgical heritage in an article devoted to the antiquity of metals is highly surprising. With people like P. C. Ray, P. Neogi in the early 20th century and the likes of Paul Craddock towards the later part of the 20th century, along with various others writing about ancient India's expertise in metals and metallic formulations, available in the public domain, not referring to them amounts to incomplete survey of the available literature and thereby suppressing India's scientific past (Ray, 1902; Neogi, 1918; Craddock et al., 1998).

Hence it becomes all the more important and timely to take cognizance of India's expertise in handling zinc and how it monopolized the supply of the pure metal to the entire world.

There are some significant misses by researchers who have written on the antiquity of brass or zinc or brass in India. Linguistics could partly be a reason because both brass and zinc have been referred to in the ancient Indian literature with multiple names. For instance, the following are the various allusions to zinc that appear in different literature:

- *Yashada* – This term is commonly used in Ayurvedic texts to refer to zinc.
- *Tuttha* – This term is used to refer to zinc or zinc oxide.
- *Kharpara* – This is used to refer to calamine
- *Jashada/Jasada* – This term is used to refer to a type of zinc ore and/or zinc itself.
- *Jasta* – This term is used to refer to a type of zinc ore and/or zinc itself.

These terms may be used interchangeably in different texts. It is interesting to note that several international languages have a word for zinc or its compound that is either derived from or is cognate with the word *tuttha*. Here are some cognates of the Sanskrit word *tuttha* in various languages:

- Persian: *Tutiya,*
- Arabic: *Tootiya,*
- Turkish: *Tutiya,*
- Latin: *Tutia,*
- Old English: *Tutie,*
- Italian: *Tuzia,*
- Spanish: *Tucia,*
- Portuguese: *Tutia,*
- French: *Tucie.*

It must however be noted that the traditional Sanskrit thesaurus *Amarakosha* translates the word *tuttha* to copper sulfate which is how is it understood in contemporary times. But in the *Arthashaastra*, the word *tuttha* is taken to be zinc ore by commentators. There is credence to this interpretation because the traditional Tamil word for zinc or its oxide is called *thutthanaagam* and even in the Chinese language, it is called *tutenage.*

Corroborating this with the above-mentioned cognates from various foreign languages supports the attribution of *tuttha* to zinc. There is a possibility that modern-day translators and scholars might have missed this word for zinc in texts like *Arthashaastra* and its contemporary texts by considering only the meaning of copper sulfate for the word *tuttha.*

Similarly, in the case of brass too, linguistics might have evaded the past researchers. The various words by which brass has been referred to in our ancient literature are:

Rīti or Rītikā; Pittala; Ārakūṭa or Āra; Svarṇakakṛta

Talking about the antiquity of brass, yet again the linguistic variants by which this alloy was addressed have evaded the researchers. For instance, in the *Atharvaveda* (15.6.9) we come across the following verse:

तदाहवैश्वानरंपित्तलंतदाऽश्विनाविवेशतांविवेशतामिति।
Tadā ha vaiśvānarampittalamtadāśvināviveśatāviveśatāmiti

"Then the Asvins entered the Pittala vessel with Vaishvanara, and so entered, they became a remedy for all diseases."

Where we come across an explicit mention of the word for brass which is still in vogue in Indian languages like Hindi and Tamil, *pittala.*

Interestingly in the other ancillary *Vedic* texts like the KātyayanaŚrautaSūtra (10.2.7), we once again come across direct reference to brass like in this verse pertaining to the *Agnihotra* ritual.

कृत्वायत्रतत्रदीसिकरोति।तस्यकुलालयश्चान्यश्चततःप्रवालानुबन्धीयात्।तत्रतदङ्कुरेणसर्वतः
प्रणाल्यांचोर्णयेद्रोदूमेनसवयंलेप्यगोपितंपित्तलेनैवाभिघच्छेत्।तद्रोदूमेनसंचितमवगाह्य।

Kṛtvāyatratatradīptimkaroti. Tasyakulālayaścānyaścatataḥpravālānubandhīyāt.
Tatratadaṅkureṇasarvataḥpraṇālyāmcorṇatetgodūmenasavayamlepyagopitam-
pittalenaivābhighacchet. Tadgodūmenasañcitamavagāhya.

"Having dug the pit, let him surround it with unbaked bricks. Let him then smear it with cow dung, and over the cow dung let him spread a covering of *pittala* (brass), and over the covering of *pittala* let him again smear cow dung."

Incidentally, brass was also referred to by another word called *svarṇakṛta* and we come across this once again in the KatyayanaŚrautaSūtra (22.1.21 and 24.4.10) in the section on the *Mahāvṛta* ritual as well as in the *Pravargya* ritual, the following verse mentions the use of a brass vessel for holding the ghee:

सपूर्वोनिर्दिशेत्, दक्षिणतउपस्थापयेत्, श्यामोभवेत्, निवेदयेत्, सुवर्णकृतांदधाति।

Sa pūrvenirdeśat, dakṣiṇataoopasthāpayet, śyāmobhavet, nivedayet, suvarṇakṛtāmdashāti

"He who knows this takes the ladle with its end turned towards the north, the Brahman with the north side of his body turned towards the north, pours the (ghee) from the Brahman's right hand into a new vessel of brass."

2.4   ZINC EXTRACTION BY TRADITIONAL METHODS

We get the first ever-explicit description of the zinc extraction process in the text called *Rasaratnasamuchaya* whose date is anywhere given between 10th and 14th centuries. Following are the exact verses that pertain to the zinc distillation in this text:

हरिद्रात्रिफलारालसिन्धुभूमैः सटङ्कणैः ।
सारूष्करैश्च पादांर्थैः साम्लैः सच्चार्धं खर्परम् ॥
लिप्त' हन्ताकमूषायां शोषयित्वा निरुध्य च ।
मूषां मूषोपरि' न्यस्य खर्परं प्रधमेत्ततः ॥
खर्परे प्रहते ज्वाला भवेन्नोला सिता यदि ।
तदा सन्दंशतो मूषां दृष्ट्वा छत्वा त्वधोमुखीम् ।
घनैरास्फालयेद्भूमौ यथा नालं न भज्यते ॥
वङ्गाभं पतितं सत्त्व' समादाय नियोजयेत् ॥ 157-161

The following excerpt is the translation given by Prof. P. C. Ray (1956):

Extraction of Zinc :—Rub calamine with turmeric, the chebulic myrobalans, resin, the salts, soot, borax, and one-fourth its weight of *Semicarpus anacardium,* and the acid juices. Smear the inside of a tubular crucible with the above mixture and dry it in the sun and close its mouth with another inverted over it, and apply heat ; when the flame issuing from the molten calamine changes from blue to white, the crucible is caught hold of by means of a pair of tongs and its mouth held downwards and it is thrown on the ground, care being taken not to break its tubular end. The essence possessing the lustre of tin, which is dropped, is collected for use.   157-161

The artistic representation of the extraction process is given in Figure 2.1.

The upper chamber is filled with the purified ore of zinc along with prescribed herbs and heated from above while its narrow mouth opens to another chamber buried beneath the ground. Once the zinc gets reduced to its metallic form, the vapors have no escape route but to travel downwards where they condense due to colder temperatures. The overall chemical reaction may be represented as:

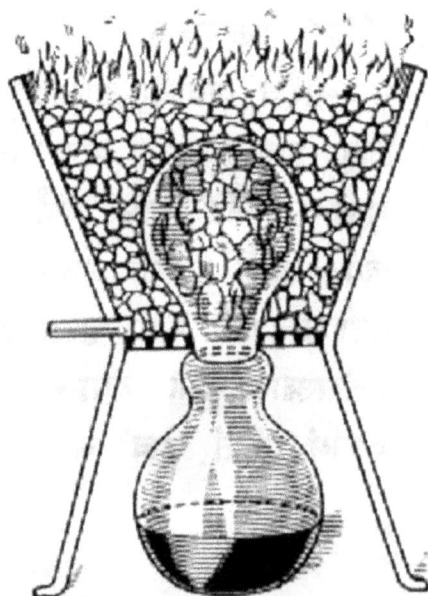

$$ZnCO_3 \rightarrow ZnO + CO_2 \quad ZnO + C \rightarrow Zn + CO \tag{2.1}$$

FIGURE 2.1 The Koshthi apparatus for zinc distillation (Image courtesy: Internet).

2.5 ZINC IN AYURVEDA

The term *yashada* or *jasadbhasma* is mentioned in several ancient Ayurvedic texts, including the *Charaka Samhita* and the *Sushruta Samhita*. These texts were written between the 2nd century BCE and the 7th century CE, making *yashada* a well-established mineral compound in Ayurvedic medicine for over two thousand years.

The use of *yashada* is described in these texts as a treatment for a variety of conditions, including skin diseases, digestive disorders, and urinary tract infections. The texts also describe the process of preparing *yashada*, which involves burning zinc in air until it is completely oxidized, and then processing the resulting ash to remove any impurities. In the *Charaka Samhita* in the *Chikitsathana* chapter verse 250, it is described thus:

"सौवीरमञ्जनंतुल्थ्यंताप्योधातुर्मनःशिला| चक्षुष्यामधुकंलोहामणयःपौष्पमञ्जनम्|| (26.250)

Undoubtedly, this could be considered as one of the earliest references of zinc in the ayurvedic texts. Interestingly, brass is also extensively mentioned in the *Charaka Samhita*. There are at least six mentions of this alloy in the text which indicates the use of zinc through brass. In the same text, the synonymous of *pittala*, i.e., *ritika* is also used once again reinforcing the prevalence of this alloy during Charaka's time or even before (सुवर्णरूप्यत्रपुताम्ररीतिकांस्यास्थिशङ्खद्रुमवेणुदन्तैः (3.7)).

The term *pittala* occurs in 11 instances in the *Sushruta Samhita* and interestingly the term *ritika* is not to be seen in this text.

Yashada known by ancient Indians and first time introduced in the *Adhamalla* commentary of *Sharangadhara* in the context of *Dhatu*. However, its ore and its alloy have been mentioned in ancient treatises including *Samhitas*. In *MadanpalNighantu, Yashad* has been introduced for the first time along with *Suvarna, Rajata, Tamra, Kamsya, Pittal,* and *Vanga*. Its synonyms and pharmacological properties are also described (Shastri, 2010).

In the *Rasayanadhikar* chapter of *Vangasen, Yashad* has been mentioned as an essential constituent of *KharparakhyaRasayana* (Gupta and Neeraj, 2010).

In *Rasa kamdhenu, Yashad* term has been mentioned as one of the synonyms of *Rasaka.*

Bhava PrakashaNighantu, Yashad has been mentioned along with *Suvarna, Rajata, Tamra, Ranga, Sisa,* and *Lauha*. Its synonyms and pharmacological properties are also described in *Dhatwadivarga* (Bhavmishra et al., 2012).

In *Ayurveda Prakash,* the author has extensively described *Ysahad* including its synonyms, source, pharmaceutical processes and pharmaco-therapeutical properties. However, he has clearly mentioned that the pharmaceutical processes of *Yashad* should be done like *Vanga* (Madhavacharya et al., 1962).

In *Rasa Tarangini*, the author has widely described *Yashad* including its synonyms, *Grahya-AgrahyaLakshanas* and different pharmaceutical processes of *Shodhana* and *Marana*. Apart from this, its pharmacological properties and indication in different diseases are listed with different *Anupana* (Sharma, 1979).

Over time, the use of *yashada* has evolved and adapted to changing medical practices and scientific knowledge. Today, *yashada* is still used in Ayurvedic medicine and is often combined with other herbs and minerals to create specific formulations for different conditions.

2.6   BRASS AND OTHER ALLOYS IN ANCIENT INDIA

Zinc forms alloys with copper in all proportions, but only those alloys containing up to about 45% zinc, and ranging in color from red through yellow to gold as the amount of zinc increases, are in commercial use as brass. Two main phases are involved: the alpha phase, which is face-centred cubic with a maximum zinc content of 39%; and the beta phase, which is body-centred cubic and occurs at 40 to 50% zinc content. Alloys composed entirely of the alpha phase are characterized by their ability to be cold-worked and are suitable for rolling, pressing, and drawing. At 40 to 45% zinc, the solidified alloys form mixed alpha and beta phases, in which hot plasticity is followed on cooling by reasonable cold-working properties. These are used in casting, hot-pressing and extruding.

Brass is an alloy of copper and zinc, with traces of other metal and varieties of it ranging from 8% to 50% of zinc. This alloy has been known to man since at least 2500 BCE and is mentioned in several contexts as occurring in most of the inhabited areas in and around India (Rajasthan), China (Shanxi), West Asia (Anatolia), and Central Asia (Armenia, Turkmenistan, Caucuses). Brass has a pleasing yellow color and is generally easier to obtain than bronze accounting for the fact that it began to replace the latter to make weapons, tools, artifacts and household articles.

In a modern context, the copper-zinc ratio in brass is found to vary depending on the distances from the surface at the micron level (being richer in copper at the surface) and this could also add to the stability of the alloy. Apart from resistance to corrosion, there were other reasons for the increased popularity of brass. Brass is more malleable (easier to hammer) than bronze and more ductile (it can be drawn out into fine wires). Ores of zinc also tend to occur with those of copper, especially in the areas in Asia mentioned above, and it is not at all difficult to conceive of a situation where the first examples of brass were obtained fortuitously when a mixed ore was accidentally smelted; this indeed must have been the origin of the cementation process, one of the two methods of producing brass.

It is also conceivable that the bright golden color associated with the common a-brass (with >8% zinc) might have excited alchemists in their quest to turn base metals into gold. We would like to mention certain aspects of the brass timeline with respect to India and surrounding areas that showcase the impressive technical advances. The entire topic of brass in ancient times has been reviewed extensively by Morton, Neogi and Craddock and it is not our intent to repeat their earlier material (Morton 2019, Craddock 1998). Neogi (1918) comments that brass coins and artifacts dating from 100 BCE have been found in north (Indus and Ganga valleys) and northwest India (Afghanistan). He states that brass might have been known even earlier in India probably as early as 300 BCE because it is mentioned in the *Charaka Samhita* along with gold, silver, copper and tin, where it is referred to as *reeti*. The usage of the word *reeti* for brass continued till 600 CE after which it was gradually replaced by *pittala* which is the word used today in many modern Indian languages.

Let us turn next to the Indus valley civilization. Mixed quaternary alloys (copper, zinc, lead, tin) containing up to 6% Zn have been found around 1500 BCE in Lothal (Gujarat), which had been settled by Harappan people from further north, nearer to sources of zinc-bearing copper ores. The Indus Valley civilization ended around 1500 BCE and by 600 BCE the Ganga plain became inhabited. Increasingly, high percentage zinc brasses have been found in copper alloys in what is modern-day Uttar Pradesh (*Atranjikhera*). It is possible that brass articles were taken by Darius (521–486 BCE) to Persia and by Alexander (356–323 BC) to Macedonia further disseminating brass technology beyond India.

Brass containing a high percentage of zinc must have been made by a process different from cementation and this new technology must have been widespread in India early on. An artifact made of high zinc brass (36% zinc) dated between 700–200 BCE was found in Senuwar, near the Bihar-Bengal border and is also pertinent to this discussion. This zinc content was extraordinarily high and it could not have been achieved unless the innovation/technology had been available on pure zinc or people were aware of some other procedure to increase the zinc content.

The boiling point of zinc is lower (907°C) and raises fundamental difficulties in the alloying process where zinc would vaporize off before the copper melts, unless there is a lower melting mixed phase. The fact that brass was made in India at least by 100 CE shows that our ancients were well aware of these complex issues and could solve metallurgical problems by maintaining high temperatures in furnaces with such a fine degree of control and great skill. Once our ancients grasped the significance of zinc distillation in the manufacture of brass, it is not difficult to understand how they subsequently came upon the process of extracting pure zinc itself.

A great deal of fine physicochemical control had now been obtained over the process. Further, the fact that Indians had mastered the *in situ* generation of zinc, meant that there must have been an inevitable curiosity to make the pure metal itself. Here, however, a formidable scientific problem arose: in the high-temperature alloying of zinc vapor with molten copper, an accumulation of zinc (vapor cum liquid) could not have occurred because the pure metal was being generated and consumed *in situ*. Given their newly found expertise in making high-zinc brasses, our ancients must have attempted to produce pure zinc from sphalerite by reduction followed by distillation and condensation as they did for brass. It is also clear that they were unsuccessful in their attempts because otherwise artifacts of pure zinc would have been recovered that share the provenance of those made of high-zinc brasses, and this is clearly not the case.

The hot zinc vapor-liquid material must have been too reactive towards oxidation or they were unable to maintain a sufficiently inert (oxygen-free) atmosphere in the reduction. Whatever be the case, a very new methodology had to be developed. This led to the discovery of downward distillation, the centerpiece of the Zawar method of preparing pure zinc, and one of the high points of ancient Hindu chemistry itself.

Kharakwal (2011) in his book on the Zinc Technology of India mentions that distillation was done for the first time in 12th century CE. Given the availability of artifacts from much earlier times, it is difficult to make out how he arrived at such a late timeline. Others have stuck to this date for zinc distillation, paying no heed to the mention of Nagarjuna in the 4th century by Craddock.

Srinivasan et al. reported contradicting dates with respect to zinc distillation. They quoted both *Rasaratnakara* of *Nagarjuna* for the *Tiryakpatanayantra* and at the same time say zinc distillation was carried out only in the 12th CE (Srinivasan and Ranganathan, 1997; Srinivasan, 2016). Both Srinivasan and Alam seem to agree with Forbes on the connection of zinc distillation with the evolution of alcohol distillation (Alam, 2020).

2.7   ZINC IN MODERN TIMES

We finally turn our attention to this interesting element in the modern era, especially with regard to aspects that concern our daily lives today, and also to India. Generally, modern means contemporary or recent, and the word has a somewhat subjective meaning and is related to the context in which it is used. Historians usually reckon the modern era as beginning around 1500 CE. This cut-off is also very suitable in the context of this chapter because it also marks the beginning of a serious decline in Hindu science which was the major or predominant contributor to science in the Indian subcontinent till then.

P. C. Ray (1956) ascribes this decline largely to a rigid stratification of the caste system by that time and a shift from the *varna* to the *jati* system where people of higher castes began to shun occupations that demanded technical expertise of the type needed in science, namely experimental work. For example, Susruta has laid down that dissection of a corpse is a *sine qua non* for a person attempting to obtain training in medicine. This did not fit well with a system where the mere touch of a corpse was considered enough to bring contamination to a person of a higher caste.

Ray points out that Europe too was not free of this menace and that scholars like Aristotle and Paracelsus had sneered at those who undertook technical work that too especially in chemistry. This tendency seems to have been especially pronounced in England and where this disdain for chemistry tended to continue till the early 19th century leading the great German chemist Liebig to comment in 1837 that "England was not the land of science" However, one must remember that a huge intellectual revolution had begun in Europe with the Renaissance and was progressing through the Age of Enlightenment through the Industrial Revolution and beyond. An age where effect was compulsorily linked to cause, deduction replaced induction and reductionism rather than holism was the principal dogma would naturally have produced a Boyle, Descartes or Newton.

The way science was done itself changed: alchemy was replaced by chemistry. The former could not hold a candle to our holistic way of doing scientific work accounting for the state-of-the-art science in our country in ancient and medieval times. In modern times, however, the latter could easily outstrip our ancient methodologies because of its speed of performance, and also because of the adverse cultural climate here that led to the decline and virtual disappearance of Hindu chemistry from this land.

It is in this background that one must understand the first successful process developed in Europe by William Champion (1709–1789) who ran a profitable brass-making business in Bristol. Champion obtained a patent for the recovery of zinc by distillation in 1738 and his system remained workable for at least a century, but there was no mistaking that he used the Zawar downward distillation process in a straightforward manner concentrating on scale-up so that he was producing 40 kg per charge of sphalerite. Champion adopted local glass furnaces to accommodate his retorts but the process of heating inverted retorts in an upper chamber and condensing the zinc in tubes leading down through holes in the floor to water-filled vessels in a cool chamber beneath, was exactly similar to the Zawar process. It was impossible that Champion was unaware of the Zawar process especially since there was increasing contact between England and India in the early 18th century. The inevitable result of his success was that the price of zinc fell in Europe much to the chagrin of the East India Company and local smelters began dealing with other metals. Champion, who had become extremely wealthy by 1770, got entangled in a web of legal problems that led to his bankruptcy and ultimate death.

However, his process was adapted and modified by continental manufacturers such as Johann Ruberg in Silesia in the late 18th century and Jean-Jacques-Daniel Dony in Liège, Belgium in the early 19th century with other retort techniques. Dony's horizontal retorts were operated in Britain as the main zinc-producing process for about 100 years starting in the mid-19th century but the process, involving relatively small-sized retorts, was physically arduous and highly environmentally unfriendly. All this led to a blast furnace process, with a much bigger charge size and vertical retorts, developed in the early 20th century in the United States which continues to be used even today despite certain problems with iron-rich zinc ores. The retort was constructed of Silicon-Carbon brick for high heat conductivity of the charge of roasted sulfide concentrate and coal was fired. Zinc vapor was removed with carbon monoxide at the top of the retort and condensed in a stirred molten-zinc bath. The heating was also done in electric furnaces in some cases. To summarize, the blast furnace route became the main pyrometallurgical means of producing zinc.

Today, the major method of recovering zinc is through electrolysis of a solution of zinc sulfide in a process that was made commercial around 1915 following a patent by Léon Letrange in 1881 and the realization that high-purity sulfate electrolyte was needed in the process.

2.8 CONCLUSION

Zinc is one of the important non-ferrous metals that is highly essential for a sustainable economy. A glance at the data from Indian Bureau of Mines Year Book (2019), we get to see that despite India having the second lowest zinc reserves in the world, it is one of the top producers of this metal (Anonymous, 2019). The long-lasting tradition and know-how of zinc distillation are indeed responsible for such a feat. Technical prowess apart, the academic landscape pertaining to the history of Indian science and technology, particularly about zinc needs an overhaul as the credit that India deserves is still due.

It is very interesting to note that even the Chinese literature openly acknowledges that they learnt the art of refining copper and making zinc from India. The Chinese verses that credit and acknowledge India for knowledge about zinc can be found in the Sui Shu or the Book of Sui, which was written during the Sui Dynasty (581–618 AD) in China. The relevant passage from the Sui Shu as translated into English is as follows (Daumas, 1969).

> The country of Tianzhu (India) produces a kind of white copper that is the color of copper but is light in weight. Therefore, people often mistake it for tin or white iron. It was not until it reached the country of Daqin (Persia) that people realized it was white copper. The method of production involves taking copper ore and refining it. It is used to cast copper coins or make copper utensils, and its profit exceeds that of copper. There is also lead ore, which, when extracted and refined, can be made into five-colored lacquer that is much admired by the people. All of these techniques were transmitted from Tianzhu (Tongdian).

In this passage, the Chinese acknowledge that the knowledge of refining zinc into brass was transmitted to them from India, which was known as the country of Tianzhu in ancient times. In Volume 31 of the "Tongdian," a Chinese encyclopedic text from the Tang Dynasty (618–907 AD), a section titled "Zhi 3: Jinmuzhuan", the following passage credits their knowledge about zinc to India (Du You, 735–812 CE).

> Zinc originates from the Western Regions and India. When it was first discovered, it was used to make bells, mirrors, and other objects.

Even the Persian records credit India for their knowledge of Zinc. Al-Biruni credits Indians for being the first to extract zinc from its ore and use it for various purposes is as follows:

> The Hindus themselves have taught us how to extract zinc from its ore, which they call *tutiya*. They heat this ore in a closed vessel until it is reduced to the metallic state; they then distil it with some salt, and the product thus obtained is called zinc. (AlBiruni1003 CE; ed. and trans. By Edward C. Sachau, 1910) (Al-Biruni, 1910).

With such confessions from our neighbors on either side, it is high time that the scientific heritage of our country with respect to zinc must be restored to its deserved heights.

REFERENCES

Alam I. 2020. The history of zinc and its use in pre-modern India. Studies in People's History 7, 23–29. 10.1177/2348448920908237.

Al-Biruni. 1910. India: An Account of the Religion, Philosophy, Literature, Geography, Chronology, Astronomy, Customs, Laws and Astrology of India About A.D. 1003. Sachau EC (Ed. and Trans.). Kegan Paul, Trench, Trübner & Co., London, 1: 152.

Anonymous. 2003. World Health Organization. Mercury: Environmental Aspects. World Health Organization, Geneva, Switzerland. who.int/foodsafety/publications/chem/mercury/en/

Anonymous. 2019. Indian Bureau of Mines Year Book.

Charaka, Agniveśa, Dridhabala. 2017. Caraka Samhitā. Chaukhamba Surbharati Prakashan, Varanasi, India.

Craddock PT. 1998. 2000 Years of Zinc and Brass. British Museum Occassional Paper No. 50, pp. 27–72.

Daumas M. 1969. A History of Technology and Invention: Progress Through the Ages. Crown Publications, New York. https://www.science.org/doi/10.1126/science.168.3932.726.b

Du You, 735–812 CE, Tongdian (通典), Vol. 23, p. 739 (Note: The "Tongdian" is a Chinese encyclopedic text compiled by Du You during the Tang Dynasty).

Gaur, RC. 1983. Excavations at Atranjikhera: Early Civilizations of the Upper Ganga Basin, Aligarh Muslim University/Motilal Banarasidas, New Delhi.

Gowland W. 1912. The Metals in Antiquity. J. Royal Anthropol. Inst. Great Britain Ireland 42, 235–287.

Gupta LN, Neeraj K. 2010. Study on Trivanga Bhasma. PhD Thesis, Department of Rasa Shastra, Faculty of Ayurveda, IMS, BHU, Varanasi.

Haynes WM. 2016–2017. The CRC Handbook of Chemistry and Physics, 97th ed., p. 4–9. CRC Press, Boca Raton, Florida. https://archive.org/details/CRCHandbookOfChemistryAndPhysics97thEdition2016

Kharakwal JS. 2011. Indian Zinc Technology in a Global Perspective. Pentagon Press 2011, India.

Madhavacharya, AP. 1962. in: Mishra GS (Ed.), Arthavidyotini and Arthaprakashini Hindi commentaries, 2nd ed. Chaukhambha Bharti Academy, Varanasi, India., 3/152–183, pp. 374–381.

Mainz VV. 2015. The metals of antiquity and their alloys, in: Seth CR (Ed.), Chemical Technology in Antiquity. ACS Symposium Series, American Chemical Society, Washington, DC. doi:10.1021/bk-2015-1211.ch005

Mason B. 1952. Principles of Geochemistry, John Wiley and Sons, New York, 276, p. 41.

Mishra B. 2012. Bhavmishra, Bhavprakash, Purvakhanda, Dhatwadivarga 33. pp 606. Chaukhambha Sanskrit Bhavan, Varanasi, India.

Morton V. 2019. Brass from the Past: Brass made, used and traded from prehistoric times to 1800. Archaeopress. 10.2307/j.ctvndv4z0.

Neogi P. 1918. Copper in Ancient India. Special Publication No. 1, IACS (http://arxiv.iacs.res.in:8080/jspui/bitstream/10821/917/1/THE%20INDIAN%20ASSOCIATION%20FOR%20THE%20CULTIVATION%20OF%20SCIENCE%20COPPER%20IN%20ANCIENT%20INDIA_P%20NEOGI_IACS_1.pdf)

Raisman G. 1998. Mercury in medicine: A history of its uses and abuses. J. Royal Soc. Med. 91, 646–649. 10.1177/014107689809101108

Ray PC. 1956. History of Chemistry in Ancient and Medieval India. Indian Chemical Society, Calcutta.

Ray PC. 1902. A History of Hindu Chemistry from the Earliest Times to the Middle of the Sixteenth Century, A.D.: With Sanskrit Texts, Variants, Translation and Illustrations. Bengal Chemical & Pharmaceutical Works, Calcutta.

Sharma S. 1979. Rasa Tarangani, in: Shastri KN (Ed.), Motilal Banarasi Das, Varanasi, 11th ed. 19/94-140, pp. 473–481.

Shastri JLN. 2010. Madanpal Nighantu. Swarnadi Varga. Chaukhambha Orientalia, 1st ed., pp. 470.

Srinivasan S. 2016. Metallurgy of zinc, high-tin bronze and gold in Indian antiquity: Methodological aspects. Indian J. Hist. Sci. 51, 22–32.

Srinivasan S, Ranganathan S. 1997. Metallurgical heritage of India, in: Golden Jubilee Souvenir, Indian Institute of Science, Bangalore. pp. 29–36,

Wainwright PR. 1910. The use of mercury in medicine. Br. Med. J. 12:2(2583), 1497–1502. doi: 10.1136/bmj.2.2583.1497. PMID: 20765547; PMCID: PMC2338581.

3 Pharmacodynamics, Pharmacokinetics and Toxic Effects of Zinc

Krishnamurthy Vinoth Kumar, B. Shyamaladevi, and Ethirajan Sukumar

3.1 INTRODUCTION

Zinc (Zn) is an essential micronutrient that the human body cannot produce or store. In 1961, the importance of Zn for humans was recognized. In the human body, 2800–3000 proteins are found to contain Zn as prosthetic groups and more than 300 enzymes rely on it for their activities. It was also found that in all the six enzyme classes *viz.* hydrolases, lyases, ligases, isomerases, oxido-reductases and transferases, zinc is an essential one for a wide range of biochemical, immunological and clinical functions. It also plays a vital role as a co-factor in enzyme action, maintenance of structures of over 300 enzymes, sensory organ adjustment, cell membrane stabilization, gene expression, cell signaling, regulation of apoptosis, metabolism of carbohydrates, proteins and lipids, physical development, reproductive health, immunological function and neuro-behavioral activity. Considering the significant and diverse roles of Zn in various biological functions, it is reasonable to presume that it is critical for human health (Praharaj et al., 2021).

Zinc deficiency was first observed in mice in 1934, and also in plants and animals. It is considered to be a significant health risk factor that can lead to death. A third of the world's population is in danger of zinc deficiency, which is especially common among children under the age of five since they require more of it for growth and development. The deficiency is linked to a variety of health issues including stunted physical growth, a weaker immune system, DNA damage, cancer, increased risk of infections, poor birth outcomes in pregnant women, etc.

3.2 PHARMACODYNAMICS OF ZINC

The study of drugs' physiological and biochemical effects in humans, animals, microorganisms or combinations of organisms, is known as pharmacodynamics. Zinc has a wide range of metabolic activities, which points to a potential physiological function. Zinc's physiological activities may be expanded by incidental effects caused by altered food intake and effects on the performance of other nutrients. Although the physiological functions of zinc are understood, their biochemistry and physiology are

DOI: 10.1201/9781003412472-3

still unknown and the precise metabolic processes that alter physiology as a result of zinc sensitivity are also unknown.

3.2.1 GROWTH

It is well known that zinc is essential for the development of many different species, from bacteria to people. Growth failure is a typical early symptom of zinc deficiency in laboratory animals. A number of mechanisms appear to be responsible for the lack of growth. However, the use of pair-fed controls revealed that zinc deficiency has effects beyond feeding behavior. Animal experiments have shown that zinc shortages result in a decrease in food intake.

There have been many documented effects of zinc deficiency on the somatotrophic axis, including a reduction in circulating Insulin-like Growth Factor-1(IGF-1) concentrations. Exogenous IGF-1 cannot cure zinc deficiency-induced growth failure. Although this is only a partial explanation, zinc in the medium is necessary for cell growth and its removal interferes with DNA synthesis and lowers thymidine kinase mRNA output. Additionally, it appears that the IGF-1 signaling pathway in cells is harmed. The healing of wounds, which is thought to be connected to other processes, also requires zinc (Levaot and Hershfinkel, 2018).

3.2.2 IMMUNE SYSTEM

The immune system is particularly impacted by zinc deficiency. A decline in cell-mediated and antibody-mediated responses, as well as lymphopenia and thymic atrophy, were noted. As with growth, it appears that several processes are in operation. In addition to having broader effects on DNA synthesis, zinc deficiency appears to promote apoptosis, which leads to the loss of Band T-cell progenitors in the bone marrow. Possible contributors include thymulin, a zinc-dependent enzyme that stimulates T-cell development in the thymus. The effects of zinc deficiency may be clinically significant because mononuclear cells produce fewer cytokines as a result. Acrodermatitis enteropathica patients are more susceptible to infections, and zinc deficiency has been linked to a number of other illnesses, such as sickle cell anemia and digestive issues (Frakeretal, 1987).

3.2.3 NERVOUS SYSTEM

The brain is particularly vulnerable to zinc deficiency during embryonic development, leading to the discovery of diseases such as abnormal neural tubes. Despite the fact that this research involved animals, it suggested that humans will have a similar association. Zinc is transported into vesicles within glutamatergic neurons by a brain-specific transporter called ZnT-3 and is found in high concentrations throughout the brain, particularly in the hippocampus. It appears to be a neurotransmission modulator and is co-secreted with the neurotransmitter. During this procedure, the synaptic cleft contained extremely high levels of zinc (4100 mM). Additionally, chemical or trauma-induced brain injury results in massive zinc release, which is thought to be the primary driver of cell death (Levaotand Hershfinkel, 2018).

3.2.4 REPRODUCTIVE SYSTEM

Zinc is abundant in seminal fluid and necessary for spermatogenesis. Prior to ejaculation, the sperm appear to acquire zinc from this source. In the initial stages of germ cell and spermatogenesis, it is involved in sperm cell development and maturation, ejaculation, liquefaction, spermatozoa proteasome binding, capacitation, and fertilization. More zinc is released into the prostate's seminal plasma following ejaculation which is essential for sperm release and motility. Malformations occur in both humans and animals as a result of deficiencies, which are also necessary for normal embryonic development. Additionally, it is crucial during the perinatal, postpartum, and neonatal periods (Harchegani et al., 2020). Additionally connected to pregnancy-related morbidity, such as premature birth, is its deficiency in mothers.

3.2.5 ANTIOXIDANT DEFENSE SYSTEM

Zinc plays a crucial role in the body's antioxidant defense system, despite not being an antioxidant itself. It can protect macromolecules by binding to protein thiol groups and preventing their oxidation. Additionally, zinc displaces redox-reactive metals like iron and copper from proteins and lipids, reducing the production of harmful hydroxyl radicals caused by these metals. Copper/Zinc superoxide dismutase, a zinc-containing antioxidant enzyme, is inhibited by low zinc levels, leading to increased oxidative stress (Prasad, 2009).

Zinc deficiency decreases the activity of ROS and RNS scavengers like glutathione, ascorbic acid, uric acid, and tocopherol. On the other hand, zinc's role in inducing metallothionein (MT), a protein that scavenges hydroxyl radicals, enhances its availability to other proteins when released under conditions of elevated oxidative stress. While zinc deficiency increases oxidative stress, excess zinc can also contribute to this issue, as MTs can release harmful amounts of zinc in certain situations (Krezel, 2007).

Interestingly, zinc deficiency has been associated with DNA deterioration, and the zinc-dependent tumor suppressor gene p53, frequently altered in cancerous tumors, may have an anti-carcinogenic effect. Disruption of cellular zinc homeostasis contributes to factors like oxidative stress, mitochondrial dysfunction, and the activation of apoptotic pathways (Prasad, 2009).

3.3 PHARMACOKINETICS OF ZINC

Pharmacokinetics explains how a specific xenobiotic or chemical affects the body after administration through the mechanisms of absorption and distribution, as well as the metabolic changes of the substance in the body and the results and routes of the drug's metabolites' excretion. The route of administration and the dose of a drug administered have an impact on the Pharmacokinetic properties of chemicals.

3.3.1 ABSORPTION

The process of influx into the enterocyte and transfer across the basolateral membrane in to the portal circulation can be referred to as absorption. Although both saturable and unsaturated pathways are thought to be involved, the sub-cellular mechanisms underlying exogenous zinc absorption are still not completely understood. It undergoes a carrier-mediated process to be absorbed in the small intestine. Food is broken down into free zinc ions, which are then bound to naturally occurring ligands and transported into the duodenum and jejunum enterocytes. In order for zinc to cross the cell membrane and enter the portal membrane and enter the portal circulation, certain transport proteins may be necessary. Large zinc also results in passive para-cellular absorption of zinc. The liver releases absorbed zinc into the bloodstream, where it is then distributed to various tissues by the portal system (Brown et al., 2004; Tubek, 2007).

The duodenum and upper small intestine absorb 20–80% of the zinc we eat, and albumin is the source of 70% of the zinc in our bloodstream. Since serum zinc levels makeup only 0.1% of the total amount of zinc in the body, any condition that affects serum albumin concentration may also affect serum zinc levels. In order to meet tissue demands, the circulating zinc is quickly replenished. Unabsorbed zinc is excreted in the urine and feces (2%), and it is also released into the gastro-intestinal system. Therefore, gastrointestinal absorption and excretion play a major role in controlling zinc homeostasis (King et al., 2000).

3.3.1.1 Factors Affecting Zinc Absorption

Zinc absorption is influenced by various factors, including the amount of zinc present in the intestinal lumen, dietary promoters or inhibitors (e.g., phytate and other minerals), chronic zinc status, and physiological conditions. Animal protein, even in small quantities, enhances both total zinc absorption and absorption efficiency. Zinc easily binds to soluble, low-molecular-weight organic substances such as sulfur-containing amino acids and hydroxy acids, facilitating its absorption. However, certain compounds like inositol hexaphosphates, pentaphosphates, and phytic acid form partially soluble complexes with zinc, hindering adequate absorption. The concentration of phytate negatively correlates with the fractional absorption of zinc. Plant-based foods, especially grains and legumes, contain varying amounts of phytate, and diets with high phytate-to-zinc molar ratios are believed to contribute to zinc deficiency on a global scale (Krebs, 2000).

3.3.2 Zinc Transporters and Binding Proteins

The ZnT transporter family, belonging to the Solute Carrier Family 30 A (SLC30A), exports zinc ions from the cytosol into cellular organelles or out of the cells, while the ZIP family moves metals from the extracellular or organellar lumen into the cytoplasm. Saccharomyces cerevisiae's Zrc1, COT1, and Cuprevious CzcD, as well as mammalian ZnT and ZIP transporters, play vital roles in maintaining zinc homeostasis and tolerance to high cation concentrations. These transporters are part of the cation diffusion facilitator (CDF) family, which shows sequence homology in the transmembrane (TM) region. The expression of zinc transporters varies in different organs, with the pancreas and gastrointestinal tract expressing specific transporters involved in zinc absorption and secretion.

Zinc levels strongly influence the expression and cellular distribution of these transporters. Metallothionein (MT), a 61-amino-acid peptide with sulfur-containing cysteine residues, is a crucial intracellular zinc-binding protein that also complexes with other divalent metals like copper, cadmium, and mercury. Mammals have four distinct isotypes of MT proteins, MT-1 and MT-2 being present in all cells. They regulate the detoxification of heavy metals and control zinc and copper levels, influencing immunological function and metal transport through the gut mucosa (Nzengue et al., 2011). The production of MT is influenced by factors like p53 protein, metal-responsive transcriptional factor 1, and intracellular glutathione levels, with a co-regulatory relationship between specific zinc transporter classes and MT-1 binders (Lichten and Cousins, 2009; Schicht et al., 2009; Thirumoorthy et al., 2011).

3.3.3 DISTRIBUTION

ZnT and ZIP transporters are essential for maintaining zinc homeostasis in cells and subcellular structures, regulating its absorption, distribution, storage, and efflux. These transporters facilitate the buffering and muffling of cytosolic zinc to counteract fluctuations in zinc concentration. Zinc absorption from food occurs in the small intestine, with a significant portion stored in skeletal muscle, bones, liver, skin, and other tissues like the kidney, pancreas, and brain. Serum zinc accounts for only 0.1% of the body's zinc, primarily bound to albumin and 2-macroglobulin. At the cellular level, zinc distribution is approximately 50% in the cytoplasm, 30%–40% in the nucleus, and 10% in the membrane. The cytosolic labile zinc ions are found in low concentrations in the picomolar to low nanomolar range, while zinc is bound to various proteins and stored in organelles and vesicles. Subcellular compartments, such as the mitochondria and endoplasmic reticulum, contain different levels of labile zinc, which requires further investigation (Jackson, 1989; King, 2000; Barnett et al., 2013).

3.3.4 METABOLISM

3.3.4.1 Carbohydrate Metabolism

The primary biochemical mechanism that ensures a steady supply of energy to living cells is carbohydrate metabolism. Glucose, which can be broken down by glycolysis to enter the Kreb's cycle and then be converted to ATP through oxidative phosphorylation, is the most important carbohydrate. Important pathways in carbohydrate metabolism include the pentose phosphate pathway (which turns hexose sugars into pentoses), glycogenesis (which insulin triggers), glycogenolysis (which glucagon triggers), and gluconeogenesis (which produces glucose from scratch). Thus, a complex regulatory mechanism involving a number of organs results in whole-body glucose homeostasis. Nutritional components are distributed in accordance with each organ's specific requirements thanks to inter-organ communication mediated by a number of circulating factors, including hormones and neuropeptides.

Along with the regulation and formation of insulin expression, zinc is also important for the proper operation of lipid and glucose metabolism. In numerous studies, zinc supplementation has been shown to lower blood pressure, hyperglycemia, and serum levels of LDL cholesterol (Ranasinghe et al., 2015). Knowing more about zinc's properties may help treat metabolic syndrome and lower the risk of stroke and angina pectoris.

3.3.4.2   Zinc in Insulin Secretion

Zinc plays a crucial role in glucose metabolism, with one of its most significant functions being in pancreatic cells, where insulin is preserved as zinc crystals. Zinc is essential for the production, storage, and secretion of insulin by pancreatic beta-cells and influences insulin's structural formation. Recent research has highlighted the negative impact of zinc deficiency on glucose tolerance and insulin sensitivity. Interestingly, providing zinc supplements to diabetic individuals has been shown to promote glucose homeostasis (Jayawardena et al., 2012). In adipocytes, zinc facilitates glucose transport, inhibits gluconeogenesis, and promotes glycolysis. Pancreatic beta-cells store and crystallize free cytosolic zinc and insulin in granules. In these granules, six insulin monomers combine to form a hexamer with two zinc ions at the center. This insulin-zinc crystal is stored and transported across the cell membrane during normal cell function. Glucose-stimulated insulin production inhibits zinc properties, leading to the dissociation of zinc-insulin complexes upon release, leaving only the active insulin monomer (Slepchenko et al., 2015).

3.3.4.3   Zinc in Glucose Transporter Protein Expression

Zinc supplementation has been found to maintain glycemic control in diabetic animal models, indicating its insulin-like properties. The enzyme 'Insulin Responsive Amino Peptidase' (IRAP) enhances zinc's impact on glucose absorption by cells, especially in adipose and muscle tissue. IRAP regulates the levels of glucose transporter 4 (GLUT4), facilitating glucose entry into the cells. In skeletal muscles, zinc-2-glycoprotein increases cellular GLUT4 protein and promotes AMP-activated protein kinase (AMPK) phosphorylation, leading to enhanced glucose uptake. Adipose tissue also shows increased GLUT4 expression due to zinc-2-glycoprotein, positively correlating with insulin sensitivity. Zinc enhances glucose uptake in adipocytes in a dose and insulin-dependent manner, leading to increased phosphorylation of the insulin receptor subunit (IR-ß) and inhibition of glycogen synthesis enzyme GSK-3. Moreover, zinc stimulates the production of lipolytic enzymes in adipocytes, promoting lipogenesis in response to insulin stimulation. Zinc's insulin-like effects, mediated by IRAP and other cellular mechanisms, demonstrate its importance in glycemic control and glucose metabolism (Balaz et al., 2014; Ranasinghe et al., 2015; Norouzie et al., 2018).

3.3.4.4   Lipid Metabolism

Lipid metabolism involves the synthesis and breakdown of various lipids to meet the body's metabolic needs. Adipose tissue serves as the primary lipid storage organ, and its dysfunction is linked to obesity. Zinc plays a crucial role in lipid metabolism, with studies showing that zinc supplementation can lower total cholesterol, LDL cholesterol, and triglycerides while increasing HDL cholesterol levels. Conversely, zinc deficiency can worsen hepatic lipid metabolism. Low zinc levels are associated with adipose tissue dysfunction in obesity, and zinc content in adipose tissue decreases significantly in obese individuals. Zinc's metabolic effects on obesity may involve interference with leptin synthesis. Overall, the relationship between zinc and lipid metabolism offers intriguing possibilities for further research (Charradi et al., 2013; Khan et al., 2013; Tinkov et al., 2016).

3.3.4.4.1   Zinc in Adipokines Secretion

Zinc plays a vital role in adipocyte activity and leptin production, affecting the development of leptin resistance. Zinc supplementation has shown potential in increasing serum leptin levels and improving metabolic parameters in obese leptin-resistant patients. Additionally, zinc impacts other adipokines like adiponectin and zinc-a2-glycoprotein (ZAG), which influence lipid metabolism and insulin resistance (Yang et al., 2013; Bing et al., 2012; Charradi et al., 2013). Zinc-containing proteins, such as ZNF638, Zfp423, Zfp467, and Zfp521, are crucial regulators of adipocyte differentiation and a dipogenesis (Kang et al., 2012). These proteins modulate the expression of key adipogenic markers, impacting both white and brown adipose tissue development. The interplay of zinc and these zinc-finger proteins offers promising avenues for exploring their potential therapeutic effects in obesity and metabolic syndrome (Meruvu et al., 2011; Quach et al., 2011; Wei et al., 2013; Yang et al., 2013).

3.3.4.5   Energy Metabolism

Zinc-mediated effects on energy metabolism may have as their targets Proliferator Activated Receptors (PPARs), which are involved in the production of mRNA genes, that are essential for energy metabolism (Goto et al., 2011; Fruchart, 2013). PPARs come in three different varieties: PPARa, PPARb/d, and PPARc. The latter is involved in lipid storage, regulates adipocyte differentiation, enhances adipogenesis, and improves insulin sensitivity. The first two are primarily involved in controlling fatty acid degradation. Though his colleagues have established the significance of zinc for the proper operation of PPARs in the zinc-finger protein (Zhou et al., 2015).

Numerous studies have demonstrated interactions between obesity and zinc homeostasis. It was found that blood zinc levels were significantly lower in obese people. Low nutritional zinc status exacerbates obesity-related metabolic abnormalities like insulin resistance, inflammation, and altered lipid profile. Additionally, it has been suggested that zinc deficiency may be a significant risk factor for Type-2 diabetes. Numerous studies have shown that people with diabetes have lower zinc concentrations than healthy people (Ekmekcioglu et al., 2001).

3.3.4.6   Protein Metabolism

The effects of zinc deficiency on protein metabolism in rats have been well studied. When assessed *in vivo* or *in vitro*, it was concluded that protein synthesis was decreased in the liver, muscle, thymus, and bone of zinc-deficient rats but not in the jejunal mucosa (Southon et al., 1985). In malnourished infants, zinc supplementation to food increases protein turnover during the recovery phase (Golden and Golden, 1992). A few studies have also been done on protein metabolism in adults and the elderly with zinc deficiency. The research by Briefel et al. (2000), confirmed that zinc supplements had a significant positive impact on the health of adults and elders.

3.4   EXCRETION

The body employs various pathways to eliminate zinc, with approximately half of it being lost through the gastrointestinal tract. While most of the zinc is reabsorbed, significant amounts are also released through biliary and intestinal secretions, making this a crucial step in zinc balance regulation. Under zinc-adequate conditions, zinc excretion is proportional to the body's zinc burden. However, certain unregulated zinc excretion pathways, such as integumentary tissues (hair loss, skin desquamation, perspiration), can also contribute to zinc loss. In humans, endogenous intestinal zinc is primarily detected in feces, and the amount excreted depends on dietary intake, nutrient absorption, and physiological demand. Humans can maintain zinc balance across a wide range of intakes, ranging from 107 to 231 moles per day (14 to 30 mg/kg). Homeostatic processes regulate zinc levels through changes in fractional absorption (usually 20–40%) and excretion through urine (0.5 mg/day) and the intestine (1–3 mg/day). Although zinc losses in urine are not significant, they do respond to dietary extremes to maintain homeostasis. For instance, adults consuming 10–15 mg of dietary zinc daily excrete 500 micrograms or less of zinc in urine, while severe zinc limitation can reduce renal excretion in 24 hours to 200 micrograms. Moreover, administration of zinc orally or through pharmaceutical preparations leads to a notable increase in renal excretion (Roohani et al., 2013; Kambe et al., 2015).

3.5   HOMEOSTASIS

The physiological concept of homeostasis describes how an organism controls its systems to maintain optimal balance despite ongoing changes brought on by new exposure and losses. When there is a shortage of micronutrients, the proper regulatory response kicks into increase acquisition effectiveness and reduce losses, whereas when there is a surplus, the opposite reaction is observed.

Depending on a food's or another material's underlying critical significance or toxic potential, the homeostatic mechanisms may have more or less evolutionary fine-tuning. Normal zinc homeostasis is necessary for a healthy immune system to maintain adequate antioxidant capacity, glucose homeostasis, and wound healing. Additionally, zinc functions as a co-factor for numerous enzymes, transcription factors, and replication factors. Zinc influx into and out of cells is one of the many processes that contribute to cellular zinc homeostasis. Pharmacokinetics was covered earlier in relation to the compartmentalization and storage within the cells (Kambe et al., 2015).

3.6   TOXIC EFFECTS OF ZINC AND ITS COMPOUNDS

Excess zinc intake can lead to acute toxicity with symptoms like nausea, vomiting, and abdominal cramps (Elinder, 1986). Zinc chloride toxicity causes a burning sensation, vomiting, and esophagitis. Chronic inhalation of zinc vapors causes "Metal Fume Fever" with symptoms like headache, cough, and fever (Anonymous, 1989). Chronic exposure to zinc can impair the immune system and cause anemia (Samman and Roberts, 1987). Teratogenic effects were not observed in animals exposed to zinc (Ketcheson et al., 1969). No significant carcinogenic activity has been reported, but increased tumor frequencies were observed in mice (Halme, 1961; Anonymous, 1991).

3.7   CONCLUSION

The studies done on various aspects of zinc confirmed its role as a drug. Detailed research on various metabolisms in animal and human studies has established the positive aspects of zinc and its selected compounds as drugs to treat several diseases. The toxicity studies further indicated their safety and dosage levels for various age groups. However, further investigations are needed to show the exact mechanisms of zinc's role in the biochemical and physiological processes.

REFERENCES

Anonymous. 1989. Toxicological Profile of Zinc. Agency for Toxic Substances and Disease Registry (ATSDR). U.S. Public Health Service, Atlanta, GA. p. 121. ATSDR/TP-89-25.

Anonymous. 1991. Carcinogenicity Assessment of Zinc and Compounds and oral RfD Assessment of Zinc phosphide, Integrated Risk Information System, U.S. Environment Protection Agency, Ohio, Cicinnati.

Balaz M, Vician M, Janakova Z, Kurdiova T, Surova M, Imrich R, et al. 2014. Subcutaneous adipose tissue zinc-alpha2-glycoprotein is associated with adipose tissue and whole-body insulin sensitivity. Obesity (Silver Spring, Md). 22, 1821–1829.

Barnett JP, Blindauer CA, Kassaar O, Khazaipoul S, Martin EM, Sadler PJ, Stewart AJ. 2013. Allosteric modulation of zinc speciation by fatty acids. Biochim. Biophys. Acta 1830, 5456 –5464.

Bing C, Mracek T, Gao D, Trayhurn P 2012 Zinc-a2-glycoprotein: an adipokine modulator of body fat mass? Int. J. Obes. 34, 1559–1565.

Briefel RR, Bialostosky K, Kennedy-Stephenson J, Mcdowell MA, Ervin RB, Wright JD. 2000. Zinc intake of the U.S. population: findings from the third national health and nutrition examination survey, 1988–1994. J. Nutr. 130, 1367S–73.

Brown KH, Rivera JA, Bhutta Z, Gibson RS, King JC, et al. 2004. International Zinc Nutrition Consultative Group (IZiNCG) technical document #1. Assessment of the risk of zinc deficiency in populations and options for its control. Food Nutr. Bull. 25, S99–203. [Pub. Med: 18046856].

Charradi K, Elkahoui S, Limam F, Aouani EJ 2013. High fat diet induced an oxidative stress in white adipose tissue and disturbed plasma transition metals in rat: prevention by grape seed and skin extract. J. Physiol. Sci. 63, 445–455.

Ekmekcioglu C, Prohaska C, Pomazal K, Steffan I, Schernthaner G, Marktl W. 2001. Concentrations of seven trace elements in different hematological matrices in patients with type 2 diabetes as compared to healthy controls. Biol Trace Elem Res. Mar 79, 205–219. doi: 10.1385/BTER:79:3:205.

Elinder CG. 1986. Zinc, in: Friberg L, Nordberg GF, Vouk VB (Eds.), Handbook on the Toxicology of Metals, 2nd ed., Vol. II, Specific Metals, Elsevier, New York.

Fraker PJ, Jardieu, P, Cook, J. 1987. Zinc deficiency and immune function. Arch. Dermatol. 123, 1699–1701.

Fruchart JC. 2013. Selective peroxisome proliferator-activated receptor a modulators (SPPARMa): the next generation of peroxisome proliferator-activated receptor a-agonists. Cardiovasc. Diabetol, 31, 82. doi:10.1186/1475-2840-12-82.

Golden BE, Golden MHN. 1992. Effect of zinc on lean tissue synthesis during recovery from malnutrition. Eur. J. Clin. Nutr. 46, 697–706.

Goto T, Lee JY, Teraminami A, Kim YI, Hirai S, Uemura T, Inoue H, Takahashi N, Kawada T. 2011. Activation of peroxisome proliferator-activated receptor-alpha stimulates both differentiation and fatty acid oxidation in adipocytes. J. Lipid Res. 52, 873–884.

Halme E. 1961. On the carcinogenic effect of drinking water containing zinc. (German). Vitalstoffe 6, 56–66.

Harchegani AB, Dahan, H, Tahmasbpour, E, Shahriary, A. 2020. Effects of zinc deficiency on impaired spermatogenesis and male infertility: The role of oxidative stress, inflammation and apoptosis. Hum. Fertil. 23, 5–16.

Jackson MJ. 1989. Physiology of zinc: general aspects, in: Mills CF (Ed.), Zinc in Human Biology. Springer, New York, pp. 1–14.

Jayawardena R, Ranasinghe P, Galappatthy P, Malkanthi R, Constantine G, Katulanda P. 2012. Effects of zinc supplementation on diabetes mellitus: a systematic review and meta-analysis. Diabetol. Metab. Syndr 4, 13.

Kambe T, Tsuji T, Hashimoto A, Itsumura N. 2015. The physiological, biochemical, and molecular roles of zinc transporters in zinc homeostasis and metabolism. Physiol. Rev 95, 749–784. 10.1152/physrev.00035.2014.

Kang S, Akerblad P, Kiviranta R, Gupta RK, Kajimura S, Griffin MJ, Min J, Baron R, Rosen ED. 2012. Regulation of early adipose commitment by Zfp521. PLoS Biol. 10, e1001433.

Ketcheson MR, Barron GP, Cox DH. 1969. Relationship of maternal dietary zinc during gestation and lactation to development and zinc, iron and copper content of the postnatal rat. J. Nutr. 98, 303–311.

Khan MI, Siddique KU, Ashfaq F, Ali W, Reddy HD, Mishra A. 2013. Effect of high-dose zinc supplementation with oral hypoglycemic agents on glycemic control and inflammation in type-2 diabetic nephropathy patients. J. Nat. Sci. Biol. Med. 4, 336–340.

King JC, Shames DM, Woodhouse LR. 2000. Zinc homeostasis in humans. J. Nutr. 130, 1360S–1366S.

Krebs NF. 2000. Overview of zinc absorption and excretion in the human gastrointestinal tract. J. Nutr. 130(5S Suppl), 1374S–1377S. 10.1093/jn/130.5.1374S.

Krezel A, Hao Q, Maret W. 2007. The zinc/thiolate redox biochemistry of metallothionein and the control of zinc ion fluctuations in cell signaling. Arch. Biochem. Biophys. 463, 188–200.

Levaot N, Hershfinkel M. 2018. How cellular Zn2+ signaling drives physiological functions. Cell Calcium 75, 53–63. doi: 10.1016/j.ceca.2018.08.004.

Lichten LA, Cousins RJ. 2009. Mammalian zinc transporters: nutritional and physiologic regulation. Ann. Rev. Nutr 29, 153–176.

Meruvu S, Hugendubler L, Mueller E. 2011. Regulation of adipocyte differentiation by the zinc finger protein ZNF638. J. Biol. Chem. 286, 26516–26523.

Norouzi S, Adulcikas J, Sohal SS, Myers S. 2018. Zinc stimulates glucose oxidation and glycemic control by modulating the insulin signaling pathway in human and mouse skeletal muscle cell lines. PLoS One. Jan 26;13, e0191727. doi: 10.1371/journal.pone.0191727.

Nzengue Y, Candéias SM, Sauvaigo S, Douki T, Favier A, Rachidi W, Guiraud P. 2011. The toxicity redox mechanisms of cadmium alone or together with copper and zinc homeostasis alteration: its redox biomarkers. J. Trace Elem. Med. Biol. 25, 171–180.

Praharaj S, Skalicky M, Maitra S, Bhadra P, Shankar T, Brestic M, Hejnak V, Vachova P, Hossain A. 2021. Zinc biofortification in food crops could alleviate the zinc malnutrition in human health. Molecules 26, 3509. 10.3390/molecules26123509.

Quach, JM, Walker, EC, Allan, E, Solano, M, Yokoyama, A, Kato, S, Martin, TJ (2011). Zinc finger protein 467 is a novel regulator of osteoblast and adipocyte commitment. J. Biol. Chem., 286(6), 4186–4198.

Ranasinghe P, Pigera S, Galappatthy P, Katulanda P, Constantine GR. 2015. Zinc and diabetes mellitus: understanding molecular mechanisms and clinical implications. Daru J. Pharm. Sci. 23, 44. doi: 10.1186/s40199-015-0127-4.

Prasad AS. 2009. Zinc: Role in immunity, Oxidative stress and Chronic inflammation. Curr. Opin. Clin. Nutr. Metab. Care 12, 646–652.

Ranasinghe P, Wathurapatha WS, Ishara MH, Jayawardana R, Galappatthy P, Katulanda P, Constantine GR. 2015. Effects of Zinc supplementation on serum lipids: a systematic review and meta-analysis. Nutr Metab (Lond). Aug 4;12, 26. doi: 10.1186/s12986-015-0023-4.

Roohani N, Hurrell R, Kelishadi, R, Schulin R. 2013. Zinc and its importance for human health: an integrative review. J. Res. Med. Sci. 18, 144–157.

Samman S, Roberts DCK. 1987. The effects of zinc supplements on plasma zinc and copper levels and the reported symptoms in healthy volunteers. Med. J. Aust. 146, 246–249.

Schicht O, Freisinger E. 2009. Spectroscopic characterization of Cicer arietinum metallothionein 1. Inorg. Chim. Acta 362, 714–724.

Slepchenko KG, Daniels NA, Guo A, Li YV. 2015. Autocrine effect of Zn2+ on the glucose stimulated insulin secretion. Endocrine. 50, 110–122.

Southon S, Livesey G, Gee JM, Johnson IT. 1985. Intestinal cellular proliferation and protein synthesis in zinc-deficient rats. Br. J. Nutr. 53, 595–603.

Thirumoorthy N, Shyam Sunder A, Manisenthil Kumar K, Senthil Kumar M, Ganesh G, Chatterjee M. 2011. A review of metallothionein isoforms and their role in pathophysiology. World J. Surg. Oncol. 9, 54.

Tinkov AA, Popova EV, Gatiatulina ER, Skalnaya AA, Yakovenko EN, Alchinova IB, Nikonorov AA. 2016. Decreased adipose tissue zinc content is associated with metabolic parameters in high fat fed Wistar rats. Acta Sci. Pol. Technol. Aliment 15, 99–105.

Tubek S. 2007. Selected zinc metabolism parameters in premenopausal and postmenopausal women with moderate and severe primary arterial hypertension. Biol. Trace Elem. Res. 116, 249–256.

Wei S, Zhang L, Zhou X, Du M, Jiang Z, Hausman GJ, Dodson MV. 2013. Emerging roles of zinc finger proteins in regulating adipogenesis. Cell. Mol. Life Sci. 70, 4569–4584.

Yang M, Liu R, Li S, Luo Y, Zhang Y, Zhang L, Liu D, Wang Y, Xiong Z, Boden G et al. 2013. Zinc-α2-glycoprotein is associated with insulin resistance in humans and is regulated by hyperglycemia, hyperinsulinemia or liraglutide administration: Cross-sectional and interventional studies in normal subjects, insulin-resistant subjects and subjects with newly diagnosed diabetes. Diabet. Care 36, 1074–1082.

Zhou T, Yan X, Wang G, Liu H, Gan X, Zhang T, Wang J, Liang Li L. 2015. Evolutionary pattern and regulation analysis to support why diversity functions existed within PPAR gene family members. BioMed Research International, Volume 2015, Article ID 613910, 11 pages, http://dx.doi.org/10.1155/2015/613910

4 Zinc-Containing Vegetables, Fruits and Other Food Products

Deep J. Kalita and Chintha Pradeepika

4.1 INTRODUCTION

Zinc is the most essential micronutrient next to iron for human activity and various physiological functions. As it cannot be stored, food is the only source to maintain its concentration in the body. According to the report by World Health Organization (WHO), a vegetable-rich diet reduces the risk of ischemic heart disease, stroke and cancers of colorectum, gastric, lung and esophageus, saving almost three million lives every year (WHO, 2002). Consumption of vegetables has been promoted by WHO as a global priority considering their ability to provide essential trace elements (FAO, 2003; FAO/WHO, 2004).

4.2 SOURCES OF ZINC NUTRITION

4.2.1 ZINC-CONTAINING VEGETABLES

As zinc is also essential for the proper growth of vegetables, its presence in various vegetable tissues (leaf, fruit, stem, bulb, fruits, etc.) depends on both the type of vegetable and its availability in the growing soil. Different vegetables can vary in their ability to metal migration, enrichment and transformation from the soil to the edible parts. Over the years, researchers investigated the level of zinc in different vegetables that are consumed daily in different parts of the world. Onianwa et al. (2001) studied the zinc content in vegetables collected from the market of Ibadan city, which forms part of the regular diet of people in Nigeria. Among them, bulb vegetables [garlic (17.30 mg/kg, fresh weight (FW) and onion (27.9 mg/kg)] were reported to have higher zinc content than leafy and fruit vegetables.

In the fruit vegetables, pepper (27.4) and garden egg (*Solanum melongena*) (9.67) had higher levels of it than others, such as okra (2.37) and tomato (1.00). Leafy greens such as cabbage (2.00), pumpkin leaves (*Telfairia occidentalis*) (3.53), water leaf (*Talinum triangulare*) (2.92), lettuce (2.30), bitter leaf (2.87), ewedu (*Corchorus olitorius*) (2.87), sokoyokoto (*Celosia argentea*) (3.53), tete (*Amaranthus thunbergii*) (2.87), efirin (*Ocimumgratissium*) (3.53) and root vegetables likecarrots(1.70) and ginger (8.27) were found to have moderate

DOI: 10.1201/9781003412472-4

zinc levels. In another study, the zinc content was investigated in vegetables commonly consumed in Amassoma, Nigeria. In this study, the mean zinc content (mgKg-1) in leafy vegetables [*Vernonia amygdalina* (1.058) and fluted pumpkin (*Telfaria occidentalis*) (0.497)], fruit vegetables [okra (*Hibiscus esculentus*) (0.830), eggplant (*Olanium melongena*), plantains (*Musa paradisiaca*) (0.283)], root vegetable (*Ipomoea batatas* (0.406±0.32)] were reported to be in the range of 0.1–3.0 mg kg-1 which is under the value of 3.00 mg kg-1as per WHO report for vegetables (Johnson, 2003).

A slightly lower level of mean zinc (mg kg-1) was reported for leafy vegetables [Indian basil (0.030), fluted pumpkin plant (0.10), gboma plant (0.12), plumed cockscomb (0.13), Jews mallow (0.064), bitter leaf (0.011), water leaf (0.07) and cabbage (0.05)] that were commonly consumed in the regular diet by people in Lagos, Nigeria. Bajaj et al. (1990) investigated the zinc content in eighteen varieties of eggplant (*Solanum melongena L*) that were cultivated in Punjab, India, and found to be in the range of 1.645–4.00 mg/100 g with a mean value of 2.5 mg/100 g. Praveen et al. (2003) studied the zinc content in different vegetables that are commonly available in the market of Karachi, Pakistan, and found to have it in low level (0.83 mg/kg) in onion while a higher level was found in eggplant (3.52), okra (4.36) and tomato (2.45). Other vegetables such as garlic, luffa guard, bitter guard, pumpkin, Indian squash, cucumber and chillies were reported to have zinc content of 5.13, 2.50, 1.98, 3.51 and 3.22 mg/kg (fresh weight basis), respectively.

It is worthwhile to note that the zinc content in garlic (*Allium sativum*) was much higher than in other vegetables that form part of the common diet in Spain. Remarkable consistency in zinc content was reported for the same type of vegetable from different regions. They included asparagus, spinach, green bean and green peas in which the zinc content was reported to be higher [500–1000 mg/100 g(fresh wt.)]. Other vegetables such as carrot, potato, radish, celeriac, cabbage, cauliflower, tomato and cucumber had comparatively lesser amounts. A study of zinc content (mg/g) in a few wild plants confirmed to have a moderate level of zinc in giant taro (*Alocasia indica*) (1.21), asparagus (*Asparagus officinalis*) (2.60), common purslane (*Portulaca oleracia*) (3.02), *spine guard (Momordica dioicia*) (1.34), golden-yellow eulophia *(Eulophiaochreata) (3.83),* poison berry (*Solanum indicum*) (0.95), assyrian plum (*Cordia myxa*) (0.35), spider plant (*Chlorophytum comosum*) (0.76)] (Aberoumand and Deokule, 2009). Many researchers around the world have analyzed a plethora of vegetables grown in different countries under varying soil, water and environmental conditions and reported zinc and other mineral/trace metal contents (Anwar et al., 2016; Woldetsadik et al., 2017; Li et al., 2018).

4.2.2 Zinc-Containing Fruits

Fruits as part of a daily diet are very important to prevent diseases (Valvi and Rathod, 2011) have investigated the zinc content in seven popular fruits that are common in the Kolhapur district of India and found that 100 g of shade-dried fruits such as *Cordia dichotoma* (Indian cherry), *Ziziphus rugosa* (Zunna berry), *Ficus racemosa* (Cluster fig), *Meyna laxiflora* (Muyna), *Flacourtia indica* (Indian plum), *Elaeagnus conferta* (wild olive) and *Glycosmis pentaphylla* (Ban-nimbu) contained be 3.85, 3.68, 5.0, 5.21, 2.13, 5.51 and 1.46 mg respectively. All these fruits also showed the presence of good amounts of other essential minerals such as iron, potassium, calcium, magnesium and sodium. Haeflein and Rasmssen (1977) studied the zinc content in dietary fruits appearing on 24-hour dietary recalls of 19 volunteers and found that its presence in these was in the range of 0.02 to 0.26 mg/100 g. These values were slightly higher for the dry fruits (0.16–0.50 µg/g) than the fresh ones (0.04–0.30 µg/g) mainly due to loss of moisture. In another comprehensive study, the same in raw peel and pulp of various citrus fruits (oranges, pomelo, lemon, mandarin, key lime and grapefruit) that are common in Turkey were determined (Czech et al., 2020). The results concluded that the content of macro- and micronutrients in the peel of most of the fruits far exceeded their quantity in the pulp. The zinc level was higher in the key lime fruit than in oranges and pomelos. The latter two had rich amounts of iron and copper. Along with zinc, key lime fruit was also found to be a good source of calcium, sodium and potassium.

Praveen et al. (2003) analyzed the zinc content in apple, musk melon, Chiku and papaya from the market in Karachi, Pakistan, and the values were reported to be 2.05, 2.73, 5.11 and 1.74 µg/g, respectively. Onianwa et al. (2001) studied various fruits that were purchased from the market in Ibadan city, Nigeria, and reported the zinc content (mg/kg, dry wt.) in raw plantain (9.00) was higher than mango (1.00), orange (2.20), pineapple (0.97), paw-paw (2.08), banana (1.50), apple (0.16), watermelon (7.40), coconut (4.50) and guava (2.00). However, its lower level was reported in watermelon (0.047), pineapple (0.050), orange (0.039), tangerine (0.082), grape (0.073), banana (0.046), pawpaw (0.045) and apple (0.045) that were collected from the three major markets in Lagos, Nigeria (Sobukola et al., 2010). San et al. (2009) studied the mineral composition of leaves and fruits of some promising jujube (*Ziziphus jujube*) genotypes. These fruits which are consumed worldwide in different forms (fresh, dried and processed) Dnianet al. (2001) indicated that 93% of the total mineral composition is composed of nitrogen, potassium and calcium besides an appreciable amount of zinc in the range of 0.53 to 1.27 mg/100 g (dry wt.) among the genotypes.

Tchepeleva et al. (1998) studied the mineral content in wild berries that are native to the Siberian region. Compared to cranberry, the zinc content was higher in both bileberry and blueberry and the values were around 30 mg/liter. The zinc contents in various sweet, sour and bitter-tasting fruits procured from the local markets in Bahawalpur, Pakistan showed its levels to be 4.81, 24.5 and 4.70 mg kg^{-1}, respectively. Sour-tasting fruits have the highest zinc content and are recommended as a food supplement for people with diabetes. In another study, Terrés et al. (2001) determined and compared the zinc content in 30 different fresh and nine different dry fruits that are part of the daily diet in South Spain. Among the fresh fruits, a higher level was reported for pomegranate (*Punica granatum*), orange (*Citrua aurantium var., dulcis*), strawberry (*Fragaria vesca*) and fig (*Ficus carica*) with average values of 5.7, 5.7, 4.1 and 3.6 (mg/g, fresh wt.) respectively than cherry (*Prunus avium*), apricot (*Armeniaca vulgaris*), apple (*Malus communis*), plum (*Prunus domestica*), melon (*Cucumis melo*), grape (*Vitis vinifera*), pear (*Pyrus communis*), cherimoya (*Annona cherimola*), olive (*Olea europea*), peach (*Prunus persica L.*), lemon

(*Citrus limnum*) and watermelon (*Citrullus vulgaris*) with average value of 1.4, 1.9,0.4,0.3, 0.6,0.3, 0.6, 0.8, 1.7, 1.8, 1.3 and 1.1 (mg/g, dry wt.) respectively.

Based on zinc content, different fruits were divided into three categories; fruits of low content (0–100 mg per 100 g), including apples (38 mg), grapes (49 mg), grapefruit (71 mg), lemon (99 mg), pineapple (94 mg); medium content (100–200 mg per 100 g) edible portion such as plum (104 mg), strawberry (130 mg), banana (164 mg), orange (117 mg) and high content (200–350 mg per 100 g), portion such as raspberry (337 mg), red currant (226 mg), black currant (283 mg). Among various fruits, apples produced in Germany are reported to have higher regional zinc content (60–75 mg) followed by those produced in the USA (20–30 mg) and Scandinavian (15–50 mg) countries (Scherz and Kirchhoff, 2006).

In a food survey, Cherfi et al. (2014) investigated the daily intake of zinc and other heavy metals (Cr, Pb, Cu) from strawberries and melons grown in Algeria. The results concluded that zinc and other metal contents in fruits [strawberries (0.163 μg) and melons (0.131 μg)] as per the estimated daily intake (EDI) were below the Target Hazard Quotient (THQ) threshold values. The findings concluded that daily intake of strawberries and melons could provide a safe level of zinc and other metal nutrients that are essential to the body.

4.2.3 Zinc-Containing Edible Roots and Tubers

Several studies reported the zinc contents of starchy roots and tubers that form staple food in developing countries. Onianwa et al. (2001) recorded its amount in some commonly available tubers of South West Nigeria such as yam (10.0 mg/kg, dry weight basis), cassava (7.40) and garri (cassava-based food) (5.13) which had higher content than potato (3.00). A similar study with the roots of 600 cassava genotypes collected from all over the world and cultivated in a research station in Colombia the zinc content in the range of 2.63–37.52 mg/kg (dry wt) (Chavez et al., 2005). Different traditional yam species grown in Sri Lanka such as *Dioscorea pentaphylla, D. alata* and *D. esculenta* revealed good amounts of zinc in them (37.58, 15.95 and 15.69 μg/g respectively) (Kulasinghe et al., 2018).

Zhu et al. (2007) stated that transgenic approaches for achieving zinc enrichment in cassava tubers were more effective than conventional approaches. In addition, a higher level of zinc biofortification was achieved in sweet potato cultivars through foliar application of glycine-chelated zinc (ZnGly) with spraying rates of 1.19 or 2.39 kg ha^{-1} (Xu et al., 2022).

4.2.4 Zinc-Containing Legumes and Seeds

Legumes and seeds are high-protein food types extensively consumed in many countries and is considered a suitable substitute for meat and seafood. They also have a much lower carbon footprint than meat products. The zinc level (mg/kg/dry wt) in seven different legumes [three types of chickpeas (*Cicer arietinum*) (33.5), two types of beans (*Phaseolus multiflorus*) (35.4) and two types of lentils (*Lens esculenta*) (45.1), soybean (37.0)] was studied and found to contain almost similar zinc levels in meats. Though the level of zinc was high in these, the presence of anti-nutritive substances like oxalates reduced their bioavailability and even caused its deficiency in vegetarian diets (Tait, 1988; WHO, 2002). The zinc content in common bean seed (42.7) was reported to be much higher than in soya beans (32.0) followed by melon seeds (28.2) and ground nut (14.0) (Onianwa et al., 2001).

Scherz and Kirchhoff (2006) investigated the leguminous and cereal grains and inferred that in pulses [red kidney bean (*Phaseolus vulgaris*) (1.9), shelled pea (*Pisum sativum*) (1.8) and green lentils (*Lens culinaris*) (2.3)], the average zinc contents (mg/100 g, dry wt.) were similar to those found in cereal grains. The zinc content in soybean whole seeds, seed coat and embryo of 21 genotypes that were grown in three different sites in Minnesota, USA, was investigated and found not to vary significantly between the growing locations. Higher zinc content was reported in both whole seed and embryo than in the seed coat (Wiersma and Moraghan, 2013).

4.2.5 ZINC-CONTAINING CEREALS

Cereals and cereal-based food products are described as good sources of zinc (Bates et al., 2016). As per the National Institutes of Health report (2022), daily intake of 75 g of cereal provides 2.8 mg of zinc to the body. The average zinc content (μg/g, fresh wt.) in various cereal grains [rye (33.4), barley (28.9), oat (33.9), buckwheat (26.3), wheat flour (14.3) and different common bread wheat (15.0)] was reported and the contents were found to be slightly higher than other cereal products (Oury et al., 2006). Similar to legumes, cereal products also possess the risk of zinc deficiency due to the presence of phytate-chelating agents that decrease the bioavailability of zinc (Tait, 1988). Prior to the green revolution in India, the zinc concentrations in wild and primitive wheat cultivars that resulted in high yields (such as *T. monococm, T. dicoccon* and *T. dicocoides*) were reported to be in the range from 4.5 to 190 mg/kg in germplasm. However, due to elimination of low-yield/high zinc content cultivars, its concentration in high-yield modern wheat (*T. aestivum*) cultivars generally ranged from 20 to 35 mg/kg (Cakmak et al., 1996 and 2010). Due to regional variation and soil quality, the zinc concentration is always lower than 10–15 mg/kg in wheat grown in zinc-deficient regions, such as India, Turkey, Australia and some parts of China which are below the required nutritional values (40–50 mg/kg) (Zhao et al., 2014).

4.2.6 ZINC-CONTAINING ANIMAL FOODS

Animal products are very good source of zinc and for children aged two–three years, meat or poultry-based products are the third major contributors while for those in the age group of four–16 years, in addition to the above, milk products and cereal-based Meats are a rich source of zinc and follow the order – beef (7 mg/85 g), crab (6 mg/85 g) and chicken (2.4 mg/85 g) (National Institutes of Health, 2022). Among the meats, the zinc content was found to be the highest in London Broil (6.77 mg) (Haeflein and Rasmussen, 1977). A study on some common foods of Southern Spain showed significantly higher amounts of zinc in high-protein foods (meat, fish, milk products, eggs, cereals and legumes) than those with a low protein content (vegetables, fresh fruits and drinks) (Terres et al., 2001). The analysis of zinc content in various types of meats revealed that it was more in the beef muscles (10.5–56.5 mg/g) than those of pork (14.9–36.0 mg/g) and chicken (8–15.4 mg/g). The pig liver had a significantly higher level (43.4–90.0 mg) than that of beef (36.3–45.5 mg). A similar trend was reported for pig kidneys with 22.0–30.5 mg in comparison to beef kidneys with 16.0–21.0 mg (Kuhne, 1976).

The zinc contents in 12 different types of meats [three types pork chop (20.9), two types rabbit (20.5), two types lamb (28.8), two type veal (35.8) and three types chicken breast (10.4)] were investigated and found that veal had the highest while the chicken breast had the least. Compared to other organs, liver showed the highest Zn content, irrespective of the type of animal (Satisbury et al., 1991; Sola et al., 1997).

4.2.7   ZINC-CONTAINING MARINE FOODS

Among the marine foods, oysters (74 mg/85 g) are one of the rich sources of zinc compared to crabs (6 mg/85 g) and other marine anthropods (National Institutes of Health, 2022). The same in fish products was found to vary depending on the type and geographic origin and the values. In general, shellfish (9.7 mg/g) had higher zinc content than fish (6.9 mg/g) (King and Keen, 1994; Tahan et al., 1995; Cakmak et al., 1999).

Onianwa et al. (2001) investigated the zinc content in the fishes that are common in South West Nigeria and found that it was in the order – stockfish (23.5 mg kg^{-1}), crayfish (15.5), dried fish (22.1) and frozen fish (18.5). The zinc content of various fishes and aquatic living organisms was investigated and it was found that their concentrations in saltwater animals depended on the contamination level of the water. In mollusks and crustaceans, the content was about four times higher than in saltwater fish (Watling and Watling, 1982).

4.2.8   OTHER ZINC-CONTAINING FOODS

Foods such as dairy, nuts, beverages, etc., are also equally important contributors of zinc. According to the National Institutes of Health (2022), a moderate amount of zinc could be obtained from food items such as yogurt (1.7 mg/226 g), milk (1.0 mg/230 mL), almonds (0.9 mg/28 g), cheese (0.9 mg/28 g) and cashews (1.6 mg/28 g).

   a. Dairy Products: In a study, the zinc content (mg/kg) in dairy products [milk (6.37), cheese (16.9), yoghurt (17.3) and milk powder (35.6)] obtained from South West Nigeria was reported to contain a higher level than that obtained from egg (6.87) (Onianwa et al., 2001). In the cow's milk which has undergone thermal treatment/pasteurization, the zinc content varied significantly with an average in ultra-high temperature (UHT) sterilized milk samples was significantly higher (4.0 µg/ml) than the pasteurized milk (3.7 µg/ml). When compared to UHT whole milk, the zinc content was higher in their skimmed versions (4.1–4.3 µg/ml) confirming that skimming facilitates the zinc concentration in the aqueous phase (Bylund, 1995). Compared to different kinds of pasteurized/fortified cows' milk, the zinc content (µg/g, fresh weight) in dairy products such as cheese (25.8), yoghurt (4.2), custard (4.0) and egg-custard (4.2) were significantly higher due to the processes involved such as clotting and dehydration.

b. Nuts: The average zinc content in nuts [almonds (*Amygdalus communis*) (29.8), peanuts (*Arachis hypogea*) (31.9), walnuts (*Juglansregia*) (36.6), seeds (*Helianthus annuus*) (64.6) and pistachio (*Pistacia vera*) (33.4 µg/g)] were reported to be much higher than those reported in for raw fruits (Tait, 1988). In another study, the same in Brazil nuts (*Bertholletia excela*), cashews (*Anacardium occidental*), hazelnuts (*Corylus avellana*) and walnuts (*Juglans regia*) were found to be 1.8, 2.4, 3.0, 1.5 and 1.8 µg/g (Scherz and Kirchhoff, 2006).

c. Weaning Meals: The nutrient content in weaning meals is very important as they complement human milk or infant formula. Among them, the zinc contents (mg/100 g) in four vegetable puree meals [carrot and pea (38.0), cauliflower and potatoes (30.0), green beans (0.27) and mixed garden vegetables (22.0)] were much higher than the fruit containing meals [mixed fruit puree (0.15), puree made of apple, banana and orange (0.16), puree made of apricot and apple (0.13)] (Bosscher et al. (2002).

d. Additives, Sweeteners, Oils, Fat and Spices: In a study, the zinc content in these was found to be [salt (2.13), sugar 0.06, jam (12.2) and honey (36.9); palm oil (7.53), groundnut oil (0.67), vegetable oil (0.60); margarine (1.53), sweets (0.27), biscuits (6.87) and bread (2.9) along with spices and seasonings [locust bean (56.9), curry (53.7), thyme (0.20)] and concluded that depending on the type of food the level of zinc varied significantly (Onianwa et al., 2001).

e. Drinks and Beverages: Zinc is also available in various alcoholic drinks/beverages; however, its content in distilled alcoholic beverages (whisky, gin, rum, brandy, alcoholic liquor) was reported to be significantly lower than in fermented ones. Its content in wine and beer was higher (1.3 and 0.5 µg/ml respectively).

Among the highly consumed non-alcoholic drinks, the zinc content in tea (8.4 µg/g, fresh weight) was reported to be significantly higher than that of coffee (1.2 µg/g, fresh wt.) (Diaz et al., 1997).

4.3   CONCLUSION

Based on the above information, it can be concluded that the zinc concentration and its variations mostly depend on specific foods and food groups and less on their origin. Among the food types, high protein diets (plant or animal) usually contain a higher level of zinc than those with a high level of fiber, fat, carbohydrate, etc. The level of heavy metal contamination in the environment (soil/water) has a significant influence on zinc concentration in plant and animal foods. The raw fruits have a much lower level of zinc than the dried form and based on the geographical location the content varies significantly for similar types of fruits. The level of zinc in legumes is much higher than those present in cereals. Compared to the original varieties, the content is much lower in the hybrid cereal varieties that are designed to have high crop yields. Compared to other main dietary foods, limited zinc can also be obtained from different drinks and snacks that are consumed in day-to-day life. Finally, it is very important to evaluate the level of micronutrients in the foods that are part of daily diet to provide recommended levels to the body and avoid adverse effects that arise due to their absence or excess.

REFERENCES

Aberoumand A, Deokule SS. 2009. Determination of elements profile of some wild edible plants. Food Anal. Methods. 2, 116–119. 10.1007/s12161-008-9038-z

Anwar S, Nawaz MF, Gul S, Rizwan M, Ali S, et al. 2016. Uptake and distribution of minerals and heavy metals in commonly grown leafy vegetable species irrigated with sewage water. Environ. Monit. Assess. 188, 541. 10.1007/s10661-016-5560-4

Bajaj KL, Kansal BD, Chadha ML, Kaur PP. 1990. Chemical composition of some important varieties of egg plant (Solanum melongena L). Tropical Sci. 30, 255–261.

Bates B, Cox L, Page SNP, Prentice A, Steer T, Swan G. 2016. National Diet and Nutrition Survey Results from Years 5 and 6 (combined) of the Rolling Programme (2012/2013-2013/2014). PublicHealth England and the Food Standards Agency.

Bosscher D, Cauwenbergh RV, Auwera JCV, Robberecht H, Deelstra H. 2002. Calcium, iron and zinc availability from weaning meals. Acta Paediatr. 91, 761–768. 10.1111/j.1651-2227.2002.tb03324.x

Bylund G. 1995.Building-blocks of dairy processing. In: Dairy Processing Handbook, Teknotex, AB (Ed.), . Tetra Pak Processing Systems AB S-221, Lund, Sweden.

Cakmak I, Ozkan H, Braun HJ, Welch RM, Romheld V. 2010. Zinc and iron concentrations in seeds of wild, primitive and modern wheats. Food Nutr. Bull. 21, 401–403. journals.sagepub.com/doi/pdf/10.1177/156482650002100411

Cakmak I, Kalayci M, Ekiz H, Braun HJ, Kilinc Y, Yilmaz A. 1999. Zinc deficiency as a practical problem in plant and human nutrition in Turkey: a NATO-Science for stability project. Field Crops Res. 60, 175–188. 10.1016/S0378-4290(98)00139-7.

Cakmak I, Yilmaz A, Kalayci M, Ekiz H, Torun B, Erenoglu B, Braun HJ. 1996. Zinc deficiency as a critical problem in wheat production in Central Anatolia. Plant Soil. 180, 165–172. 10.1007/BF00015299

Chavez AL, Sanchez T, Jaramillo G, Bedoya JM, Echeverry J, Bolanos EA, Ceballos H, Iglesias CA. 2005. Variation of quality traits in cassava roots evaluated in landraces and improved clones. Euphytica 14, 125–133. 10.1007/s10681-005-3057-2.

Cherfi A, Abdoun S, Gaci O. 2014. Food survey: levels and potential health risks of chromium, lead, zinc and copper content in fruits and vegetables consumed in Algeria. Food Chem.Toxicol. 70, 48–53. 10.1016/j.fct.2014.04.044.

Czech A, Zarycka E, Yanovych D, Zasadna Z, Grzegorczyk I, Kłys S. 2020. Mineral content of the pulp and peel of various citrus fruit cultivars. Biol. Trace Elem. Res. 193, 555–563. 10.1007/s12011-019-01727-1.

Diaz JP, Navarro M, Lopez MC, Lopez H. 1997. Determination of selenium levels in dairy products and drinks by hydride generation atomic absorption spectrometry: correlation with daily dietary intake. Food Addit. Contam. 14, 109–119. 10.1080/026520397093 74505.

FAO, 2003. Increasing fruit and vegetable consumption becomes a global priority. fao.org/english/newsroom/focus/2003/fruitveg1.html (26 June 2015).

FAO/WHO, 2004. Fruit and vegetables for health—report of a Joint FAO/WHO Workshop, 1–3 September 2004, Kobe, Japan, World Health Organization and Food and Agriculture Organization of the United Nations. fao.org/ag/magazine/faowho-fv.pdf (26 June 2015).

Haeflein, KA, Rasmussen, AI. 1977. Zinc content of selected foods. J. Am. Diet. Assoc. 70, 610–616. PMID: 864153.

Johnson MA, 2003. COPPER Physiology in: Caballero B (Eds.), Encyclopaedia of Food Technology and Nutrition (Second Edition), Academic Press, London, pp. 1640–1647, ISBN 9780122270550, https://doi.org/10.1016/B0-12-227055-X/00298-4

King JC, Keen CL. 1994. Zinc. in: Shils ME, Dr. Olson JA, Dr. Shike M, Dr. (Eds.), Modern Nutrition in Health and Disease, 8th edn. Lea and Febiger, London.

Kuhne D. 1976. ZumMineralstoffgehaltverschiedenerMuskeln von Schwein und Rind. Fleischwirtschaft. 5, 570–573.

Kulasinghe WMAA, Wimalasiri KMS, Samarasinghe G, Silva R, Madhujith T. 2018. Macronutrient composition of selected traditional yams grown in Sri Lanka. Trop. Agric. Res. 29, 113–122. 10.4038/tar.v29i2.8282

Li X, Li Z, Lin CJ, Bi X, Liu J, Feng X, Zhang H, Chen J, Wu T. 2018. Health risks of heavy metal exposure through vegetable consumption near a large-scale Pb/Zn smelter in central China. Ecotoxicol. Environ. Saf. 161, 99–110. 10.1016/j.ecoenv.2018.05.080

National Institutes of Health. Zinc-Fact Sheet for Health Professionals. 2022. Available online: ods.od.nih.gov/factsheets/Zinc-HealthProfessional/ (accessed on26 December 2023).

Onianwa PC, Adeyemo AO, Idowu OE, Ogabiela EE. 2001. Copper and Zinc contents of Nigerian foods and estimates of the adult dietary intakes. Food Chem. 72, 89–95. 10.1016/S0308-8146(00)214-4.

Oury FX, Leenhardt F, Rémésy C, Chanliaud E, Duperrier B, Balfourier F, Charmet G. 2006. Genetic variability and stability of grain magnesium, zinc and iron concentrations in bread wheat. Eur. J. Agronom. 25, 177–185.

Parveen Z, Khuhro MI, Rafiq N. 2003. Market basket survey for lead, cadmium, copper, chromium, nickel, and zinc in fruits and vegetables. Bull Environ Contam Toxicol. 71, 1260–1264.

Salisbury CD, Chan W, Saschenbrecker PW. 1991. Multi-element concentrations in liver and kidney tissues from five species of Canadian slaughter animals. J. Assoc. Anal. Chem. 74, 587–591. 10.1093/jaoac/74.4.587

San B, Yildirim AN, Pola TM, Yildirim F. 2009. Mineral composition of leaves and fruits of some promising Jujube (Zizyphus jujuba Miller) genotypes. Asian. J. Chem. 21, 2898–2902. www.researchgate.net/publication/331564205_mineral_composition_of_ripe_and_unripe_selected_edible_fruits.

Scherz H, Kirchhoff E. 2006. Trace elements in foods: Zinc contents of raw foods—A comparison of data originating from different geographical regions of the world. J. Food Compos. Anal. 19, 420–433. 10.1016/j.jfca.2005.10.004.

Sobukola OP, Adeniran OM, Odedairo AA, Kajihausa OE. 2010. Heavy metal levels in some fruits and leafy vegetables from selected markets in Lagos, Nigeria. Afr. J. Food Sci. 4, 383–393.

Sola S, Barrio T, Martin A. 1997. Essential elements (Mn, Fe, Cu, Zn) in pork and duck liver paste produced in Spain. Food Addit. Contam. 14, 135–141. 10.1080/02652039709374508.

Tahan JE, Sanchez JM, Granadillo VA, Cubillan HS, Romero RA. 1995. Concentration of total Al, Cr, Cu, Fe, Hg, Na, Pb, and Zn in commercial canned seafood determined by atomic spectrometric means after mineralization by microwave heating. J. Agri. Food Chem. 43, 910–915. 10.1021/jf00052a012

Tait SJF. 1988. Zinc in human nutrition. Nutr. Res. Rev. 1, 23–37. 10.1079/NRR19880005

Tchepeleva GG, Gardienko GP, Polovinkina NI, Efremov AA. 1998. Vitamins and mineral substances in fruit and nuts of wild Siberian plants. Int. J. Appl. Biol. 2(1), 392–396.

Terrés C, Navarro M, Martín-Lagos F, Giménez R, López H, López MC. 2001. Zinc levels in foods from southeastern Spain: relationship to daily dietary intake. Food Addit Contamin. 18, 687–695. doi: 10.1080/02652030121584.

Valvi SR, Rathod VS. 2011. Mineral composition of some wild edible fruits from Kolhapur district. Int. J. Appl. Biol. Pharm. Technol. 2(1), 392–396.

Watling HR, Watling RJ. 1982. Metal concentration in oysters from South African Coast. Bull. Environ. Contam. Toxicol. 28, 460–466.

WHO, Office of World Health Reporting. 2002. The World health report: 2002: reducing risks, promoting healthy life: overview. World HealthOrganization. https://apps.who.int/iris/handle/10665/67454

Wiersma JV, Moraghan JT. 2013. Within seed distribution of selected mineral elements among soybean genotypes that vary in iron efficiency. Crop Sci. 53(5), 2051–2062. 10.2135/cropsci2012.10.0599

Woldetsadik D, Drechsel P, Keraita B, Itanna F, Gebrekidan H. 2017. Heavy metal accumulation and health risk assessment in wastewater-irrigated urban vegetable farming sites of Addis Ababa, Ethiopia. Int. J. Food Contam. 4, 9. 10.1186/s40550-017-0053-y

Xu M, Liu M, Ma Q, Wu L. 2022. Glycine-chelated zinc lowered foliar phytotoxicity than excess zinc sulfate and improved zinc use efficiency in two sweet potato cultivars. Sci Hortic. 295, 110880. 10.1016/j.scienta.2022.110880

Zhao AQ, Tian XH, Cao YX, Lu XC, Liu T. 2014. Comparison of soil and foliar zinc application for enhancing grain zinc content of wheat when grown on potentially zinc-deficient calcareous soils. J. Sci. Food Agri. 94 (10), 2016–2022. 10.1002/jsfa.6518

Zhu C, Naqvi S, Gomez-Galera S, Pelacho AM, Capell T, Christou P. 2007. Transgenic strategies for the nutritional enhancement of plants. Trend Plant Sci. 12, 548–555. 10.1016/j.tplants.2007.09.007

5 Zinc in Marine Ecology and Environment

K. Altaff and R. Vijayaraj

5.1 INTRODUCTION

Zinc is an essential trace element for all living organisms and it is the second most abundant trace element, after Fe, in most vertebrates. Zn ensures the integrity of biological components like DNA and biological structures like membranes and ribosomes as a component of more than 200 metalloenzymes and other metabolic chemicals (Bellotti et al., 2021; Luchinat et al., 2022). Subsequently, the majority of Zn supplied to aquatic environments partitions into the sediments. Zn is used in numerous enzyme systems involved with a variety of metabolic processes (Luchinat et al., 2022). In the ocean, total dissolved Zn concentration has a nutrient-like vertical profile with a particularly strong correlation with silicate (Wong et al., 2021). High dissolved oxygen, low salinity, and low pH promote the release of zinc from sediments. Typically, dissolved zinc is composed of different chemical and inorganic compounds as well as the hazardous aquo ion $[Zn (H_2O)_6]_2^+$ (Gogoi et al., 2020; Kicińska et al., 2022). Aquo ions and other hazardous species affect aquatic organisms most severely in environments with low pH, low alkalinity, low dissolved oxygen levels, and high temperatures (Gogoi et al., 2020; Kiciska et al., 2022). Zn regulates RNA and DNA through Zn-dependent enzymes, which are the main metabolic targets of Zn (Gaffney-Stomberg, 2019). Metallothioneins, low-molecular-weight proteins, are crucial for maintaining mammalian Zn homeostasis and preventing Zn toxicity (Na-Phatthalung et al., 2021; Xiao et al., 2022); Zn is a potent inducer of metallothioneins (Na-Phatthalung et al., 2021; Xiao et al., 2022). The pancreas and bone seem to be primary targets of Zn intoxication in birds and mammals; gill epithelium is the primary target site in fish.

5.2 CHEMICAL PROPERTIES OF Zn IN MARINE ENVIRONMENTS

According to Azizi et al. (2018), Zn mobility in marine ecosystems depends on the composition of suspended and bed sediments, concentrations of dissolved and particulate iron and manganese, pH, salinity, concentrations of complexing ligands, and Zn concentration. According to Kalavathy and Bhaskar (2019) and Zhou et al. (2021), zinc is most soluble in marine water at low pH and low alkalinity (10 mg Zn/L of solution at pH 6 that decreases to 6.5 mg Zn/L at pH 7 and 0.65 mg Zn/L at pH 8). In brackishwater, dissolved Zn concentrations rarely exceed 40 µg/L; higher levels are typically linked to anthropogenic activities and deposits of Zn-enriched ore (Geddie

DOI: 10.1201/9781003412472-5

and Hall, 2019a). Marine waters typically have more than 10 µg/L of the most strongly adhering suspended particles, whereas saturated saltwater can have between 1.2 and 2.5 mg/L of zinc. The conditions where Zn is most bioavailable and also detrimental to aquatic species are low pH, low alkalinity, low dissolved oxygen, and high temperatures (Roshan et al., 2018). Chemically soluble forms of zinc are the most toxic and bioavailable, and the aquo ion dominates other dissolved species and is thought to be the most toxic (Boguta and Sokoowska, 2020). However, aquo ion concentrations fall under conditions of high alkalinity, at pH > 7.5, and increasing salinity. The most prevalent species are $ZnHCO_3^+$, Zn_2^+, and $ZnCO_3$ with high alkalinity and pH 6.5; at low alkalinity and an enhanced pH 8.0, the most prevalent species are $Zn_2^{+,}$ $ZnCO_3$, Zn humic acid, $ZnOH^+$, and $ZnHCO_3^+$ in that order. Saltwater contains Zn, which is dissolved, precipitated as a solid, and adsorbed on particle surfaces. According to Koschinsky and Hein (2003) and Salomons and Förstner (2012), soluble Zn can be found in seawater as free (hydrated), inorganic (the predominant form in the open sea), or organic compounds. The four primary species of soluble Zn in saltwater at pH 8.1 are Zn hydroxide (62%), free ion (17%), monochloride ion (6.4%), and Zn carbonate (5.8%). At pH 7, 50% of the dissolved zinc is present as a free ion, whereas the majority of the dissolved zinc is present as organo zinc complexes in the presence of dissolved organic molecules (Zirino and Yamamoto, 1972; Fru, 2020). In estuaries and other marine environments, the relative abundance of Zn species varies with salinity. Low salinities have higher concentrations of $ZnSO_4$ and $ZnCl^+$ (Salomons and Förstner, 2012). The aquo ion dominates with greater salinities. However, as salinity falls, the concentration of the free Zn ion rises and the concentration of Zn chloro complexes falls, leading to an increase in the bioavailability of the free metal ion and an increase in the bioconcentration by marine organisms. Before eventually partitioning into the sediments, the majority of Zn that is released into aquatic ecosystems absorbed into hydrous iron and manganese oxides, clay minerals, and organic components (Mossa et al., 2020; Moiseenko and Gashkina, 2020). Likewise, Insoluble organic complexes, insoluble sulphides, precipitated Zn hydroxide, ferric and manganic oxyhydroxide precipitates, and various forms of Zn are all found in sediments. Soluble Zn is mobilised and released as sediments transition from a reduced to an oxidised state, although the bioavailability of various sediment Zn forms varies greatly, and the transfer processes are not well understood. The pH >7 resulted in significant sorption to sediments; however, at pH ≤7, sorption was minimal (Mossa et al., 2020; Moiseenko and Gashkina, 2020). Due to the displacement of adsorbed Zn ions by alkali and alkaline earth cations, which are plentiful in brackish waters, Zn is dissolved from sediments at low salinities. When Zn is present in high concentrations, precipitation of the hydroxide, carbonate, or sulphate may take place. In reducing environments, sulphide precipitation in sediments significantly reduces Zn mobility.

5.3   MARINE ECOLOGY OF Zn

The distribution of Zn in marine waters, animals, and sediments has not been investigated in coastal waters and estuaries, despite reports of Zn concentrations in some offshore places in the Indian Ocean (Goto et al., 2021; Yap and Al-Mutairi,

2021). This assumes significance in these regions due to the potential for greater elemental exchange between sediments and waters as well as the fact that Zn is crucial for phytoplankton growth (Wang et al., 2019; Müsing et al., 2022). This biochemical functionality would offer a beautiful explanation of the relationships with the major nutrients in seawater (Nowinski and Moran, 2021; Müsing et al., 2022). The typical Zn concentration in Indian saltwater ranges from 0.6 to 5 ppb (Jha et al., 2019). Zn occurs 20–700 ppm in marine algae (Geddie and Hall, 2019a), However, estuaries often have Zn concentrations of 5 to 10 ppb, 3–25 ppm in marine fish and shells, 100–900 ppm in oysters and 7–50 ppm in lobsters (Geddie and Hall, 2019b). In saltwater, Zn concentrations range from 0.6 to 5 ppb on average. The values for the Andaman Sea were 0.4–0.9 nM and 9.6–11.4, compared to 0.5 nM and 10.0 for the northern Indian Ocean. Various sources of Zn complexing ligands might be derived in the Andaman Sea (Kim et al., 2015a; Kim et al., 2015b; Kim et al. 2015c). Furthermore, reported levels of Zn in seawater fluctuate considerably, most probably as a result of contamination as well as variation in the levels that are truly there. Because of saltwater contamination, many reported quantities of Zn in seawater are abnormally high (Stewart et al., 2021). Typically, Zn concentrations in estuaries and coastal waters are substantially greater than those in the ocean, with levels occasionally reaching 25 µg/L and regularly reaching 4 µg/L. Estuaries receiving discharge from metal mining or smelting operations may have Zn concentrations of greater than 1,000 µg/L (Barletta et al., 2019; Birch et al., 2020). The concentration of free Zn ions in surface waters of the western Pacific ranges from 0.0001 to 0.0009 g/L due to the high complex formation of more than 98% of the dissolved Zn with dissolved complex compounds. In the North Atlantic Ocean's surface waters, between 96 and 99% of the Zn has been complexed with an organic ligand. In surface waters, the Zn-binding ligand can be found in concentrations ranging from 0.4 to 2.5 nM. With depth, the concentration hardly changes (Lohan, 2003). The majority of the dissolved Zn in saltwater is complex due to the strong affinity of the organic ligand for Zn, leaving very little as free ionic Zn (Zn^+_2) in the solution (Krężel and Maret, 2016). According to Ellwood and van den Berg (2000), the amount of free Zn^+_2 in the surface waters of the open North Atlantic Ocean ranges from 0.0004 to 0.001 g/L; in surface coastal and deep offshore seas, the amount rises to roughly 0.01 g/L. Approximately 26% of the total Zn in the Irish Sea's surface waters and the surface waters in the southern North Sea contain 54% of their total Zn as adsorbed to suspended particles. Surface waters of the North Atlantic Ocean having 0.032 to 0.447 mg/L suspended particle matter have a Zn content of 0.004 to 0.003 g/L.

5.4 BIOACCUMULATION OF Zn IN MARINE ORGANISMS

Zn is an essential element; many marine species seem to be able to control tissue Zn levels in seawater (Pajarillo et al., 2021). In contrast to organically complexed Zn, dissolved, free ionic Zn from seawater can be bioaccumulated by the polychaete worm *Neanthes arenaceodentata* (Mason et al., 1988). The digestive gland of the deposit-feeding clam *Scrobicularia plana* has the highest Zn contents, proving that ingested sediments are the main source of Zn. (Cheggour et al., 2005. In the Bou

Regreg Estuary in Morocco, *Scrobicularia plana* and *Neries diversicolor* easily collect Zn from sediments contaminated with sewage treatment plant effluents. In the haemolymph of shore crabs (*Carcinus maenas*), Zn is retained and has a significant relationship with organic components in the blood (Martin and Rainbow, 1998). The tissues of barnacles (*Balanus amphitrite*) have a dry weight Zn content of up to 16,000 µg/g dry weight. The Zn is absorbed by the barnacles from their meal extremely effectively, and they release it into the environment very gradually (Wang et al., 2000). The midgut epithelium stores the majority of the Zn as Zn phosphate granules. The liver of squirrelfish contains significant levels of Zn. Female squirrelfish liver can accumulate more than 5,000 µg/g dry weight. Complexes with the Zn that is stored in the liver are formed by the two Zn binding proteins, metallothionein in the nuclei of hepatocytes and female-specific Zn binding protein in the cytoplasm of liver cells (Hogstrand and Wood, 1996). Additionally, squirrelfish accumulate large amounts of Zn in their ovaries. Zn concentrations in the liver and ovaries change in response to oestrogen administration, suggesting that Zn is stored in the liver by females for use in egg development and ovulation. In sediments with Zn concentrations as high as 10,000 µg/g, *Polynices sordidus* retain an internal Zn content that is largely stable. Crab *Carcinus maenas* and barnacle *Elminius modestus* both retain rather constant body Zn residues (83.2 19.4 µg /g of dry weight) in the presence of dissolved Zn concentrations up to about 400 g/L (Martin and Rainbow, 1998). Zn absorption rates for several marine animal species fed various natural meals vary from 16 to 93%. The green mussel *Perna viridis* assimilates 21 to 36 percent of the Zn it consumes, but the clam *Ruditapes philippinarum* assimilates 29 to 59% of the Zn from the same five kinds of algae (Chong and Wang, 2000). Shrimp *Lysmata seticaudata* and crab *Carcinus maenas* receive the same amount of Zn from water alone as they do from water plus food after three months of exposure to (65) Zn in water or water and food. Gobia sp., a fish, acquires 2.5 times more Zn from the water pathway alone than from the food pathway mixed with water (Boothe and Knauer, 1972). *Pugettia producta*, a type of crab, can absorb more than 65% of the zinc present in its macroalgal diet. The skin and gills of the bottom-feeding fish *Platichthys flesus* contain the highest concentrations of Zn, indicating uptake from saltwater via surface adsorption, presumably (Singh et al., 1992). *Pleuronectes platessa* juveniles more effectively collect Zn from food sources like *Artemia* nauplii or polycheates or *Nereis diversicolor* than from water. Only 10% of the Zn found in the tissues comes from water (Pentreath, 1973; Milner, 1982). A Zn-binding metallothionein is easily generated in intestinal cells of the winter flounder *Pleuronectes americanus*, a flatfish that is closely related to it.

5.5   TOXICITY OF Zn IN MARINE ORGANISMS

Human activities increasingly cause negative impacts on the marine environment, frequently having a negative impact on the productivity and health of marine ecosystems (Smale et al., 2019). Plastic waste has already been discovered in all investigated marine areas, and plastic waste is officially acknowledged as a serious anthropogenic pollution (Davison et al., 2021). Common ocean plastics can operate as a source and a vector for a variety of organic contaminants as well as a variety of metals, including zinc (Zn) and copper (Cu). Zn leaching from both new and old plastic and car tyre rubber has been reported to occur at considerable levels in numerous studies examining the impact of plastic pollution on marine species (Capolupo et al., 2020; Capolupo et al., 2021). Zn is a metal that is commonly utilised in the creation of various plastic and rubber products (Mostoni et al., 2019). Other man-made pollutants have also been noted as potential contributors to high Zn in the marine environment, such as antifouling chemicals and atmospheric aerosols. These pollutants are in addition to plastic pollutants in the environment (Sarker et al., 2021). Given the increasing likelihood that marine environments will experience elevated Zn levels, there is now a need to consider the effect this may have on key marine organisms (Sarker et al., 2021). Zn homeostasis is critical to many organisms, including microbes, with cells required to maintain minute concentrations for enzymatic function and cell growth whilst avoiding excess accumulation due to the risk of toxicity at higher levels (Sarker et al., 2021). In aquatic microorganisms, Zn toxicity has been shown to affect different cellular processes, but such investigations have primarily focused on freshwater organisms. Relatively few marine microorganisms have been characterised regarding their Zn toxicity response, as Zn levels in marine waters are typically low. In the surface waters of the North Atlantic and North Pacific Oceans, the total dissolved Zn concentration is expected to be around 0.3 nM (Mishra et al., 2019; Sarker et al., 2021). Very low amounts (1–20 pM) of free Zn ($Zn2+$) are present in the surface waters under study since the vast majority of this Zn (98%) is bound to unidentified organic ligands (Sarker et al., 2021). Although most marine life would never come into contact with Zn at dangerous levels due to these normally low levels, the rising load of anthropogenic contaminants in marine habitats suggests that it is now appropriate to conduct research on the toxicity of Zn in these organisms (Sarker et al., 2021). Free Zn concentrations have been recorded at levels as low as 3.27 µg l^{-1} in India and as high as 154.71 µg l^{-1} with an average of 32.15 µg l^{-1} in Port Jackson estuary, Australia, 6.50–28 µg l^{-1} in Mersey estuary, UK, and so on (Sarker et al., 2021). The accumulation of plastic debris within open ocean gyres has the potential to cause some Zn transport, even though high concentrations of Zn are not anticipated to be present in these environments on a large scale (Sarker et al., 2021). This is because studies show that plastic pollution may reach these areas relatively quickly and within the timeframe that additive Zn may still be leaching from plastic debris. Zn is only moderately hazardous to some marine creatures because they can control tissue residues of Zn over vast ranges of Zn concentrations in the ambient water, sediments, and diet. The only form of zinc that is harmful is the free ion, which makes up a very small portion of the overall amount of zinc in untreated saltwater (Sarker et al., 2021). Total Zn concentrations in solutions that are acutely fatal typically vary from 100 to 50,000 µg/L (Sarker et al., 2021). The most resilient organisms include fish, which are also more sensitive than phytoplankton and some larval crustaceans and molluscs (Sarker et al., 2021). Zn is an essential element for microalgae, as shown by the inhibition of growth of coastal strains of the microalgae *Thalassiosira pseudonana* and *T. weissflogii* at concentrations of dissolved ionic Zn lower than 0.00065 g/L (Sunda and Huntsman, 1995). Between 0.00065 µg/L and 0.65 µg/L, growth is at its maximum; at greater concentrations, it is suppressed. However, *Emiliania huxleyi's* growth is unaffected by ionic Zn concentrations up to at least 6.5 µg/L (Bryan and Langston, 1992). Except in significantly polluted estuaries like the Fal Estuary in England, the majority of the Zn in saltwater is complexed or adsorbed, keeping free ionic Zn concentrations below this level. Zn ion concentrations may be too low to allow for phytoplankton development in open ocean surface waters, but not in coastal waters. *Nitzschia closterium* diatom growth is inhibited at 20 µg/L total dissolved Zn concentrations. Except in severely polluted estuaries like the Fal Estuary in England, the majority of the Zn in saltwater is complexed or adsorbed, keeping free ionic Zn concentrations below this level. Zn ion concentrations may be too low to allow for phytoplankton development in open ocean surface waters, but not in coastal waters. Expansion of diatoms At 20 g/L of total dissolved Zn, Nitzschia closterium is inhibited. At a concentration of total dissolved Zn of 15 g/L, the carbon fixation of mixed wild populations of marine phytoplankton is hindered (Stauber and Florence, 1990). Phytoplankton is slightly more responsive to Zn than marine macroalgae (Altaff and Vijayaraj, 2021a). Six Brazilian seaweed species experience growth inhibition at a dose of 20 g/L (Filho et al., 1997). *Ulva latuca, Enteromorpha flexuosa* and *Hypnea musciformis* all perished at a Zn concentration of 5,000 µg/L, 1,000 µg/L and 100 µg/L, respectively (Ojaveer et al., 1980; Verriopoulos and Hardouvelis, 1988; Hunt and Anderson, 1989). When it comes to some phytoplankton and invertebrate eggs, especially, the Zn concentrations in contaminated estuarine and coastal waters can occasionally be higher than those that have been found to have harmful effects on sensitive species (Langston, 2018). So it's probable that some estuarine and marine habitats close to the coast are being harmed by Zn pollution. Zn is less accessible and hazardous than free Zn ion in coastal and estuary waters because it is complexed with inorganic and organic ligands or adsorbed to suspended particles. The frequency with which Zn poses a risk to marine habitats is therefore undetermined.

5.6   Zn SOURCED FROM FISH FEED PRODUCTS

The aquaculture industry is shifting further toward other sources of finfish feed due to its existing unsustainable reliance on fishmeal and fish oil products (Altaff and Vijayaraj et al., 2021a). To maintain a balanced diet, certain trace metals must be present in feed items for fish because they are necessary for their growth and development (Altaff and Vijayaraj et al., 2021a; Altaff and Vijayaraj et al., 2021b). Zn, a crucial micronutrient for the prevention of cataract formation and other health issues, is present in the feed (Lall and Kaushik, 2021). Zn levels in salmon diet, waste products, and related sediments were measured by Dean et al. in 2007. Based on the concentration of Zn in salmon (196.4 mg/kg in feed to 364.52 mg/kg in faeces) and feed loss to the environment, it is possible to directly attribute 87% of the Zn found in sediments to the feed products. The amount of feed input also varies according to the salmon's growth and harvesting cycles. The fact that feed products are the primary source of Zn pollution in farms is reiterated by a correlation tendency between sediment Zn contents and these cycles. The Marlborough Sound sites' salmon feed products have a Zn content of about 100 mg/kg. As a point of comparison, the Zn content in diets for Atlantic salmon ranges from 6 to 240 mg/kg of feed. Recent decreases in zinc supplements translate to current usage levels of about 160 mg/kg (100 mg/kg of supplemented zinc plus roughly 60 mg/kg of naturally occurring zinc in fishmeal). The majority of Zn supplements are currently given to feed products as Zn sulphate, an inorganic compound. Fish can more effectively absorb zinc when it is in an organic form that is coupled to a protein or amino acid, according to research by Mahboub et al. (2020). By using this organic form, the overall Zn level in feed can be reduced to 90 mg/kg. Other feed groups are beginning to switch to a zinc supplement known as zinc methionine, which is a more accessible form of zinc. By lowering the amount of Zn in finfish feed to the lowest levels necessary for fish health and ensuring more efficient uptake by the fish, the amount of Zn discharged into the marine environment is anticipated to decrease by around 50%.

5.7 Zn AS AN ESSENTIAL MICROELEMENT FOR ALGAL GROWTH

Zn is a crucial mineral nutrient for the growth of plants and algae and is a part of several enzymes, such as superoxide dismutase, carboxypeptidase, carbonic anhydrase, and a number of dehydrogenases (Fariduddin et al., 2022). Electron transport and phosphorylation are significantly reduced in plants with low Zn levels (Fru, 2020; Kim et al. 2015c; Kolber et al., 2000; Read et al., 2021; Singh et al., 2005; Yanko et al. 1999). Zn is a micronutrient but is frequently present in high concentrations in algal biomass due to accumulation and sequestration in polyphosphate bodies in cell sectors (El-Agawany and Kaamoush, 2022). A wide variety of the roles of Zn in protein and carbohydrate metabolism has been made clear through studies with numerous highly purified enzymes (Banaszak et al., 2021). It can have a stimulatory or an inhibitory effect, depending on the amount of Zn present (Banaszak et al., 2021). While high Zn concentrations hindered development and reduced cell division in certain algae, low Zn concentrations promoted growth. Additionally, in regions with the highest Zn content, some hardy blue-green algal species and green algae may coexist (Banaszak et al., 2021). According to numerous research, dead algae absorbs far more Zn_2^+ than living algae. Zn is a necessary component for the growth of all living things, but excessive amounts might limit growth and endanger ecosystems (Banaszak et al., 2021; El-Agawany and Kaamoush, 2022). Zn toxicity in algae is influenced by a variety of environmental factors, including pH, particles and complexing agents, the presence of other elements and ions like Ca_2^+, Mg_2^+, P_2^+, C_2^+, and NO_3 or other metals, sulphur compounds, and amino acids, ionic strength, cell concentrations, temperature, salinity, light intensity, medium aeration, exchange reactions between suspended sediments, and water quality. After entering the mammalian body through the respiratory and digestive systems, ZnO causes toxicological harm to animals as well as DNA damage by producing oxidative stress (Sonwani et al., 2021). After entering the mammalian body through the respiratory and digestive systems, ZnO causes toxicological harm to animals as well as DNA damage by producing oxidative stress (Sonwani et al., 2021). For the purpose of estimating metal toxicity, it is crucial to determine the half-maximal effective concentration of a heavy metal (EC50). This can be done by using the EC50 values of some heavy metals, such as Cu, Ni, and Zn, which have been found to have a greater toxic effect on algae than Ni or Zn (Filová et al., 2021). Metal tolerance in algae may be based on a number of mechanisms that are not all fully understood. The majority of the internal Zn2+ was thought to be kept in vacuoles in the large cells of *Chara corallina* (Wherrett, 2006). According to De-Filippis (1981), Cyanophyceae are more susceptible to Cu, Cd, and Zn metals than other algae examined for photosynthetic activity due to photosystem II suppression and/or decreased activity of CO2 fixation enzymes (El-Agawany and Kaamoush, 2022). Higher Zn_2^+ concentrations reduced cell mobility, carotenoid/chlorophyll ratio, total chlorophyll content, and cell division (El-Agawany and Kaamoush, 2022). Omega-3 fatty acids, which guard against chronic illnesses such as coronary heart disease, diabetes, and cancer, have been found to be abundant in microalgae.

5.8 CONCLUSIONS

Metal-containing waste is frequently dumped into the ocean, posing a serious health risk to aquatic life. Trace metals, especially Zn, have a significant effect on ocean biota, impacting biological productivity and altering coastal ecosystems. Zn can start causing different levels of toxicity in marine biota, depending on their physicochemical characteristics. In addition, it acts as a micronutrient regulating communities of phytoplankton and associated microorganisms. However, recent advances in oceanographic research have led to an improved understanding of marine biogeochemistry of Zn elements and also appear to mediate primary productivity/ phytoplankton community structure. Some laboratory studies have shown the importance of several metals for phytoplankton growth and the mechanisms they develop to maintain their optimal requirements. In the same way, the importance of Zn for marine microorganisms is well established through its different biological roles. Zn is one of the essential micro-nutrients for human beings obtained from the diet and plays an important role in many physiological functions such as promoting normal growth, maintenance and regulation of the immune system and immune response. Due to its structural, catalytic and intercellular-signalling component, Zn transporters in conjunction with proteins control the homeostasis of the body at tissue, cellular, and subcellular levels. Hence it is imperative to monitor Zn levels for maintaining a healthy food chain in the marine ecosystem.

REFERENCES

Altaff K, Vijayaraj R. 2021a. Micro-algal diet for copepod culture with reference to their nutritive value – A Review. Int. J. Cur. Res. Rev.l 13, 86–96.
Altaff K, Vijayaraj R. 2021b. Influence of heat shock protein (HSP-70) enhancing compound from red alga (*Porphyridium cruentum*) for augmenting egg production in copepod culture–A new in silico report. Mar. Sci. Technol. Bull. 10, 186–192.
Amado Filho GM, Karez CS, Andrade LR, Yoneshigue-Valentin Y, Pfeiffer WC. 1997. Effects on growth and accumulation of Zn in six seaweed species. Ecotoxicology and environmental safety 37(3), 223–228.

Azizi G, Akodad M, Baghour M, Layachi M, Moumen A. 2018. The use of *Mytilus spp.* mussels as bioindicators of heavy metal pollution in the coastal environment. A review. J. Mater. Environ. Sci. 9, 1170–1181.

Banaszak M, Górna I, Przysławski J. 2021. Zn and the Innovative Zn-α2-Glycoprotein Adipokine Play an Important Role in Lipid Metabolism: A Critical Review. Nutrients 13(6), 2023.

Barletta M, Lima AR, Costa MF. 2019. Distribution, sources and consequences of nutrients, persistent organic pollutants, metals and microplastics in South American estuaries. Science of the Total environment 651, 1199–1218.

Bellotti, D, Rowińska-Żyrek, M, Remelli, M (2021). How Zinc-Binding Systems, Expressed by Human Pathogens, Acquire Zinc from the Colonized Host Environment: A Critical Review on Zincophores. Current Medicinal Chemistry, 28(35), 7312.

Birch GF, Lee JH, Tanner E, Fortune J, Munksgaard N, Whitehead J, ... Steinberg P. 2020. Sediment metal enrichment and ecological risk assessment of ten ports and estuaries in the World Harbours Project. Marine Pollution Bulletin 155, 111129.

Boguta P, Sokołowska Z. 2020. Zn binding to fulvic acids: Assessing the impact of pH, metal concentrations and chemical properties of fulvic acids on the mechanism and stability of formed soluble complexes. Molecules 25(6), 1297.

Boothe PN, Knauer GA. 1972. The Possible Importance Of Fecal Material In The Biological Amplification Of Trace And Heavy Metals 1. Limnology and Oceanography 17(2), 270–274.

Bryan GW, Langston WJ. 1992. Bioavailability, accumulation and effects of heavy metals in sediments with special reference to United Kingdom estuaries: a review. Environmental pollution 76(2), 89–131.

Capolupo M, Gunaalan K, Booth AM, Sørensen L, Valbonesi P, Fabbri E. 2021. The sublethal impact of plastic and tire rubber leachates on the Mediterranean mussel Mytilus galloprovincialis. Environmental Pollution 283, 117081.

Capolupo M, Sørensen L, Jayasena KDR, Booth AM, Fabbri E. 2020. Chemical composition and ecotoxicity of plastic and car tire rubber leachates to aquatic organisms. Water Research 169, 115270.

Cheggour M, Chafik A, Fisher NS, Benbrahim S. 2005. Metal concentrations in sediments and clams in four Moroccan estuaries. Marine Environmental Research 59(2), 119–137.

Chong K, Wang WX. 2000. Assimilation of cadmium, chromium, and Zn by the green mussel Perna viridis and the clam Ruditapes philippinarum. Environmental Toxicology and Chemistry: An International Journal 19(6), 1660–1667.

Davison SM, White MP, Pahl S, Taylor T, Fielding K, Roberts BR, ... Fleming LE. 2021. Public concern about, and desire for research into, the human health effects of marine plastic pollution: Results from a 15-country survey across Europe and Australia. Global Environmental Change 69, 102309.

De Filippis LF, Hampp R, Ziegler H. 1981. The effects of sublethal concentrations of Zn, cadmium and mercury on Euglena. Archives of Microbiology 128(4), 407–411.

Dean RJ, Shimmield TM, Black KD. 2007. Copper, Zn and cadmium in marine cage fish farm sediments: an extensive survey. Environmental Pollution 145(1), 84–95.

El-Agawany NI, Kaamoush MI. 2022. Role of Zn as an essential microelement for algal growth and concerns about its potential environmental risks. Environmental Science and Pollution Research, 1–12.

Ellwood MJ, Van den Berg CM. 2000. Zn speciation in the northeastern Atlantic Ocean. Marine Chemistry 68(4), 295–306.

Fariduddin Q, Saleem M, Khan TA, Hayat S. 2022. Zn as a Versatile Element in Plants: An Overview on Its Uptake, Translocation, Assimilatory Roles, Deficiency and Toxicity Symptoms. Microbial Biofertilizers and Micronutrient Availability, 137–158.

Filová A, Fargašová A, Molnárová M. 2021. Cu, Ni, and Zn effects on basic physiological and stress parameters of Raphidocelis subcapitata algae. Environmental Science and Pollution Research 28(41), 58426–58441.

Fru W. 2020. Copper and Zn in water, sediment and gastropods in the harbours of the Cape Town Metropole, South Africa (Doctoral dissertation, Cape Peninsula University of Technology).

Gaffney-Stomberg E. 2019. The impact of trace minerals on bone metabolism. Biological trace element research 188(1), 26–34.

Geddie AW, Hall SG. 2019a. An introduction to copper and Zn pollution in macroalgae: for use in remediation and nutritional applications. Journal of Applied Phycology 31(1), 691–708.

Geddie AW, Hall SG. 2019b. The effect of salinity and alkalinity on growth and the accumulation of copper and Zn in the Chlorophyta Ulva fasciata. Ecotoxicology and Environmental Safety 172, 203–209.

Gogoi A, Taki K, Kumar M. 2020. Seasonal dynamics of metal phase distributions in the perennial tropical (Brahmaputra) river: Environmental fate and transport perspective. Environmental research 183, 109265.

Goto KT, Sekine Y, Ito T, Suzuki K, Anbar AD, Gordon GW, …Kiyokawa S. 2021. Progressive ocean oxygenation at~ 2.2 Ga inferred from geochemistry and molybdenum isotopes of the Nsuta Mn deposit, Ghana. Chemical Geology 567, 120116.

Hogstrand C, Wood CM. 1996. The physiology and toxicology of Zn. Fish Toxicology of Aquatic Pollution, 61–84.

Hunt JW, Anderson BS. 1989. Sublethal effects of Zn and municipal effluents on larvae of the red abalone Haliotis rufescens. Marine Biology 101(4), 545–552.

Jha DK, Ratnam K, Rajaguru S, Dharani G, Devi MP, Kirubagaran R. 2019. Evaluation of trace metals in seawater, sediments, and bivalves of Nellore, southeast coast of India, by using multivariate and ecological tool. Marine pollution bulletin 146, 1–10.

Kalavathy G, Baskar G. 2019. Synergism of clay with Zn oxide as nanocatalyst for production of biodiesel from marine Ulva lactuca. Bioresource technology 281, 234–238.

Kicińska A, Pomykała R, Izquierdo-Diaz M. 2022. Changes in soil pH and mobility of heavy metals in contaminated soils. European Journal of Soil Science 73(1), e13203.

Kim T, Obata H, Gamo T. 2015a. Dissolved Zn and its speciation in the northeastern Indian Ocean and the Andaman Sea. Frontiers in Marine Science 2, 60.

Kim T, Obata H, Gamo T, Nishioka J. 2015b. Sampling and onboard analytical methods for determining subnanomolar concentrations of Zn in seawater. Limnology and Oceanography: Methods 13(1), 30–39.

Kim T, Obata H, Kondo Y, Ogawa H, Gamo T. 2015c. Distribution and speciation of dissolved Zn in the western North Pacific and its adjacent seas. Marine Chemistry 173, 330–341.

Kolber ZS, Van Dover CL, Niederman RA, Falkowski PG. 2000. Bacterial photosynthesis in surface waters of the open ocean. Nature 407(6801), 177–179.

Koschinsky A, Hein JR. 2003. Uptake of elements from seawater by ferromanganese crusts: solid-phase associations and seawater speciation. Marine Geology, 198(3-4), 331–351.

Krężel A, Maret W. 2016. The biological inorganic chemistry of Zn ions. Archives of biochemistry and biophysics 611, 3–19.

Lall SP, Kaushik SJ. 2021. Nutrition and metabolism of minerals in fish. Animals 11(09), 2711.

Langston WJ. 2018. Toxic effects of metals and the incidence of metal pollution in marine ecosystems. Heavy metals in the marine environment, 101–120.

Lohan MC. 2003. Studies on the biogeochemistry of Zn in the subartic North Pacific (Doctoral dissertation, University of Southampton).

Luchinat E, Cremonini M, Banci L. 2022. Radio Signals from Live Cells: The Coming of Age of In-Cell Solution NMR. Chemical Reviews 122(10), 9267–9306.

Mahboub HH, Shahin K, Zaglool AW, Roushdy EM, Ahmed SA. 2020. Efficacy of nano Zn oxide dietary supplements on growth performance, immunomodulation and disease resistance of African catfish Clarias gariepinus. Diseases of Aquatic Organisms 142, 147–160.

Martin DJ, Rainbow PS. 1998. Haemocyanin and the binding of cadmium and Zn in the haemolymph of the shore crab Carcinus maenas (L.). Science of the Total Environment 214(1-3), 133–152.

Mason AZ, Jenkins KD, Sullivan PA. 1988. Mechanisms of trace metal accumulation in the polychaete Neanthes arenaceodentata. Journal of the Marine Biological Association of the United Kingdom 68(1), 61–80.

Milner NJ. 1982. The accumulation of Zn by O-group plaice, Pleuronectes platessa (L.), from high concentrations in sea water and food. Journal of Fish Biology 21(3), 325–336.

Mishra S, Bharagava RN, More N, Yadav A, Zainith S, Mani S, Chowdhary P. 2019. Heavy metal contamination: an alarming threat to environment and human health. In Environmental biotechnology: For sustainable future (pp. 103–125). Springer, Singapore.

Moiseenko TI, Gashkina NA. 2020. Distribution and bioaccumulation of heavy metals (Hg, Cd and Pb) in fish: Influence of the aquatic environment and climate. Environmental Research Letters 15(11), 115013.

Mossa AW, Young SD, Crout NM. 2020. Zn uptake and phyto-toxicity: Comparing intensity-and capacity-based drivers. Science of the Total Environment 699, 134314.

Mostoni S, Milana P, Di Credico B, D'Arienzo M, Scotti R. 2019. Zn-based curing activators: new trends for reducing Zn content in rubber vulcanization process. Catalysts 9(8), 664.

Müsing K, Clarkson MO, Vance D. 2022. The meaning of carbonate Zn isotope records: Constraints from a detailed geochemical and isotope study of bulk deep-sea carbonates. Geochimica et Cosmochimica Acta 324, 26–43.

Na-Phatthalung P, Min J, Wang F. 2021. Macrophage-mediated defensive mechanisms involving Zn homeostasis in bacterial infection. Infectious Microbes & Diseases 3(4), 175–182.

Nowinski B, Moran MA. 2021. Niche dimensions of a marine bacterium are identified using invasion studies in coastal seawater. Nature Microbiology 6(4), 524–532.

Ojaveer E, Annist J, Jankowski H, Palm T, Raid T. 1980. On effect of copper, cadmium and Zn on the embryonic development of Baltic spring spawning herring. Finnish Marine Research 247, 135–140.

Pajarillo EAB, Lee E, Kang DK. 2021. Trace metals and animal health: Interplay of the gut microbiota with iron, manganese, Zn, and copper. Animal Nutrition 7(3), 750–761.

Pentreath RJ (19734. The roles of food and water in the accumulation of radionuclides by marine teleost and elasmobranch fishes. In Radioactive Contamination of the Marine Environment. Proceedings of the Symposium on the interaction of Radioactive Contaminants with the constituents of the Marine Environment, Seattle, 1972, pp. 421-436.

Read TL, Doolette CL, Howell NR, Kopittke PM, Cresswell T, Lombi E. 2021. Zn accumulates in the nodes of wheat following the foliar application of 65Zn oxide nano- and microparticles. Environmental Science & Technology 55(20), 13523–13531.

Roshan S, DeVries T, Wu J, Chen G. 2018. The internal cycling of Zn in the ocean. Global Biogeochemical Cycles 32(12), 1833–1849.

Salomons W, Förstner U. 2012. Metals in the Hydrocycle. Springer Science & Business Media.

Sarker I, Moore LR, Tetu SG. 2021. Investigating Zn toxicity responses in marine Prochlorococcus and Synechococcus. Microbiology 167(6).

Singh KP, Zaidi SI, Raisuddin S, Saxena AK, Murthy RC, Ray PK. 1992. Effect of zinc on immune functions and host resistance against infection and tumor challenge. Immunopharmacology and Immunotoxicology 14(4), 813–840.

Singh B, Natesan SKA, Singh BK, Usha K. 2005. Improving Zn efficiency of cereals under Zn deficiency. Current Science, 36–44.

Smale DA, Wernberg T, Oliver EC, Thomsen M, Harvey BP, Straub SC, Burrows MT, Alexander LV, Benthuysen JA, Donat MG, Feng M. 2019. Marine heatwaves threaten global biodiversity and the provision of ecosystem services. Nature Climate Change 9(4), 306–312.

Sonwani S, Madaan S, Arora J, Suryanarayan S, Rangra D, Mongia N, … Saxena P. 2021. Inhalation exposure to atmospheric nanoparticles and its associated impacts on human health: a review. Frontiers in Sustainable Cities 3, 690444.

Stauber JL, Florence TM. 1990. Mechanism of toxicity of Zn to the marine diatomNitzschia closterium. Marine Biology 105(3), 519–524.

Stewart BD, Jenkins SR, Boig C, Sinfield C, Kennington K, Brand AR, Lart W Kröger R, … Kröger R. 2021. Metal pollution as a potential threat to shell strength and survival in marine bivalves. Science of the Total Environment 755, 143019.

Sunda WG, Huntsman SA. 1995. Cobalt and Zn interreplacement in marine phytoplankton: Biological and geochemical implications. Limnology and Oceanography 40(8), 1404–1417.

Verriopoulos G, Hardouvelis D. 1988. Effects of sublethal concentration of Zn on survival and fertility in four successive generations of Tisbe. Marine pollution bulletin 19(4), 162–166.

Wang RM, Archer C, Bowie AR, Vance D. 2019. Zn and nickel isotopes in seawater from the Indian Sector of the Southern Ocean: the impact of natural iron fertilization versus Southern Ocean hydrography and biogeochemistry. Chemical Geology 511, 452–464.

Wang WX, Rainbow PS. 2000. Dietary uptake of Cd, Cr, and Zn by the barnacle Balanus trigonus: influence of diet composition. Marine Ecology Progress Series 204, 159–168.

Wherrett T. 2006. Mechanisms of aluminium toxicity, tolerance and amelioration in wheat (Doctoral dissertation, University of Tasmania).

Wong KH, Obata H, Ikhsani IY, Muhammad R. 2021. Controls on the Distributions of Dissolved Cd, Cu, Zn, and Cu-Binding Organic Ligands in the East China Sea. Journal of Geophysical Research: Oceans 126(6), e2020JC016997.

Xiao C, Kong L, Pan X, Zhu Q, Song Z, Everaert N. 2022. High Temperature-Induced Oxidative Stress Affects Systemic Zn Homeostasis in Broilers by Regulating Zn Transporters and Metallothionein in the Liver and Jejunum. Oxidative Medicine and Cellular Longevity, 2022.

Yanko V, Arnold AJ, Parker WC. 1999. Effects of marine pollution on benthic foraminifera. In Modern foraminifera (pp. 217–235). Springer, Dordrecht.

Yap CK, Al-Mutairi KA. 2021. Ecological-health risk assessments of heavy metals (Cu, Pb, and Zn) in aquatic sediments from the ASEAN-5 emerging developing countries: A review and synthesis. Biology 11(1), 7.

Zhou W, Wang Y, Ni C, Yu L. 2021. Preparation and evaluation of natural rosin-based Zn resins for marine antifouling. Progress in Organic Coatings 157, 106270.

Zirino A, Yamamoto S. 1972. A pH-dependent model for the chemical speciation of copper, Zn, cadmium, and lead in seawater. Limnology and oceanography 17(5), 661–671.

6 Zinc

A Promising Micronutrient for Probiotic Absorption

D. Srinivasan, Krishnamurthy Vinoth Kumar,
B. Shyamaladevi, and Ethirajan Sukumar

6.1 INTRODUCTION

A bacteria is defined as a single or multiple culture of living things that when administered to animals or people, improve the host's natural microflora. This definition emphasizes the value of live microorganisms that are found in the mouth, gastrointestinal tract (GIT), and upper respiratory or urogenital tracts and help to improve human health (Havenaar et al., 1992). Probiotics should be able to thrive in the human intestine and be able to survive the GIT, which involves exposure to bile and hydrochloric acid in the small intestine and the stomach, respectively. Foods containing probiotics, such as those found in milk, yoghurt, and cheese, are referred to as functional foods. This also includes any fresh or processed foods that go beyond their purely nutritional roles to make health-promoting or disease-prevention claims. The species *Lactobacillus* and *Bifidobacterium* make for the best candidates for adding healthy bacteria to foods intended for human consumption. These microorganisms are known GIT residents and have a number of characteristics in common, including tolerance to bile and acid, capacity for adhesion to intestinal cells, and GRAS status (Dunne et al., 2001).

6.2 PREBIOTICS

Prebiotics are unique plant fibers that promote the growth of good bacteria in the gut. The digestive system functions better as a result. Prebiotics and probiotics are both beneficial to the digestive system, but they work in different ways. In actuality, prebiotics provide food for the beneficial bacteria in the gut. They are indigestible carbohydrates that pass through the lower digestive tract and act as food for the good bacteria, encouraging their growth. The combination of both is referred to as a "synbiotic," which is defined as "combination of a probiotic and a prebiotic that benefits the host by improving the survival and the implantation of live microbial dietary supplements in the GIT in addition to selectively stimulating the growth

DOI: 10.1201/9781003412472-6

and/or by activating the metabolism of one or more health-promoting bacteria (Roberfroid et al., 1998; Roberfroid et al., 2000; Roberfroid et al., 2007).

6.3   CLASSIFICATION OF PROBIOTICS

Probiotics are classified into three groups Mono probiotics that contain bacteria of one species and strain. Examples include Coli drugs, Lactobacilli drugs, Bacillary drugs, and Bifido drugs. Polyprobiotics which contain Coderzhat bacteria of one species but different strains. Examples: Bacillary drugs, Bifido drugs, and Lactobacilli drugs. Combined probiotics that have bacteria of different types. Examples: Lactobacilli + Bifido and Coli + Bifido and Lactobacilli + Enterococcus + Bifido.

6.4   BENEFITS OF PROBIOTICS

Consuming live probiotic bacteria is vital for their positive effects, requiring survival through digestion, coagulation in the intestines, and adherence to gut epithelium for in vivo impact. These microorganisms generate bacteriocin, hydrogen peroxide, and biosurfactant for intestinal survival, while also elevating mucin-coding genes, promoting mucus production for protection. Effective probiotics complement healthy flora rather than replace it. *Lactobacillus rhamnosus* GGLGG is a well-studied probiotic, showing resilience and adherence in the gastrointestinal tract, even persisting up to a week post-oral administration, especially in infants with rotavirus diarrhea (Franchi et al., 2006).

The human intestinal tract harbors around 100 trillion microbes, primarily in the colon, reaching a density of 1011–1012 cells per milliliter Whitman et al. (1998). Germ-free infants acquire their microbiota from various sources, such as maternal delivery, diet, genetics, or the environment (Mandar and Michailsaar, 1996). The human gastrointestinal tract offers a diverse source of potential probiotic microorganisms, including facultative and obligate anaerobes (Naidu et al., 1999). Probiotics exhibit varied effects on intestinal microbiota. *Bifidobacterium animalis*, for instance, alters microbiota metabolism, boosting carbohydrate and nucleotide production while reducing lipid and amino acid metabolism (McNulty et al., 2011).

6.5   PROBIOTICS IN ENHANCING IMMUNITY

Zinc greatly influences both the innate and evolved immune systems, with severe deficiency leading to adverse effects on thymic size, lymphocytes, immune responses, and even mortality (Prasad et al., 1988). Zinc deficiency increases the risk of childhood diarrhea and related dysbiosis in the gut microbiota, impacting nervous, reproductive, and digestive systems (Walker et al., 2013; Tompkins et al., 2007). Studies by Reed et al. (2015) reveal low zinc is associated with reduced Clostridiales and Verrucomicrobia, while Enterobacteriaceae and Enterococcus populations rise. This imbalance can lead to the growth of pathogenic bacteria at the expense of beneficial ones like Akkermansiamuciniphila, which plays a role in lipid and carbohydrate metabolism and reduces inflammation (Dao et al., 2016).

Probiotics wield immune-stimulating and immunomodulatory effects against GIT pathogens. Intestinal antibiotics enhance barrier defenses, induce cytokines (TNF, IL, TGFb, IL-10), and maintain mucosal integrity (Villamil et al., 2014). Consumption of multi-strain probiotics containing *Lactococcus* and *Streptococcus* species produces IL-10 and IL-12 (Daudelin et al., 2011). In pigs infected with ETEC F4, Pediococcus acids and *Saccharomyces cerevisiae* subsp. boulardii stimulate pro-inflammatory cytokines (IL-6, IL-8, TNF) (Ragland and Criss, 2017). Probiotic consumption activates intraepithelial leucocytes, bolsters mucus production, tightens junctions, and stimulates interferon synthesis by natural killer cells and Th1-lymphocytes (Fasina et al., 2008). Interestingly, a multi-strain probiotic containing *Lactobacillus* and *Enterococcus* species reduces interferon levels (Chen et al., 2011).

6.6   IMMUNO MODULATING EFFECTS OF PROBIOTICS

Probiotics primarily modulate dendritic cells and T lymphatic cell regulation, holding promise for diverse healthcare applications. Species-specific effects will guide probiotic selection. Bacterial cultures induce regulatory cell production and development, evidenced in animal models and cytokine studies. *Lactobacillus* strains suppress T-cell proliferation, encourage IL-10 and TGF-b production, and influence Th1 and Th2 cytokines in autoimmune inflammatory diseases. Certain lactic strains promote mixed Th1 and Th2 responses or elevate Th2 levels, while others stimulate Th1 cytokines. Bifidobacterium strains may trigger varied responses, including distinct cytokine patterns. Probiotics enhance IgA immune response for parenteral and oral vaccines (He et al., 2001; Mullie et al., 2004).

6.7   ZINC AS A PREBIOTIC

Zinc is an essential element for growth, vital for over 80 enzymes as coenzymes, impacting nucleic acid metabolism, transcription, translation, and RNA processing (Vallee, 1977). Prebiotics selectively enhance specific bacteria in the colon, with zinc sources gaining significant attention (Gibson and Roberfroid, 1995). Organic zinc boosts growth rate and feed conversion in broiler chickens (Cao et al., 2000). Consistent daily zinc intake is crucial due to the absence of a dedicated storage system in mammals. Zinc forms, like sulfate, gluconate, organic-enriched grain, or microorganisms, have varying bioavailability (Allen, 1998). *Lactobacilli* and *Bifidobacteria* can enhance zinc bioavailability (Mogna et al., 2012). ZnONP administration reduced harmful bacteria and increased *Lactobacillus* and *Bifidobacterium* in a colitis mice model (Li et al., 2017).

6.8   ABSORPTION OF ZINC IN INTESTINES

Mammalian intestines house a diverse ecosystem with numerous species occupying distinct niches. Resource availability and competition among species impact population structure (Hornung et al., 2018). Oxygen availability during inflammation disrupts the microbiota, favoring Enterobacteriaceae expansion. Essential nutrient-rich foods, like zinc, influence microbiota development. Zinc affects metal absorption, potentially influencing bacterial distribution. Specific transporters, such as Zip4 and Znt1, facilitate zinc absorption in different parts of the small intestine, with the jejunum showing the highest absorption rate (Lee et al., 1989).

6.9   BIOAVAILABILITY OF ZINC

Certain microbes like yeast, lactobacilli, and spirulina strains, have been used to convert inorganic zinc to a more bioavailable organic form in order to increase zinc bioavailability (Slavik et al., 2008). Due to its structure and cellular membrane composition, the microbial cell is a natural adsorbent for metal ions (Blackwell et al., 1995). Some organisms, particularly lactobacilli, can more easily soak up these elements when inorganic metal ions are added to the cultivation medium. The inorganic zinc-bio transforming *Lactobacillus fermentum* has the highest concentration.

According to reports, certain *Lactobacillus* strains can effectively bind and adsorb metals like zinc, and the Zn-enrichment process carried out by these strains of *Lactobacilli* and *Bifidobacteria* can increase the bioavailability of this vital metal. A strain of *Lactobacillus* that has been Zn-enriched may have a synergistic effect on gut health. Different metal transport pathways are used by strains and the various surface functional groups involved in Zn binding (Zoghi et al., 2014).

6.10   BIOABSORPTION OF ZINC

Microorganisms effectively absorb metal ions, influenced by metal chemistry, microbial surface properties, physiology, and environmental factors. Sorption, a passive process, involves adsorption, ion exchange, complexation, chelation, and microprecipitation, often followed by slower bioaccumulation binding (Brady and Duncan, 1994). Lactic acid bacteria (LAB) enriched with zinc serve as safe, nonpathogenic food sources and can enhance human absorption of zinc organic compounds (Dobrzan and Jamroz, 2003). Zinc uptake across the cytoplasmic membrane is facilitated by transporters like ZnuACB and ZupT. ZupT is high-affinity and ZnuACB is low-affinity. ZnuA encodes the zinc-binding part, znub the trans-membrane, and znuc the ATPase subunit. Dimeric Znuc hydrolyzes ATP to move Zn^{2+} through Znub pore during Znu system-mediated zinc uptake (Vallee, 1977).

6.11   ZINC SYNERGISM WITH PROBIOTIC

The combined benefits of probiotics and Zn for gut health have been underexplored. Zn-enriched probiotics have shown promise in reducing *Escherichia coli* infection (Ren et al., 2011) and protecting against ulcerative colic (UC) (Duranti et al., 2016), a condition linked to Zn deficiency. Oral Zn supplementation has been effective in UC treatment (Siva et al., 2016). Probiotics, particularly *Lactobacilli* and *Bifidobacteria*, play a key role in heavy metal absorption from various sources (Hu et al. (2009); Wang and Chen, 2009; Kirillova et al., 2017; Ameen et al., 2020).

The negatively charged bacterial cell wall, composed of peptidoglycan and teichoic acid, interacts electrostatically with heavy metals, facilitating their removal from water (Abbas et al., 2014; Halttunen et al., 2007). *Lactobacillus acidophilus* and *Bifidobacterium reuteri* are adept at absorbing non-essential heavy metals like chromium and mercury (Huët and Puchooa, 2017). While limited, studies indicate the potential of certain *Lactobacillus* strains to bioremediate heavy metals, including Cu^{2+} and Zn^{2+}. Such metal contamination negatively impacts biological processes, underscoring the need to assess *Lactobacillus* species' capacity for absorbing high copper and zinc concentrations (Vallee, 1977).

6.12 CONCLUSION

Zinc's prebiotic action improves probiotic absorption through synergistic effects. Prebiotics alter the flora of the intestinal tract and work with the host's immune system to protect against particular pathogens. Thus, zinc becomes crucial for improving immune function, intestinal permeability, epithelial and enzymatic functions, and the transport of electrolytes, in addition to its role as a nutrient supplement.

REFERENCES

Abbas SH, Ismail IM, Mostafa TM, Sulaymon AH. 2014. Biosorption of heavymeta ls: A Review, J. Chem. Sci. Technol, 3, 74–102.

Allen, LH. 1998. Zinc and micronutrient supplements for children. Am J Clin Nutr, 68(2 Suppl), 495S–498S.

Ameen FA, Hamdan AH, El-Naggar MY. 2020. Assessment of the heavy metal bioremediation efficiency of the novel marine lactic acid bacterium, lactobacillus plantarum MF042018. Sci. Rep. 10, 314.

Blackwell KJ, Singleton I, Tobin JM. 1995. Metal cation uptake by yeast: a review Appl Microbiol Biotechnol. 43, 579–584.

Brady D, Duncan JR. 1994. Bioaccumulation of metal cations by Saccharomyces cerevisiae. Appl. Microbiol. Biotechnol. 41, 149–154.

Cao J, Henry PR, Guo R, Holwerda RA, Toth JP, Littell RC, Miles RD, Ammerman CB. 2000. Chemical characteristics and relative bioavailability of supplemental organic zinc sources for poultry and ruminants. J. Anim. Sci. 78, 2039–2054.

Chen L, Lin YL, Peng G, Li F. 2011. Structural basis for multifunctional roles of mammalian aminopeptidase N. Proc. Natl. Acad. Sci. USA. 109, 17966–17971.

Dao MC, Everard A, Aron-Wisnewsky J, Sokolovska N, Prifti E, Verger EO, Kayser BD, Levenez F, Chilloux J, Hoyles L, Dumas ME, Rizkalla SW, Doré J, Cani PD, Clément K. 2016. Akkermansiamuciniphila and improved metabolic health during a dietary intervention in obesity: relationship with gut microbiome richness and ecology. Gut. 65, 426–436.

Daudelin JF, Lessard M, Beaudoin F, Nadeau E, Bissonnette N, Boutin Y, Brousseau JP, Lauzon K, Fairbrother JM. 2011. Administration of probiotics influences F4 (K88)-positive enterotoxigenic Escherichia coli attachment and intestinal cytokine expression in weaned pigs. Vet. Res. 42, 69.

Dobrzan Z, Jamroz D. 2003. Bioavailability of selenium and zinc supplied to the feed for laying hens in organic and inorganic form. Electron. J. Pol. Agric. Univ. 6, 1–8.

Dunne C, O'Mahony L, Murphy L, Thornton G, Morrissey D, O'Halloran S, Feeney M, Flynn S, Fitzgerald G, Daly C, Kiely B, O'Sullivan GC, Shanahan F, Collins JK. 2001. In vitro selection criteria for probiotic bacteria of human origin: Correlation with in vivo findings. Am. J. Clin. Nutr. Feb; 73(2 Suppl), 386S–392S.

Duranti S, Gaiani F, Mancabelli L et al. 2016. Elucidating gut microbiome of ulcerative colitis: Bifidobacteria as novel microbial biomarkers. FEMS Microbiol. Ecol. 92(12): fiv 191.

Fasina YO, Holt PS, Moran ET, Moore RW, Conner DE, McKee SR. 2008. Intestinal cytokine response of commercial source broiler chicks to Salmonella typhimurium infection. Poult Sci. 87, 1335–1346.

Franchi L, McDonald C, Kanneganti TD, Amer A, Núñez G. 2006. Nucleotide-binding oligomerization domain-like receptors: Intracellular pattern recognition molecules for pathogen detection and host defense. J. Immunol. 177, 3507–3513.

Gibson R, Roberfroid MB. 1995. Dietary modulation of the human colonic microbiota: Introducing the concept of prebiotics. J. Nutr. 125, 1401–1412.

Halttunen T, Salminen S, Tahvonen R 2007. Rapid removal of lead and cadmium from water by specifc lactic acid bacteria. Int J Food Microbiol, 114, 30–35.

Havenaar R, Huisin't Veld JHJ. 1992. Probiotics: A General View, in the Lactic Acid Bacteria: The Lactic Acid Bacteria in Health and Disease, Wood BJB (Ed.), Chapman and Hall, New York, pp. 209–248.

He F, Tuomola E, Arvilommi H, Salminen S. 2001. Modulation of the human humoral immune response through orally administered bovine colostrum. FEMS Immunol. Med. Microbiol. 31, 93–96.

Hornung B, Martins Dos Santos VAP, Smidt H, Schaap PJ. 2018. Studying microbial functionality within the gut ecosystem by systems biology. Genes. Nutr. 13, 5.

Hu X, Cook S, Wang P, Hwang HM. 2009. In vitro evaluation of cytotoxicity of engineered metal oxide nanoparticles. Sci. Total. Environ. 407, 3070–3072.

Huët MAL, Puchooa D. 2017. Bioremediation of heavy metals from aquatic environment through microbial processes: A potential role for probiotics J. App. Biol. Biotech. 5, 14–23.

Jones N, Ray B, Ranjit KT, Manna AC. 2008. Antibacterial activity of ZnO nanoparticle suspensions on a broad spectrum of microorganisms. FEMS Microbiol. Lett. 279, 71–76.

Kirillova AV, Danilushkina AA, Irisov DS, Bruslik NL, Fakhrullin RF, Zakharov YA, Bukhmin VS, Yarullina DR. 2017. Assessment of Resistance and Bioremediation Ability of Lactobacillus Strains to Lead and Cadmium Int J Microbiol, Art. ID 9869145. https://doi.org/10.1155/2017/9869145

Lee HH, Prasad AS, Brewer GJ, Owyang C. 1989. Zinc absorption in human small intestine. Am J Physiol, 256(1 Pt 1), G87–G91.

Li J, Chen H, Wang B, Cai C, Yang X, Chai Z, Feng W. 2017. ZnO nanoparticles act as supportive therapy in DSS-induced ulcerative colitis in mice by maintaining gut homeostasis and activating Nrf2 signaling. Scientific Reports. 7, 43126. doi: 10.1038/srep43126.

McNulty NP, Yatsunenko T, Hsiao A, Faith JJ, Muegge BD, Goodman AL, Henrissat B, Oozeer R, Cools-Portier S, Gobert G, Chervaux C, Knights D, Lozupone CA, Knight R, Duncan AE, Bain JR, Muehlbauer MJ, Newgard CB, Heath AC, Gordon JI. 2011. The impact of a consortium of fermented milk strains on the gut microbiome of gnotobiotic mice and monozygotic twins. Sci. Transl. Med. 3, 106ra106.

Mogna L, Nicola S, Pane M, Lorenzini P, Strozzi G, Mogna G. 2012. Selenium and zinc internalized by Lactobacillus buchneri Lb26 (DSM 16341) and Bifidobacteriumlactis Bb1 (DSM 17850): improved bioavailability using a new biological approach. J. Clin. Gastroenterol. 46, Suppl: S41-5.

Mullie C, Yazourh A, Thibault H, Odou MF, Singer E, Kalach N, Kremp O, Romond MB. 2004. Increased poliovirus-specific intestinal antibody response coincides with the promotion of Bifidobacteriumlongum-infant is and Bifidobacterium breve in infants: A randomized, double-blind, placebo-controlled trial. Pediatr. Res. 56, 791–795.

Mändar R, Mikelsaar M. 1996. Transmission of the mother's microflora to the newborn at birth. Biol. Neonate. 69, 30–35.

Naidu AS, Bidlack WR, Clemens RA. 1999. Probiotic spectra of lactic acid bacteria (LAB). Crit. Rev. Food Sci. Nutr. 39, 13–126.

Prasad AS, Meftah S, Abdallah J, Kaplan J, Brewer GJ, Bach JF, Dardenne M. 1988. Serum thymulin in human zinc deficiency. J. Clin. Invest. 82, 1202–1210.

Ragland SA, Criss AK. 2017. From bacterial killing to immune modulation: Recent insights into the functions of lysozyme. PLoS Pathog. 13, e1006512.

Reed S, Neuman H, Moscovich S, Glahn RP, Koren O, Tako E. 2015. Chronic zinc deficiency alters chick gut microbiota composition and function. Nutrients. 7, 9768–9784.

Ren Z, Zhao Z, Wang Y, Huang K. 2011. Preparation of selenium/zinc-enriched probiotics and their effect on blood selenium and zinc concentrations, antioxidant capacities, and intestinal microflora in canine. Biol Trace Elem Res, 141, 170–183. https://doi.org/10.1007/s12011-010-8734-x

Roberfroid M. 2007. Prebiotics: The concept revisited. J. Nutr. 137(3 Suppl 2), 830S–837SS.

Roberfroid M, Slavin J. 2000. Nondigestible oligosaccharides. Crit. Rev. Food Sci. Nutr. 40, 461–480.

Roberfroid MB. 1998. Prebiotics and synbiotics: Concepts and nutritional properties. Br. J. Nutr. 80, S197–S202.

Siva S, Rubin DT, Gulotta G, Wroblewski K, Pekow J. 2016. Zinc Deficiency is Associated with Poor Clinical Outcomes in Patients with Inflammatory Bowel Disease Inflamm Bowel Dis, 23, 152–157.

Slavik P, Illek J, Brix M, Hlavicova J, Rajmon R, Jilek F. 2008. Influence of organic versus inorganic dietary selenium supplementation on the concentration of selenium in colostrum, milk and blood of beef cows. Acta Vet Scand, 50(1), 43. https://doi.org/10.1186/1751-0147-50-43.

Tompkins TA, Renard NE, Kiuchi A. 2007. Clinical evaluation of the bioavailability of zinc-enriched yeast and zinc gluconate in healthy volunteers. Biol. Trace Elem. Res. 120, 28–35.

Vallee BL. 1977. Zinc biochemistry in normal and neoplastic growth processes. Experientia. 33, 600–601.

Villamil, L, Reyes, C, Martínez-Silva, M A (2014). *In vivo* and*in vitro* assessment of*Lactobacillus acidophilus*as probiotic for tilapia (*Oreochromis niloticus*, Perciformes: Cichlidae) culture improvement. Aquaculture Research, 45, 1116–1125.

Walker CLF, Rudan I, Liu L, Nair H, Theodoratou E, Bhutta ZA, O'Brien KL, Campbell H, Black RE. 2013. Global burden of childhood pneumonia and diarrhea. Lancet. 381, 1405–1416.

Wang J, Chen C. 2009. Biosorbents for heavy metals removal and their future. Biotechnol Adv, 27, 195–225.

Whitman WB, Coleman DC, Wiebe WJ. 1998. Prokaryotes: The unseen majority. Proc. Natl. Acad. Sci. USA. 95, 6578–6583.

Zoghi A, Khosravi-Darani K, Sohrabvandi S. 2014. Surface binding of toxins and heavy metals by probiotics. Mini. Rev. Med. Chem. 14, 84–98.

7 Zinc in Oral Health and Dental Diseases

Zoha Abdullah, M. G. Nithin, and Sana Siddiqui

7.1 INTRODUCTION

Trace elements are essential micronutrients that support regenerative processes, manage oxidative stress and maintain immunity. They play a crucial role in maintaining physiological and metabolic processes in living tissues. Zinc, classified as an essential trace element, is abundant in organisms and present in all enzyme classes. (Bhattacharya et al., 2016) The recommended daily intake of zinc is 15 mg, with the human body containing about 2–4 g of zinc distributed throughout various tissues (Lynch, 2011). Blood plasma and leukocytes contain significant amounts of zinc, while serum contains around 100 µg per deciliter bound to α-2 macroglobulin and albumin (Fatima, 2016). Zinc is vital for growth, development, metabolism of enzymes and proteins, and psychosocial functioning. Inadequate zinc levels increase the risk of infections and degenerative pathologies (Fatima, 2016; Uwitonze et al., 2020).

7.2 ZINC AND ORAL ANATOMY/ORAL TISSUES

7.2.1 ENAMEL

Tooth enamel is a highly mineralized and avascular tissue covering the tooth surface. It is primarily composed of 95% inorganic material, mainly calcium hydroxyapatite crystals, along with 2% organic material (including proteins) and 3% water. Trace elements contribute negligibly to the inorganic portion of enamel (Dogan, 2018).

Zinc concentration in enamel mirrors the pattern of fluoride. It is higher in the outer layer and lower in the subsurface. Zinc content ranges from 430 to 2100 ppm across tooth layers. Post-eruptive phase shows increased zinc uptake, similar to fluoride, possibly from oral fluids (Fatima, 2016).

7.2.2 DENTIN

Dentin consists of 70% inorganic material, including hydroxyapatite crystals and trace elements, 18% organic material, and 12% water (Dogan, 2018). It has lower crystallinity and varying trace element levels compared to enamel, making it more

DOI: 10.1201/9781003412472-7

susceptible to acidic dissolution. Zinc's astringent property influences dentin's organic components during demineralization and remineralization (Fatima, 2016). Dentin extracts contain enzymes like alkaline phosphatase, cation-binding proteins from serum, and other molecules (Goldberg et al., 2011).

7.2.3 CEMENTUM

Cementum is a unique connective tissue that covers the outermost layer of the calcite matrix on the root surface, connecting the periodontal ligament to the adjacent alveolar bone. Cementum is generally less mineralized than root dentin. Cementum contains a number of trace elements in concentrations detectable by electron microprobe analysis, in particular Cu^{+2}, Zn^{+2} and Na^{+2}; however, their distribution and significance do not seem to have been studied in detail (Dogan, 2018).

7.2.4 DENTAL PULP

No studies found on the mineral content of dental pulp.

7.2.4.1 Saliva and Dental Plaque

Zinc is omnipresent in the body, therefore it is naturally present in saliva and teeth (Lynch, 2011; Fatima, 2016). It is also found naturally in dental plaque, Lynch et al. (2011) summarized the values of zinc concentrations in plaque reported by several authors (Figure 7.1).

The comparison between the reported values was not straightforward, as concentrations for 'wet' or 'dry' plaque were reported. However, assuming that drying increases the apparent concentration sevenfold, values were broadly similar.

Studies have reported that zinc is taken up by the salivary pellicle which makes it likely that the oral mucosa is the most important oral reservoir, though insufficient data exist to support this proposition conclusively (Lynch, 2011).

7.3 ZINC IN ORAL DISEASES

7.3.1 ZINC AND DENTAL CARIES

The role of zinc in dental caries is debated. Hussein et al. (2013) found higher zinc levels in children with carious lesions, while Ückardes et al. observed that zinc in saliva reduced the likelihood of dental caries. Another study comparing saliva zinc concentration between non-caries and multiple caries groups showed significant variation, with higher levels in children without caries (Zahir and Sarkar, 2006; Dogan, 2018).

It has been established that zinc is important for the mineralization of the enamel and is known to reduce the susceptibility to dental caries. Deficiency of Zinc leads to demineralization of the enamel causing dental caries (Jayapal et al., 2018).

(a)

(b)

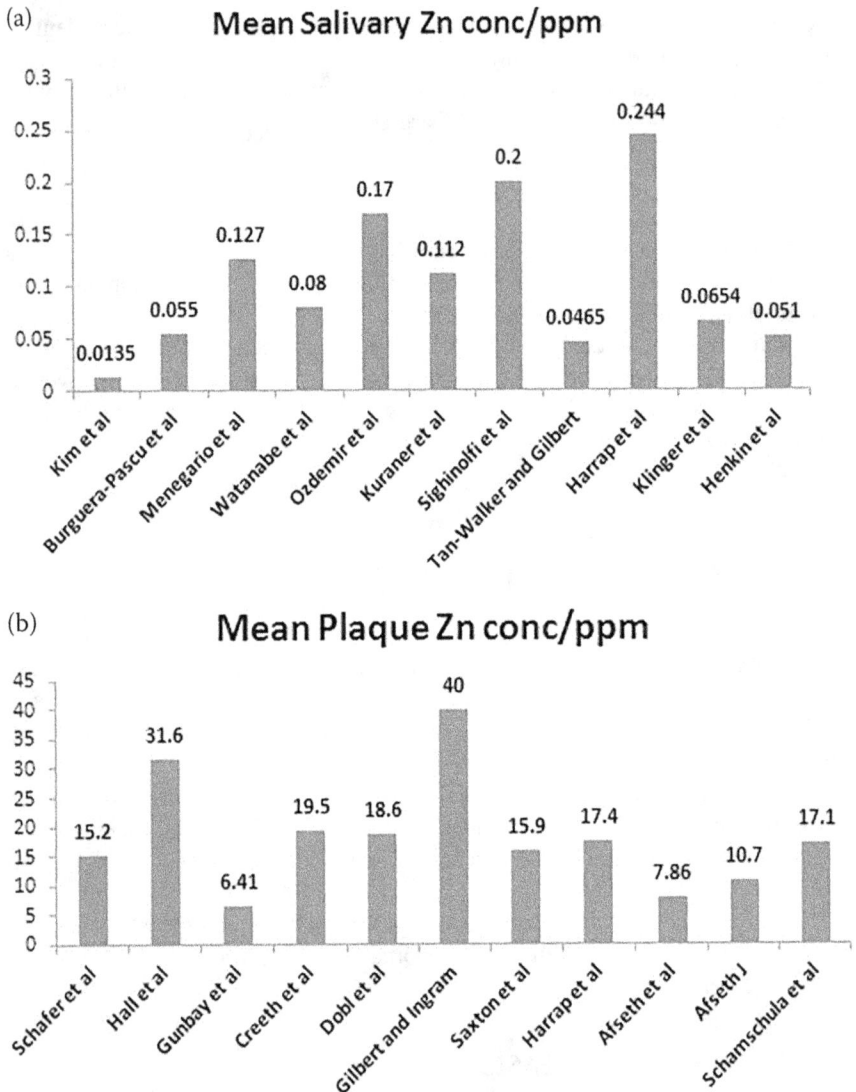

FIGURE 7.1 Reported values in the literature (Mean±SD) for the concentrations of zinc in saliva and plaque.

Increased caries incidence was reported when rats were fed a zinc-deficient diet. However, there are a few differences that warrant caution when these findings are extrapolated to humans. It should be noted that prolonged chronic zinc deficiency in humans is rare, the salivary pH in rodents is higher than humans, and concentrations of calcium and phosphate may also differ (Lynch, 2011).

7.3.2 Zinc, a Marker in Oral Sub-mucous Fibrosis, Leukoplakia and Oral Squamous Cell Carcinoma

Research on the role of trace elements in premalignant and malignant conditions of the oral cavity has shown that there was a significant depletion in the serum zinc concentration. Research comparing the serum zinc levels between healthy individuals and those with oral submucous fibrosis (OSMF) revealed a significant reduction in the serum zinc levels of the OSMF patients.

Yadav et al. (2015) attributed the low levels of serum zinc in patients with OSMF patients to the following:

a. The high demand for zinc by malignant cells can lead to a depletion of zinc levels in the serum.
b. The increase in copper levels among patients with OSMF may also explain the subsequent reduction in zinc levels, as there is a negative interaction between copper and zinc.
c. The elevated levels of copper in the serum can be attributed to the high copper content (302 nmol/ng) of areca nut, which is the primary etiological factor of OSMF in Indian patients.

Zinc levels may serve as a supplementary biomarker for diagnosing and predicting oral submucous fibrosis and other precancerous lesions (Yadav et al., 2015). Low serum zinc levels could also indicate potentially premalignant conditions like oral leukoplakia (Bhattacharya et al., 2016; Fatima, 2016; Dogan, 2018). Monitoring zinc levels in patients with suspicious lesions may aid in early detection and treatment of oral squamous cell carcinoma, improving prognosis.

Upon examining multiple studies on zinc and other trace elements, most of the literature reviews and meta-analyses concluded that individuals with OSMF had decreased levels of zinc in their serum, plasma, or saliva, which corresponded with the advancement of the condition. (Sachdev et al., 2018). However, a few studies included did not report lower Zn levels in OSF patients (Ayinampudi and Narsimhan, 2012; Khanna et al., 2013; Kode and Karjodkar, 2013).

A study on albino rats found that tumor growth in the submandibular salivary glands was slowed down when their drinking water contained 250 ppm of zinc (Ciapparelli et al., 1972). The induced tumor growth was modified, leading to a decrease in carcinomatous epithelium and increased inflammatory response. The impact of diet on zinc deficiency in oral submucous fibrosis (OSMF) patients remains unclear, but trace element supplementation has shown potential in correcting zinc deficiency (Sachdev et al., 2018; Uwitonze et al., 2020). While most studies emphasize the protective and anticarcinogenic effects of zinc, limited literature suggests a possible carcinogenic effect (Mulware, 2013).

7.3.3 Zinc in Periodontal Diseases

Optimal zinc levels are essential for periodontal tissue growth and development. Erythrocyte Cu-Zn superoxide dismutase in the periodontal ligament protects against free radicals. Zinc deficiency in rats harms oral mucosa, leading to poorer periodontal health, increased infection susceptibility, and reduced osteoblastic activity and collagen synthesis. It also causes hyperkeratosis, mouth floor ulcers, and higher plaque and gingival index scores (Meunier et al., 2005; Orbak et al., 2007).

Thomas et al. compared serum zinc levels in patients with diabetes and periodontitis, systemically healthy patients with periodontitis, and healthy control subjects. They found significantly lower zinc levels in periodontitis patients, with and without diabetes, compared to healthy individuals. Zinc absorption decreases in diabetic patients, and hyperglycemia leads to increased urinary losses of zinc, affecting insulin production. Adequate zinc levels are crucial for immune protection and bone regeneration processes, preventing periodontitis progression.

Thomas et al. (2013) and Aziz et al. (2021) confirm lower serum zinc levels in chronic periodontitis patients. Low zinc is linked to increased alveolar bone resorption, altered collagen metabolism, reduced synthesis and turnover, and decreased alkaline phosphatase activity.

7.3.4 ZINC IN PARAKERATOTIC CHANGES OF TONGUE, CHEEK AND ESOPHAGUS

Zinc deficiency can be indicated by parakeratotic changes observed in the cheek, tongue, and esophagus, and thickening of the buccal mucosa. The loss of filiform papillae is also a common manifestation of zinc deficiency (Arbabi-kalati et al., 2012).

7.3.5 ZINC AND TASTE DISORDERS

Zinc plays a vital role in taste perception, impacting taste buds, nerve transmission, and the brain. It maintains cell membrane integrity and enzyme function. Zinc deficiency can lead to taste disorders, reduced taste sensitivity, lingual trigeminal nerve sensitivities, and decreased salivary flow. In a case series, zinc sulfate supplements improved symptoms of oral dysgeusia. Correcting zinc deficiency is essential in treating patients with taste imbalances (Guan, 2018; Uwitonze et al., 2020).

7.3.6 ZINC AND ORAL MUCOSITIS

Oral mucositis is an acute inflammation of the oral mucosa caused by chemotherapy and/or radiotherapy, resulting in pain and difficulty eating, drinking, and swallowing. Zinc sulfate has been studied as a potential treatment to prevent or alleviate symptoms of oral mucositis (Martinez et al., 2014).

Tian et al. (2018) conducted a meta-analysis of five randomized controlled trials examining the impact of oral zinc sulfate on chemotherapy-induced oral mucositis. They found no significant change in the incidence or severity of oral mucositis across most studies. However, Mehdipour et al. reported significantly lower severity scores in patients treated with zinc sulfate compared to those receiving chlorhexidine gluconate, particularly at the second- and third-week intervals (Mehdipour et al., 2011).

A systematic review by Shuai et al. (2019) found that zinc sulfate may not provide preventive benefits for radiation-induced oral mucositis in head and neck cancer patients. Gholizadeh et al. (2017) reported no association between zinc sulfate therapy and decreased pain intensity. Mansouri et al. (2012) conducted a randomized controlled trial showing no significant association between zinc sulfate and the prevention or reduction of intensity and duration of oral mucositis in patients undergoing high-dose chemotherapy.

In a randomized controlled trial, the group given 50 mg zinc sulfate had a 2.1 times lower incidence of chemotherapy-induced oral mucositis and reduced severity compared to the control group (Rambod et al., 2018). Zinc has potential as an approach to delay the onset and reduce the severity of chemotherapy-induced oral mucositis (de Menêses et al., 2020).

7.3.7 Role of Zinc Deficiency in Cleft Lip, Cleft Palate and Salivary Glands

Studies have shown that higher plasma zinc levels in women of reproductive age and adequate zinc exposure during pregnancy are associated with a reduced risk of cleft lip/palate in children (Tamura et al., 2005). Conversely, low zinc intake during pregnancy has been linked to orofacial clefts (Hozyasz et al., 2009).

In an animal study, increased zinc intake protected the sublingual gland against oxidative damage caused by chronic cadmium exposure (Kostecka-Sochon et al., 2018). However, monitoring increased zinc intake is crucial as zinc toxicity can adversely affect saliva quantity and quality, potentially through neurological pathway changes or impact on acinar cells (Mizari et al., 2012).

7.3.8 Other Conditions Related to Zinc Deficiency

Zinc stimulates IL-1, IL-6, and TNF-α production in blood cells. Zinc deficiency disrupts cytokine production, and low serum zinc levels correlate with reduced Th1-type cytokines. Zinc deficiency may contribute to recurrent aphthous stomatitis by triggering abnormal cytokine cascades and tissue damage in the early stages of ulcer formation. Animal studies show aphthous ulcers have a greater impact on alveolar mucosa (Orbak et al., 2007; Ozler, 2014; Seyedmajidi et al., 2014).

Zinc mouthwash and zinc sulfate therapy have shown efficacy in treating recurrent aphthous stomatitis (Mehdipour et al., 2011). Zinc deficiency should be considered in patients with recurrent aphthous stomatitis and other oral mucosal diseases (Yildirmyan et al., 2019).

Although human studies have shown an association between zinc deficiency and aphthous ulcers, it is important to note that whether this association is causal or consequential requires further experimental validation (Uwitonze et al., 2020).

COVID-19 patients have lower serum zinc concentrations, and levels below 50 µg/dL at admission are linked to worse outcomes and higher mortality. Zinc deficiency is considered a causative factor for COVID-19-related gustatory dysfunction and may contribute to xerostomia through hyposalivation (Vogel-González et al., 2021).

7.4   ZINC SUPPLEMENTATION

Zinc supplementation is effective against oral diseases like gingivitis, periodontitis, and halitosis, while zinc deficiency is associated with suboptimal oral health (Fatima, 2016; Rosing et al., 2017). Zinc fights bacteria causing gingivitis, inhibits proteases and has a lasting effect on aphthous ulcers (Fatima, 2016; Uwitonze et al., 2020).

A randomized controlled trial on primary school children from a low socioeconomic background showed that a daily 15 mg zinc supplement for ten weeks significantly improved the plaque index, suggesting that zinc supplementation may help prevent dental caries (Üçkardeş et al., 2009).

7.5   ANTIMICROBIAL EFFECT OF ZINC

A systematic review of seventeen studies found that zinc exhibited antibacterial properties against *Streptococcus* mutans, even at low concentrations (Almoudi et al., 2018). However, Bradshaw et al. reported an ineffectiveness of zinc citrate, possibly due to the use of complex media components (Bradshaw et al., 1993).

Zinc oxide nanoparticles (ZnO-NPs) have shown excellent antimicrobial properties and biocompatibility, with minimal toxicity to humans (Allaker, 2010). ZnO-NPs exhibit significant antibacterial activity against various species (Ahrari et al., 2015; Ramazanzadeh et al., 2015; Sirelkhatim et al., 2015). However, higher concentrations may induce cytotoxicity and genotoxicity, and the potential systemic effects and inflammation caused by oral exposure are of concern (Meyer et al., 2011).

Zinc mouthwashes inhibit bacterial growth associated with halitosis and peri-implantitis by suppressing various bacteria (Suzuki et al., 2018). Zinc supplementation fights gingivitis-causing bacteria and disrupts bacterial cell membranes (Fatima, 2016; Uwitonze et al., 2020).

There are several mechanisms proposed that may be responsible for the antimicrobial effect of zinc compounds (Kim, 2007):

- Inhibition of oxidation of essential thiol groups of bacterial enzymes
- Displacement of Mg ion essential for enzymatic processes
- Non-specific reaction to (-) charged proteins in bacteria
- Inhibition of protein profile in bacteria
- Reduction of bacterial growth by reducing metabolic activity

Sirelkhatim et al. (2015) proposed that the antimicrobial mechanism of ZnO-NPs could be due to:

- Leaching of Zn2+ into the growth media,
- Disruption of bacterial cell wall integrity by direct contact with ZnO-NP,
- Interference of Zn2+ ions with the enzyme systems of the bacteria by displacing with magnesium ions which is crucial for bacterial enzymatic activities.

However, it should be highlighted that a definite antimicrobial mechanism of zinc is not established clearly.

7.6 USES OF ZINC IN DENTISTRY

7.6.1 In Mouth Rinses

Zinc is employed in mouth rinses as an antibacterial agent to control the growth of dental plaque, calculus, bleeding gums, and malodour. An in-vivo study has demonstrated that zinc citrate has an inhibitory effect on the formation of dental plaque (Addy, Richards and Williams, 1980).

Triclosan combined with zinc citrate in mouthwashes enhances the inhibition of gram-negative periodontopathogens (*F. nucleatum* and *P. gingivalis*) (Bradshaw et al., 1993). Fluocinolone-containing mouth rinse with zinc is more effective in treating oral lichen planus (Mehdipour et al., 2010; Gholizadeh et al., 2017). Zinc may have anti-inflammatory properties beneficial for managing Recurrent Aphthous Stomatitis (Belenguer–Guallar, Jimenez–Soriano, and Claramunt–Lozano, 2014).

Recent laboratory studies showed that toothpastes containing zinc or stannous and mouthwash formulas with cetylpyridinium chloride (CPC) neutralize the virus that causes COVID-19 by 99.9% (*British Dental Journal*, Product News, December 2020) (Toothpaste and mouthwash inactivate 99.9% of the virus that causes COVID-19, 2020).

7.6.2 Nano-Therapeutics

Nano-therapeutics using biocidal nanoparticles, such as ZnO nanoparticles (NPs), have gained interest in controlling oral biofilm growth. ZnO NPs offer high selectivity, cytotoxicity, biocompatibility, and are effective in reducing biofilm formation (Fatima, 2016).

Various uses of zinc nanoparticles in dentistry

- Used as Orthodontic adhesives to control the oral biofilm and decrease the demineralization around the brackets has attracted much attention.
- Used to prevent failure of dental implant caused due to plaque accumulation in the oral cavity.
- Addition of NPs in composite resins and glass ionomers boosts the mechanical and antibacterial activity.
- Research in recent years has exhibited that addition of antimicrobial NPs plays a crucial role in the prevention of primary and persistent endodontic infections after treating or recolonizing the filled canal system.

Research has indicated that zinc oxide nanoparticles exhibit a wide range of antibacterial effects, covering both Gram-positive and Gram-negative bacteria, which includes significant foodborne pathogens, like *Escherichia coli*, *Listeria monocytogenes*, *Salmonella typhii*, and *Staphylococcus aureus* (Liu et al., 2009).

7.6.3 ZINC IN TOOTHPASTES

Zinc, including zinc citrate, zinc oxide (ZnO), and zinc chloride, exhibits antibacterial properties and controls plaque and calculus formation. Zinc salts inhibit acid production by oral streptococci and the protease of *P. gingivalis*. Combining triclosan and zinc citrate may have additive or synergistic effects. Further research is needed to explore zinc citrate's role in controlling gingivitis (Marsh, 1992).

7.6.4 OTHER EFFECTS OF TRICLOSAN/ZINC CITRATE IN THE ORAL CAVITY INCLUDE (MARSH, 1992)

- Prevention of gingivitis
- Reduction of supragingival plaque and calculus formation (Stephen et al., 1990)
- Suppression of the enrichment in plaque of anaerobic bacteria and Actinomyces species.
- Inhibition of growth of relevant plaque bacteria (Marsh, 1992) (Figure 7.2)

Zinc exhibits good oral substantivity, remaining in the mouth after administration. A clinical trial found that a toothpaste containing zinc reduced calculus formation by 30% compared to a control toothpaste. No side effects were observed, supporting zinc's role as an anti-calculus agent (Creeth et al., 1993).

Percentage inhibition* by Triclosan + zinc citrate

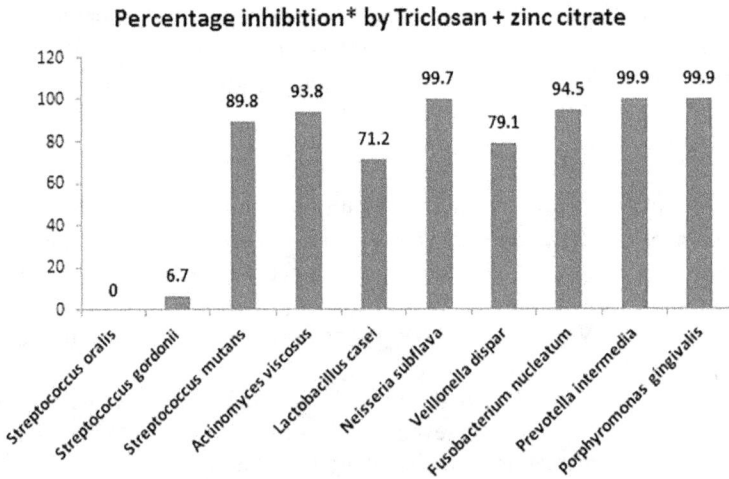

* Percentage inhibition is defined as the reduction in viable count of each species 24 h after a pulse of inhibitor (data taken from Bradshaw et al., 1991)

FIGURE 7.2 Effects of pulses of triclosan (10 Mg/L) and zinc citrate (25 mg/L) on a mixed culture of oral bacteria grown In continuous culture.

A six-month randomized clinical study comparing dentifrices containing fluoride with and without zinc found that the dentifrice with zinc significantly reduced both gingivitis and dental plaque (Zhang et al., 2015).

7.7 ZINC IN DENTAL RESTORATIVE MATERIALS

There are several materials that include zinc in their formulations.

7.7.1 ZINC OXIDE EUGENOL

The first proprietary zinc oxide-eugenol (ZOE) cement was introduced in 1894 by Wessler in Sweden. It was claimed to be "non-irritating, a poor conductor of heat, a powerful antiseptic and anodyne" (Brauer, 1964).

Zinc oxide-eugenol cements have been widely used in the dental field. They are available in powder-liquid form or in a ready-made paste of zinc oxide and eugenol.

Uses

- Temporary fillings
- Sedative bases
- Cementing media for crown and bridge work
- As a soft tissue pack in oral surgery, periodontics and certain phases of restorative dentistry
- As root canal sealers in endodontics
- And with modifying agents as impression pastes

Disadvantages

- High solubility
- Low mechanical properties

To overcome the disadvantages of ZoE, a number of additives (polymer or resins) have been incorporated to obtain cements showing improved physical properties and are called 'Modified Zinc Oxide Eugenol Type Cements'.

7.8 ZINC PHOSPHATE CEMENT

Zinc phosphate cement, the oldest luting cement in dentistry, was developed in the 18th century. Augustin and Charles Rostaing created a cement using zinc oxide and phosphoric acid in 1858. Dr. Otto Hoffman later refined it and officially traded the high-quality zinc phosphate dental cement in 1892.

Applications

- Permanent cementation of crowns (luting agent)
- Cementation of orthodontic bands
- As a base under permanent fillings
- Occasionally for Temporary filling of teeth

7.9   ZINC OXIDE NANOPARTICLES INCORPORATED RESIN

Adding ZnO nanoparticles (ZnO-NPs) to composite resin can enhance its mechanical properties. Research suggests that combining different nanoparticles with resin composite improves its mechanical characteristics up to a certain threshold. Furthermore, the inclusion of ZnO-NPs in composite resin significantly inhibits the growth of Streptococcus mutans (Tavassoli Hojati et al., 2013).

7.10   OTHER APPLICATIONS OF ZINC

Khajuria et al. (2018) developed a chitosan-based dental film with risedronate/zinc hydroxyapatite, which effectively reduced alveolar bone loss in an animal model of periodontitis. Photodynamic inactivation using quaternized Zn (II) phthalocyanines and light irradiation showed efficacy against drug-resistant periodontal bacteria (Uwitonze et al., 2020).

In a study on adult patients with chronic periodontitis, a zinc and octenidine-containing chemical device significantly reduced bacterial levels, including Treponema Denticola and Treponema Forsythia. The device improved patient compliance, pocket access, and required lower antimicrobial agent dosage (Lauritano et al., 2018).

7.11   SIGNIFICANCE OF ZINC IN IMMUNO-COMPROMISED INDIVIDUALS

Zinc deficiency can impair wound healing and weaken immunity, affecting T-lymphocytes and immunologic-based disorders like Lichen Planus. Zinc plays a crucial role in epithelial growth and its deficiency may hinder immune response (Haase and Rink, 2009).

REFERENCES

Addy M, Richards J, Williams G. 1980. Effects of a zinc citrate mouthwash on dental plaque and salivary bacteria. J. Clin. Periodontol. 7(4), 309–315. Available at: doi.org/10.1111/j.1600-051x.1980.tb01973.x.

Ahrari F, et al. 2015. The antimicrobial sensitivity of Streptococcus mutans and Streptococcus sangius to colloidal solutions of different nanoparticles applied as mouthwashes. J. Dent. Res. 12(1), 44–49. Available at: doi.org/10.4103/1735-3327.150330.

Allaker RP. 2010. The use of nanoparticles to control oral biofilm formation. J. Dent. Res. 89(11), 1175–1186. Available at doi.org/10.1177/0022034510377794.

Almoudi MM, et al. 2018. A systematic review on antibacterial activity of zinc against Streptococcus mutans. Saudi Dent. J. 30(4), 283–291. Available at: doi.org/10.1016/j.sdentj.2018.06.003.

Arbabi-kalati, F, Arbabi-kalati, F, Deghatipour, M, & nsari Moghadam A, A (2012). Evaluation of the efficacy of zinc sulfate in the prevention of chemotherapy-induced mucositis: a double-blind randomized clinical trial. Arch Iran Med, 15(7), 413–7.

Ayinampudi B, Narsimhan M. 2012. Salivary copper and zinc levels in oral pre-malignant and malignant lesions. J. Oral Maxillofac. Pathol. 16(2), 178. Available at: doi.org/10.4103/0973-029X.98452.

Aziz J, Rahman MT, Vaithilingam RD. 2021. Dysregulation of metallothionein and zinc aggravates periodontal diseases. J. Trace Elem. Med. Biol. 66, 126754. Available at: doi.org/10.1016/j.jtemb.2021.126754.

Belenguer-Guallar I, Jiménez-Soriano Y, Claramunt-Lozano A. 2014. Treatment of recurrent aphthous stomatitis. A literature review. J. Clin. Exp. Dent. 6(2), e168–e174. Available at: doi.org/10.4317/jced.51401.

Bhattacharya PT, Misra SR, Hussain M. 2016. Nutritional aspects of essential trace elements in oral health and disease: An extensive review. Scientifica 2016, 1–12. Available at: doi.org/10.1155/2016/5464373.

Bradshaw DJ, et al. 1993. The effects of triclosan and zinc citrate, alone and in combination, on a community of oral bacteria grown in vitro. J. Dent. Res. 72, 25–30. Available at: doi.org/10.1177/00220345930720010301.

Brauer GM. 1964. A review of zinc oxide-eugenol type filling materials and cements. NBS RPT 8647. Gaithersburg, MD: National Bureau of Standards, p. NBS RPT 8647. Available at: https://doi.org/10.6028/NBS.RPT.8647.

Ciapparelli L, Retief DH, Fatti LP. 1972. The effect of zinc on 9, 10-dimethyl-1, 2-benzanthracene (DMBA) induced salivary gland tumours in the albino rat—A preliminary study. S. Afr. Med. J. 37, 85–90.

Creeth JE, et al. 1993. Oral delivery and clearance of antiplaque agents from Triclosan-containing dentifrices. Int. Dent. J. 43(4 Suppl 1), 387– 397.

de Menêses AG, et al. 2020. Effects of oral supplementation in the management of oral mucositis in cancer patients: A meta-analysis of randomized clinical trials. Journal of Oral Pathology & Medicine: Official Publication of the International Association of Oral Pathologists and the American Academy of Oral Pathology 49, 117–125. Available at: doi.org/10.1111/jop.12901.

Doğan MS. 2018. Relation of trace elements on dental health, trace elements – human health and environment. IntechOpen. Available at: doi.org/10.5772/intechopen.75899.

Fatima T. 2016. Zinc: A precious trace element for oral health care? J. Pak. Med. Assoc. 66(8), 5.

Gholizadeh N, et al. 2017. The effect of orally-administered zinc in the prevention of chemotherapy-induced oral mucositis in patients with acute myeloid leukemia. Int. J. Cancer Manag. 10(8). Available at: doi.org/10.5812/ijcm.9252.

Goldberg M, et al. 2011. Dentin: Structure, composition and mineralization. Front. Biosci. (Elite Edition) 3, 711–735.

Guan G. 2018. A case series: Zinc deficiency as a potential contributor to oral dysgeusia. Modern Approaches in Dentistry and Oral Health Care 2. Available at: doi.org/10.32474/MADOHC.2018.02.000146.

Haase H, Rink L. 2009. Functional significance of zinc-related signaling pathways in immune cells. Annu. Rev. Nutr. 29, 133–152. Available at: doi.org/10.1146/annurev-nutr-080508-141119.

Hozyasz KK, et al. 2009. Relation between the concentration of zinc in maternal whole blood and the risk of an infant being born with an orofacial cleft. Br. J. Oral Maxillofac. Surg. 47, 466–469. Available at: doi.org/10.1016/j.bjoms.2009.06.005.

Hussein AS, et al. 2013. Salivary trace elements in relation to dental caries in a group of multi-ethnic school children in Shah Alam, Malaysia. Eur. J. Paediatr. Dent. 14(2), 113–118.

Jayapal N, et al. 2018. Role of zinc in oral health: A review. Int. J. Multidiscip. Res. Dev. 5, 3864–3867.

Khajuria DK, Zahra SF, Razdan R. 2018. Effect of locally administered novel biodegradable chitosan based risedronate/zinc-hydroxyapatite intra-pocket dental film on alveolar bone density in rat model of periodontitis. J. Biomater. Sci. Polym. 29, 74–91. Available at: doi.org/10.1080/09205063.2017.1400145.

Khanna S, et al. 2013. Trace elements (copper, zinc, selenium and molybdenum) as markers in oral sub mucous fibrosis and oral squamous cell carcinoma. Journal of Trace Elements in Medicine and Biology: Organ of the Society for Minerals and Trace Elements (GMS) 27, 307–311. Available at: doi.org/10.1016/j.jtemb.2013.04.003.

Kim YJ. 2007. Effects of zinc on oral bacteria and volatile sulfur compound (VSC) In oral cavity. 대한구강내과학회지 32(3), 273–281.

Kode, Manasi Ankolekar, Karjodkar Freny Rashmiraj (2013). Estimation of the Serum and the Salivary Trace Elements in OSMF Patients. JOURNAL OF CLINICAL AND DIAGNOSTIC RESEARCH. 10.7860/jcdr/2013/5207.3023

Kostecka-Sochoń P, Onopiuk BM, Dąbrowska E. 2018. Protective effect of increased zinc supply against oxidative damage of sublingual gland in chronic exposure to cadmium: Experimental study on rats. Oxid. Med. Cell. Longev. 2018, 3732842. Available at: doi.org/10.1155/2018/3732842.

Lauritano D, et al. 2018. Zinc plus octenidine: A new formulation for treating periodontal pathogens. A single blind study. J. Biol. Regul. Homeost. Agents 32, 231–236.

Liu Y, et al. 2009. Antibacterial activities of zinc oxide nanoparticles against Escherichia coli O157:H7. J. Appl. Microbiol. 107, 1193–1201. Available at: doi.org/10.1111/j.1365-2672.2009.04303.x.

Lynch RJM. 2011. Zinc in the mouth, its interactions with dental enamel and possible effects on caries: A review of the literature. Int. Dent. J. 61, 46–54. Available at: doi.org/10.1111/j.1875-595X.2011.00049.x.

Mansouri A, et al. 2012. The effect of zinc sulfate in the prevention of high-dose chemotherapy-induced mucositis: A double-blind, randomized, placebo-controlled study. Hematol. Oncol. 30, 22–26. Available at: doi.org/10.1002/hon.999.

Marsh PD. 1992. Microbiological aspects of the chemical control of plaque and gingivitis. J. Dent. Res. 71, 1431–1438. Available at: doi.org/10.1177/00220345920710071501.

Martinez JM, et al. 2014. Mucositis care in acute leukemia and non-Hodgkin lymphoma patients undergoing high-dose chemotherapy. Supportive Care in Cancer: Official Journal of the Multinational Association of Supportive Care in Cancer 22, 2563–2569. Available at: doi.org/10.1007/s00520-014-2199-y.

Mehdipour M, et al. 2010. Comparison of the effect of mouthwashes with and without zinc and fluocinolone on the healing process of erosive oral lichen planus. J. Dent. Res. Dent. 4, 25–28. Available at: doi.org/10.5681/joddd.2010.007.

Mehdipour M, et al. 2011. A comparison between zinc sulfate and chlorhexidine gluconate mouthwashes in the prevention of chemotherapy-induced oral mucositis. DARU J. Pharm. Sci. 19, 71–73.

Meunier N, et al. 2005. Importance of zinc in the elderly: The ZENITH study. Eur. J. Clin. Nutr. 59 Suppl 2, S1–S4. Available at: doi.org/10.1038/sj.ejcn.1602286.

Meyer K, et al. 2011. ZnO nanoparticles induce apoptosis in human dermal fibroblasts via p53 and p38 pathways.In Vitro Toxicol. 25(8), 1721–1726. Available at: doi.org/10.1016/j.tiv.2011.08.011.

Mizari N, et al. 2012. Effect of subchronic zinc toxicity on rat salivary glands and serum composition. Toxicol. Ind. Health 28, 917–922. Available at: doi.org/10.1177/0748233711427052.

Mulware SJ. 2013. Trace elements and carcinogenicity: A subject in review. 3 Biotech 3, 85–96. Available at: doi.org/10.1007/s13205-012-0072-6.

Orbak R, et al. 2007. Effects of zinc deficiency on oral and periodontal diseases in rats. J. Periodontal Res. 42, 138–143. Available at: doi.org/10.1111/j.1600-0765.2006.00939.x.

Ozler, GS (2014). Zinc deficiency in patients with recurrent aphthous stomatitis: a pilot study. J Laryngol Otol, 128(6), 531–3. 10.1017/S0022215114001078.

Ramazanzadeh B, et al. 2015. Comparison of antibacterial effects of ZnO and CuO nanoparticles coated brackets against streptococcus mutans. J. Dent. (Shiraz, Iran) 16, 200–205.

Rambod M, Pasyar N, Ramzi M. 2018. The effect of zinc sulfate on prevention, incidence, and severity of mucositis in leukemia patients undergoing chemotherapy. European Journal of Oncology Nursing: The Official Journal of European Oncology Nursing Society 33, 14–21. Available at: doi.org/10.1016/j.ejon.2018.01.007.

Rösing CK, et al. 2017. Efficacy of two mouthwashes with cetylpyridinium chloride: A controlled randomized clinical trial. Braz. Oral Res. 31, e47. Available at: doi.org/ 10.1590/1807-3107BOR-2017.vol31.0047.

Sachdev PK, et al. 2018. Zinc, copper, and iron in oral submucous fibrosis: A meta-analysis. Int. J. Dent. 2018, 3472087. Available at: doi.org/10.1155/2018/3472087.

Shuai T, et al. 2019. Prophylaxis with oral zinc sulfate against radiation induced oral mucositis in patients with head and neck cancers: A systematic review and meta-analysis of four randomized controlled trials. Front. Oncol. 9, 165. Available at: doi.org/10.3389/fonc.2019.00165.

Sirelkhatim A, et al. 2015. Review on zinc oxide nanoparticles: Antibacterial activity and toxicity mechanism. Nanomicro Lett. 7, 219–242. Available at: doi.org/10.1007/s4082 0-015-0040-x.

Stephen KW, et al. 1990. Control of gingivitis and calculus by a dentifrice containing a zinc salt and triclosan. J. Periodontol. 61, 674–679. Available at: doi.org/10.1902/jop. 1990.61.11.674.

Suzuki N, et al. 2018. Two mechanisms of oral malodor inhibition by zinc ions. J. Appl. Oral Sci. 26. Available at: doi.org/10.1590/1678-7757-2017-0161.

Tamura, T, Munger , RG, orcoran C, C, Bacayao , JY, Nepomuceno , B, & Solon , F (2005). Plasma zinc concentrations of mothers and the risk of nonsyndromic oral clefts in their children: a case-control study in the Philippines. Birth Defects Res A Clin Mol Teratol, 73(9), 612–6. 10.1002/bdra.20179.

Tavassoli Hojati, S, Alaghemand, H, Hamze, F, Ahmadian Babaki, F, ajab-Nia R, R, Rezvani, MB, Kaviani, M, & Atai, M (2013). Antibacterial, physical and mechanical properties of flowable resin composites containing zinc oxide nanoparticles. Dent Mater, 29(5), 495–505. 10.1016/j.dental.2013.03.011.

Thomas B, et al. 2013. Evaluation of micronutrient (zinc, copper and iron) levels in periodontitis patients with and without diabetes mellitus type 2: A biochemical study. Indian J. Dent. Res. 24, 468. Available at: doi.org/10.4103/0970-9290. 118400.

Tian X, et al. 2018. Oral zinc sulfate for prevention and treatment of chemotherapy-induced oral mucositis: A meta-analysis of five randomized controlled trials. Front. Oncol. 8, 484. Available at: doi.org/10.3389/fonc.2018.00484.

Toothpaste and mouthwash inactivate 99.9% of the virus that causes COVID-19. 2020. Br. Dent. J. 229, 753–753. Available at: doi.org/10.1038/s41415-020-2476-8.

Üçkardeş Y, et al. 2009. The effect of systemic zinc supplementation on oral health in low socioeconomic level children. Turk. J. Pediatr. 51, 5.

Uwitonze AM, et al. 2020. Zinc adequacy is essential for the maintenance of optimal oral health. Nutrients 12, 949. Available at: doi.org/10.3390/nu12040949.

Vogel-González M, et al. 2021. Low zinc levels at admission associates with poor clinical outcomes in SARS-CoV-2 infection. Nutrients 13, 562. Available at: doi.org/10.3390/ nu13020562.

Yadav A, et al. 2015. Estimation of serum zinc, copper, and iron in the patients of oral submucous fibrosis. Natl. J. Maxillofac. Surg. 6(2), 190–193. Available at: doi.org/ 10.4103/0975-5950.183851.

Yıldırımyan, N, zalp Ö, Ö, Şatır, S, Altay, MA, & Sindel, A (2019). Recurrent Aphthous Stomatitis as a Result of Zinc Deficiency. Oral Health Prev Dent, 17(5), 465–468. 10. 3290/j.ohpd.a42736.

Zahir S, Sarkar S. 2006. Study of trace elements in mixed saliva of caries free and caries active children. J. Indian Soc. Pedod. Prev. Dent. 24, 27–29. Available at: doi.org/ 10.4103/0970-4388.22832.

Zhang, J, Park, YD, Bae, WJ, El-Fiqi, A, Shin, SH, Lee, EJ, Kim, HW, & Kim, EC (2015). Effects of bioactive cements incorporating zinc-bioglass nanoparticles on odontogenic and angiogenic potential of human dental pulp cells. J Biomater Appl, 29(7), 954–64. 10.1177/0885328214550896.

8 Zinc in Child Health

Judie Arulappan and Sophia Cyril Vincent

8.1 INTRODUCTION

Zinc is a micronutrient essential for maintaining a constant cellular function in all forms of life. Almost all cells in our body contain Zn and it is a vital nutrient for growth, development, immune function, wound healing, blood clotting, thyroid function and clear vision. It also acts as a co-factor for numerous enzymes. It functions as a structural component of many proteins, hormones, hormone receptors and neuropeptides (Roohani et al., 2013). It is required for cognitive development in children during infancy and adolescence, the periods of rapid growth and development. It is also required for better neuropsychological functioning in childhood (Peland et al., 1997). Zn is not synthesized within the human body and hence should be taken from outside to maintain adequate levels. Its requirements are high in pregnant and lactating women (Narvaez-Caicedo et al., 2018) also in preterm infants as their stores are inadequate due to their higher metabolic rate and decreased gut absorption (Dao et al., 2017). Zinc is the key modulator of intracellular and intercellular neuronal signalling, and it is essential for the activity of a large number of metalloenzymes and cellular functions including DNA and RNA synthesis, cellular growth, differentiation and metabolism (Ramamoorthi et al., 2014).

8.2 ZINC DEFICIENCY

Zn deficiency in humans was first reported in 1961 and is considered a global health problem and one of the significant contributing factors to the burden of diseases in developing countries (Khalid et al., 2014; Roohani et al., 2013). It is found even in developed countries and occurs due to malnutrition, aging and chronic illnesses. More than 25% of elders living in rural America were reported to have Zn deficiency (Plum et al., 2010). Estimates of the World Health Organization (WHO) state that Zn deficiency affects 31% of the world's population and has affected 58% of Africans. The prevalence rates range from 4–73% in various regions of the world (Caulfield and Black, 2004).

8.3 ZINC DEFICIENCY AND CHILDREN

Zn deficiency affects children more as during infancy it could cause impaired growth and cognitive development in them (Chavez, 2022). Low plasma Zinc is

DOI: 10.1201/9781003412472-8

linked with decreased height for age. Zn-deficient children tend to be heavy and short (Hammill, 1985). In school age, boys are more vulnerable to its deficiency than girls (Gibson et al., 1989). In Ethopia, 47% of death in children occurs due to Zn deficiency (Walker et al., 2009) and about 44% of children were found to be stunted (IZiNCG, 2004). High prevalence of Zn deficiency is also reported in Iran (Salimi et al., 2004), Kenya (Mitheko et al., 2013), Nepal (Jiang et al., 2005), India (Pathak et al., 2008), Cameroon (Engle-Stone et al., 2014) and Sudan (Bushra et al., 2010). In Southeast Asia, nearly 95% of the under-five population is at risk of Zn deficiency (Kapil and Jain, 2011). The cognitive development in children is affected during infancy and adolescence, which are periods of rapid growth and development. Children from middle-income families were found to have moderate Zn deficiency during infancy (Skinner et al., 1997).

Zinc-deprived infants who had enhanced zinc supplements showed a greater linear growth and it was higher in girls. The infants who received Zn had better scores in motor development (Friel et al., 1993) and school-age children showed a positive impact by better performing in neuropsychologic tests (Penland et al., 1997).

8.4   ZINC DEFICIENCY AND PREGNANCY

Zn deficiency during pregnancy has detrimental effects on the mother and fetus. It leads to adverse birth outcomes such as impaired glucose tolerance, growth retardation, cognitive impairment, delayed development of the immune system, congenital malformations, low birth weight, increased risk of abortion, pregnancy-induced hypertension, preterm labor, miscarriages, stillbirths and prolonged labor (Khalid et al., 2014).

8.4.1   FINDINGS OF ANIMAL RESEARCH

Experimental studies in animals have shown that, during the early stages of brain development, deficiency of Zn caused brain abnormalities, reduction in the size of the cerebellum and altered zinc homeostasis. Deficiency during the later stages also impairs brain development and leads to learning and working memory deficits (Miranda et al., 2022). Zn deficiency in rats during infancy led to increased emotionality to stress (Halas et al., 1983). In young rhesus monkeys, it caused decreased motor activity, reduced performance in attention and short-term memory (Golub et al., 1994), and while in monkeys during pre-pubertal period, resulted in decreased activity and accuracy related to inhibitory control (Goulb et al., 1996).

8.5   ETIOLOGY OF ZINC DEFICIENCY

Zn deficiency is linked to deficits in cognitive and motor functioning, especially in premature children and those with chronic illness/nutritional issues (Prasad, 2013). Severe deficiency impairs emotional and behavioral responses by causing abnormal cerebellar function (Henkin et al., 1975). Zn deficiency in children occurs mostly due to malnutrition, aging and chronic illnesses. Acquired Zn deficiency may be attributed to excessive intake of phytates (present in seeds, legumes, soy products and whole grains), lack of meat intake, and consumption of oxalates (found in okra, spinach, tea and nuts). Zn absorption is affected due to calcium, phosphate and phytates (Stamoulis et al., 2007).

Inadequate Zn intake occurs due to Crohn's disease, short bowel syndrome, hookworm infestation, small bowel, malabsorption and pancreatic insufficiency (Prasad, 2009). It may also be due to some medications besides haemodialysis, haemolysis, burns, diarrhoea, urinary loss, diuretics and alcohol (Prasad, 2009).

8.6 COMMON ZINC DEFICIENCY DISORDERS IN CHILDREN

The most severe form of Zn deficiency is *Acrodermatitis enteropathica*, a rare autosomal recessive metabolic disorder resulting from the mutation in the intestinal Zip4 transporter. Acquired deficiency has been observed in children receiving total parental nutrition without supplementation of Zn (Black and Hurley, 2007).

8.6.1 ACRODERMATITIS ENTEROPATHICA

This is a recessively inherited disorder that arises due to a partial defect in intestinal Zn absorption and mutation in the SLC39A4 gene, which encodes a protein involved in Zn transport (Riveros et al., 2002; Wang et al., 2001). The affected infants develop diarrhea, vesiculobullous and erythematous dermatitis, alopecia, corneal scarring, retinal degeneration, cataract formation and optic atrophy, delayed sexual maturation, severe growth retardation, frequent infections and neuro-psychiatric manifestations (Cameron and McClain, 1986).

8.6.2 CROHN'S DISEASE AND ULCERATIVE COLITIS

Crohn's disease results in low plasma Zn concentration (Naber, 1998), and symptoms include dermatitis similar to *Acrodermatitis enteropathica*. These children also had eczematous changes and alopecia that responded to administration of Zn (Krasovecand Frenk, 1996). In addition, growth retardation, hypogonadism and taste abnormalities have also been reported in some cases (Krasovec and Frenk, 1996).

8.6.3 CYSTIC FIBROSIS

In infants with this disorder, dermatitis is the presenting feature (Wenk et al., 2010) which resembles *Acrodermatitis enteropathica* and the symptoms do not respond to Zn supplementation (Darmstadt et al., 1992). However, the same is suggested to these children to manage growth retardation, delayed sexual maturation, increased susceptibility to infections and anorexia (Turck et al., 2016).

8.6.4 SICKLE CELL DISEASE

In children with sickle cell disease, low Zn levels occur and are associated with poor and delayed growth (Leonard et al., 1998; Phebus et al., 1988). Evidence showed that the sickle cell crisis decreases with Zn supplementation (Swe et al., 2013).

8.6.5   LIVER DISEASE

Low plasma Zn levels have been reported in children and adults with chronic liver disease arising out of reduced intake, hypoalbuminemia, and increased urinary excretion which could be corrected with liver transplantation (Narkewicz et al., 1999; Pescovitz et al., 1996).

8.6.6   INTESTINAL FAILURE

This occurs due to insufficient Zn concentration in the parenteral nutrition and such children have the risk of Zn deficiency due to malabsorption (Balay et al., 2012).

8.6.7   RENAL DISEASE

Nephrotic syndrome in children can be complicated due to Zn deficiency secondary to increased urinary excretion (Perrone et al., 1990). Likewise, uremic patients experience Zn deficiency secondary to Zn malabsorption, reduced dietary intake (Mahajan et al., 1989).

8.7   CLINICAL MANIFESTATIONS OF ZINC DEFICIENCY IN CHILDREN

Children with mild Zn deficiency experience impaired taste and smell, depressed immunity, onset of night blindness and psoriasis. Severe deficiency showed the symptoms of dermatitis including scaly, erythematous, vesiculobullous and pustular lesions in the perioral and perianal areas. Other symptoms include frequent infections, diarrhoea, alopecia, and a depressed immune system (Shankar et al., 1998).

8.8   DIAGNOSIS AND MANAGEMENT OF ZINC DEFICIENCY IN CHILDREN

Plasma Zn concentration measurement is the most frequently used test for Zn deficiency (Kiliç et al., 1998). Serum Zn concentrations can be measured along with C-reactive protein (King et al., 2015). Low plasma Zn refers to less than 60 micrograms/dL (Hambidge et al., 1986). Zn levels may also be measured in neutrophils, lymphocytes and erythrocytes (Ruz et al., 1992).

If the deficiency occurs due to inadequate intake, oral replacement dose of 1–2 mg/kg/day is recommended. In some cases, it is necessary to take copper also to ensure that its deficiency does not occur due to high Zn administration. Similar treatment is applicable to cystic fibrosis, Crohn's disease, liver disease or sickle cell disease also (American Academy of Pediatrics Committee on Nutrition, 2019). Likewise, infants with intestinal failure may need additional Zn along with their parenteral nutrition and during therapy, its levels should be monitored closely (Yang et al., 2011). In children affected with *Acrodermatitis enteropathica*, doses to the tune of 3 mg/kg/day are required and their levels are monitored for three to six months (Maverakis et al., 2007).

8.9   CHILDHOOD DISEASES AND ZINC: WHO RECOMMENDATIONS

The WHO recommends Zn supplementation for infants and children with acute or persistent diarrhea (WHO, 2005). Children of less than six months age shall be given 10 mg/day for 10–14 days while those above six months are given 20 mg/day for 10–14 days (Lazzerini and Wanzira, 2016; Lukacik et al., 2008). Oral zinc supplementation reduces the incidence of diarrhea and related mortality in children (McDonald et al., 2015). Many studies reported that Zn supplementation reduced the frequency, decreased the incidence and provided a protective effect on the risk of diarrhea (McDonald et al., 2015; Mayo-Wilson et al., 2014; Aggarwal et al., 2007; Bayley, 1993; Sazawal et al., 1995). Pregnant women who received Zn supplements during pregnancy had their infants with less frequency of diarrhea (Iannotti et al., 2010). Children between two months and five years of age who received Zn supplementation showed a reduced incidence of pneumonia by 20% (Acevedo-Murillo et al., 2019).

8.10   DIETARY SOURCES OF ZINC AND RECOMMENDED INTAKE

Many foods are rich in zinc, the most prominent being oysters. Poultry and red meat are also important sources (Institute of Medicine, 2001). Other foods include beans, nuts, whole grains, almonds, ginger root, pecan nuts, whole-wheat grain, pumpkin seeds, crab, lobster, peas and turnip (Lim, 2013). Foods that contain high amounts of proteins are rich sources of Zn. The human body contains 2–3 g of Zn and the recommended daily intake is 11 mg/day for men and 8 mg/day for women to maintain regular homeostasis. Inappropriate consumption may lead to morbidity, especially in young children. The recommended dietary allowances are shown in Table 8.1.

8.11   CONCLUSION

Zinc is an essential micronutrient associated with more than 300 biological functions. Marginal Zn deficiency is common in children due to its poor intake or high consumption of phytates that compromise its intake. Children with Zn deficiency are

TABLE 8.1
Recommended Dietary Intake of Zinc

Subjects	Recommended Intake (Per Day)
0–6 months	2 mg
7 months - 3 years	3 mg
4–8 years	5 mg
9–13 years	8 mg
Females (14 to18 years)	9 mg
Males (14 years and above)	11 mg
Lactating Women & Pregnant	25 mg

at increased risk of morbidity due to respiratory infection and diarrhoea. The WHO recommends that all children with acute diarrhoeal illness be treated with Zn regardless of the etiology.

REFERENCES

Acevedo-Murillo JA, García León ML, Firo-Reyes V, et al. 2019. Zinc supplementation promotes a TH1 response and improves clinical symptoms in fewer hours in children with pneumonia younger than 5 years old. A randomized controlled clinical trial. Front. Pediatr. 7, 431.

Aggarwal R, Sentz J, Miller MA. 2007. Role of zinc administration in prevention of childhood diarrhea and respiratory illnesses: A meta-analysis. Pediatrics 119, 1120.

American Academy of Pediatrics Committee on Nutrition. 2019. Trace elements, in: Kleinman RE. Greer, FR (Eds.), Pediatric Nutrition, Eighth ed. American Academy of Pediatrics, Itasca, IL, p. 591.

Balay KS, Hawthorne KM, Hicks PD, et al. 2012. Low zinc status and absorption exist in infants with jejunostomies or ileostomies which persists after intestinal repair. Nutrients 4, 1273.

Bayley N. 1993. Bayley Scales of Infant Development, Third ed. The Psychological Corporation, San Antonio, TX.

Black MM, Hurley KM. 2013. Helping Children Develop Healthy Eating Habits. In: Tremblay, RE, Boivin, M, Peters, RDeV eds. Encyclopedia on Early Childhood Development [online]. https://www.child-encyclopedia.com/child-nutrition/according-experts/helping-children-develop-healthy-eating-habits

Bushra M, Elhassan EM, Ali NI, Osman E, Bakheit KH, Adam II. 2010. Anaemia, zinc and copper deficiencies among pregnant women in central Sudan. Biol. Trace. Elem. Res. 137, 255–261.

Cameron JD, McClain CJ. 1986. Ocular histopathology of acrodermatitisenteropathica. Br. J. Ophthalmol. 70, 662.

Caulfield LE, Black RE. 2004. Zinc deficiency. Comparative quantification of health risks: Global and regional burden of disease attributable to selected major risk factors. World Health Org. 1, 257–280.

Chavez M. 2022. Zinc deficiency in cognitive development and mental health in children from Latin America. CISLA Senior Integrative Projects. 45. https://digitalcommons.conncoll.edu/sip/45

Dao DT, Anez-Bustillos L, Cho BS, Li Z, Puder M, Gura KM. 2017. Assessment of micronutrient status in critically ill children: Challenges and opportunities. Nutrients Oct 28; 9(11), 1185–1210, Doi: 10.3390/nu9111185.

Darmstadt GL, Schmidt CP, Wechsler DS, et al. 1992. Dermatitis as a presenting sign of cystic fibrosis. Arch. Dermatol. 128, 1358.

Engle-Stone R, Ndjebayi AO, Nankap M, Killilea DW, Brown KH. 2014. Stunting prevalence, plasma zinc concentrations, and dietary zinc intakes in a nationally representative sample suggest a high risk of zinc deficiency among women and young children in Cameroon. J. Nutr. 144, 382–391.

Friel JK, Andrews WL, Matthew JD, Long DR, Cornel AM, Cox M, Zerbe GO. 1993. Zinc supplementation in very-low-birth-weight infants. J. Pediatr. Gastroenterol. Nutr. 17, 97–104.

Gibson RS, Vanderkooy PD, MacDonald AC, Goldman A, Ryan B, Berry M. 1989. A growth-limiting, mild zinc-deficiency syndrome in some southern Ontario boys with low height percentiles. AJCN. 49, 1266–1273.

Golub MS, Takeuchi PT, Keen CL, Gershwin ME, Hendrickx AG, Lonnerdal B. 1994. Modulation of behavioral performance of prepubertal monkeys by moderate dietary zinc deprivation. AJCN. 60, 238–243.

Golub MS, Takeuchi PT, Keen CL, Hendrickx AG, Gershwin ME. 1996. Activity and attention in zinc-deprived adolescent monkeys. AJCN. 64, 908–915.

Halas ES, Eberhardt MJ, Diers MA, Sandstead HH. 1983. Learning and memory impairment in adult rats due to severe zinc deficiency during lactation. Physiol. Behav. 30, 371–381.

Hambidge KM, Casey CE, Krebs NF. 1986. Zinc. In: Mertz, W (Ed), Trace elements in human and animal nutrition, Academic Press, Orlando. p.1.

Hammill DD. 1985. Detroit tests of learning aptitude. Austin, TX: Pro-ed.

Henkin RI, Patten BM, Re PK, Bronzert DA. 1975. A syndrome of acute zinc loss: Cerebellar dysfunction, mental changes, anorexia and taste and smell dysfunction. Arch. Neurol. 32, 745–751.

Iannotti LL, Zavaleta N, León Z, et al. 2010. Maternal zinc supplementation reduces diarrheal morbidity in peruvian infants. J. Pediatr. 156, 960.

Institute of Medicine. 2001. Food and Nutrition Board. Dietary Reference Intakes for Vitamin A, Vitamin K, Arsenic, Boron, Chromium, Copper, Iodine, Iron, Manganese, Molybdenum, Nickel, Silicon, Vanadium, and Zinc. National Academy Press.

IZiNCG. 2004. Assessment of the risk of zinc deficiency in populations and options for its control. International Zinc Nutrition Consultative Group Technical document #1, Washington, DC.

Jiang T, Christian P, Khatry SK, Wu L, West KP. 2005. Micronutrient deficiencies in early pregnancy are common, concurrent, and vary by season among rural Nepali pregnant women. J. Nutr. 135, 1106–1112.

Kapil U, Jain K. 2011. Magnitude of zinc deficiency amongst under five children in India. Indian J. Pediatr. 78, 1069–1072.

Khalid N, Ahmed A, Bhatti MS, Randhawa MA, Ahmad A, Rafaqat R. 2014. A question mark on zinc deficiency in 185 million people in Pakistan—Possible way out. Crit. Rev. Food Sci. Nutr. 54, 1222–1240.

Kiliç I, Ozalp I, Coŝkun T, et al. 1998. The effect of zinc-supplemented bread consumption on school children with asymptomatic zinc deficiency. J Pediatr. Gastroenterol. Nutr. 26, 167.

King JC, Brown KH, Gibson RS, et al. 2015. Biomarkers of nutrition for development (BOND)-zinc review. J. Nutr. 146, 858S.

Krasovec M, Frenk E. 1996. Acrodermatitis enteropathica secondary to Crohn's disease. Dermatology 193, 361.

Lazzerini M, Wanzira H. 2016. Oral zinc for treating diarrhoea in children. Cochrane Database Syst. Rev. 12, CD005436.

Leonard MB, Zemel BS, Kawchak DA, et al. 1998. Plasma zinc status, growth, and maturation in children with sickle cell disease. J. Pediatr. 132, 467.

Lim KH, Riddell LJ, Nowson CA, Booth AO, Szymlek-Gay EA. 2013. Iron and zinc nutrition in the economically-developed world: A review. Nutrients 5, 3184–3211.

Lukacik M, Thomas RL, Aranda JV. 2008. A meta-analysis of the effects of oral zinc in the treatment of acute and persistent diarrhea. Pediatrics 121, 326.

Mahajan SK, Bowersox EM, Rye DL, et al. 1989. Factors underlying abnormal zinc metabolism in uremia. Kidney Int. Suppl. 27, S269.

Maverakis E, Fung MA, Lynch PJ, et al. 2007. Acrodermatitis enteropathica and an overview of zinc metabolism. J. Am. Acad. Dermatol. 56, 116.

Mayo-Wilson E, Junior JA, Imdad A, et al. 2014. Zinc supplementation for preventing mortality, morbidity, and growth failure in children aged 6 months to 12 years of age. Cochrane Database Syst. Rev. CD009384. doi: 10.1002/14651858.CD009384.pub2.

McDonald CM, Manji KP, Kisenge R, et al. 2015. Daily zinc but not multivitamin supplementation reduces diarrhea and upper respiratory infections in Tanzanian infants: A randomized, double-blind, placebo-controlled clinical trial. J. Nutr. 145, 2153.

Miranda CTDOF, Vermeulen-Serpa KM, Pedro ACC, Brandão-Neto J, de Lima Vale SH, Figueiredo MS. 2022. Zinc in sickle cell disease: A narrative review. J. Trace Elem. Med. Biol. 126980. doi: 10.1016/j.jtemb.2022.126980.

Mitheko A, Kimiywe J, Njeru PN. 2013. Dietary, Socio-economic and Demographic Factors Influencing Serum Zinc Levels of Pregnant Women at Naivasha Level 4 Hospital Nakuru County, Kenya. University of Nairobi.

Naber THJ, Van Den Hamer CJA, Baadenhuysen H, Jansen JBMJ. 1998. The value of methods to determine zinc deficiency in patients with Crohn's disease. Scand. J. Gastroenterol. 33, 514–523.

Narkewicz MR, Krebs N, Karrer F, Orban-Eller K, Sokol RJ. 1999. Correction of hypozincemia following liver transplantation in children is associated with reduced urinary zinc loss. Hepatology. Mar; 29(3), 830–833.

Narváez-Caicedo C, Moreano G, Sandoval BA, Jara-Palacios MÁ. 2018. Zinc deficiency among lactating mothers from a Peri-urban community of the Ecuadorian andean region: An initial approach to the need of zinc supplementation. Nutrients. Jul 05; 10, 869–877. doi: 10.3390/nu10070869.

Pathak PKU, Dwivedi SN, Singh R. 2008. Serum zinc levels amongst pregnant women in a rural block of Haryana state, India. Asia Pac. J. Clin. Nutr. 17, 2.

Perafán Riveros C, França LFS, Alves ACF, Sanches JJA. 2002. Acrodermatitis enteropathica: Case report and review of the literature. Pediatr. Dermatol. 19, 426–431.

Perrone L, Gialanella G, Giordano V, et al. 1990. Impaired zinc metabolic status in children affected by idiopathic nephrotic syndrome. Eur. J. Pediatr. 149, 438.

Penland JG, Sandstead HH, Alcock NW, Dayal HH, Chen XC, Li JS, Yang JJ. 1997. A preliminary report: Effects of zinc and micronutrient repletion on growth and neuropsychological function of urban Chinese children. J. Am. Coll. Nutr. 16, 268–272.

Pescovitz MD, Mehta PL, Jindal RM, et al. 1996. Zinc deficiency and its repletion following liver transplantation in humans. Clin. Transplant 10, 256.

Phebus CK, Maciak BJ, Gloninger MF, Paul HS. 1988. Zinc status of children with sickle cell disease: Relationship to poor growth. Am. J. Hematol. 29, 67.

Plum LM, Rink L, Haase H. 2010. The essential toxin: Impact of zinc on human health. Int. J. Environ. Res. Public Health 7, 1342–1365.

Prasad AS. 2013. Discovery of human zinc deficiency: Its impact on human health and disease. Ad. Nut. An. Int. Rev. J. 4, 176–190.

Prasad AS. 2009. Impact of the discovery of human zinc deficiency on health. J. Am. Coll. Nutr. Jun. 28, 257–265.

Ramamoorthi R, Graef KM, Krattiger A, Dent JC. 2014. WIPO Re: Search: Catalyzing collaborations to accelerate product development for diseases of poverty. Chem. Rev. 114, 11272–11279.

Roohani N, Hurrell R, Kelishadi R, Schulin R. 2013. Zinc and its importance for human health: An integrative review. Journal of Research in Medical Sciences: The Official Journal of Isfahan University of Medical Sciences 18, 144.

Ruz M, Cavan KR, Bettger WJ, Gibson RS. 1992. Erythrocytes, erythrocyte membranes, neutrophils and platelets as biopsy materials for the assessment of zinc status in humans. Br. J. Nutr. 68, 515.

Salimi S, Yaghmaei M, Joshaghani H, Mansourian A. 2004. Study of zinc deficiency in pregnant women. Iranian J. Pub. Health. 33, 15–18.

Sazawal S, Black RE, Bhan MK. 1995. Zinc supplementation in young children with acute diarrhea in India. N. Engl. J. Med. 333, 839.

Shankar AH, Prasad AS. 1998. Zinc and immune function: The biological basis of altered resistance to infection. Am. J. Clin. Nutr. 68, 447S.

Skinner JD, Carruth BR, Houck KS, Coletta F, Cotter ROTTD, McLEOD MAX. 1997. Longitudinal study of nutrient and food intakes of infants aged 2 to 24 months. J. Am. Diet. Assoc. 97, 496–504.

Stamoulis I, Kouraklis G, Theocharis S. 2007. Zinc and the liver: An active interaction. Dig. Dis. Sci. Jul; 52, 1595–1612.

Swe KMM, Abas ABL, Bhardwaj A, Barua A, Nair NS. 2013. Zinc supplements for treating thalassaemia and sickle cell disease. Cochrane Database Syst. Rev. 6, CD009415. 10. 1002/14651858.

Turck D, Braegger CP, Colombo C, et al. 2016. ESPEN-ESPGHAN-ECFS guidelines on nutrition care for infants, children, and adults with cystic fibrosis. Clin. Nutr. 35, 557.

Walker CF, Ezzati M, Black R. 2009. Global and regional child mortality and burden of disease attributable to zinc deficiency. Eur. J. Clin. Nutr. 63, 591–597.

Wang K, Pugh EW, Griffen S, Doheny KF, Mostafa WZ, al-Aboosi MM, Gitschier, J. 2001. Homozygosity mapping places the acrodermatitisenteropathica gene on chromosomal region 8q24. 3. Am. J. Hum. Genet. 68, 1055–1060.

Wenk KS, Higgins KB, Greer KE. 2010. Cystic fibrosis presenting with dermatitis. Arch. Dermatol. 146, 171.

World Health Organization: The treatment of diarrhoea: A manual for physicians and other senior health workers. 2005. Available at: https://www.who.int/maternal_child_ adolescent/documents/9241593180/en/ (Accessed on March 05, 2020).

Yang CF, Duro D, Zurakowski D, et al. 2011. High prevalence of multiple micronutrient deficiencies in children with intestinal failure: A longitudinal study. J. Pediatr. 159, 39.

9 Role of Zinc in Cancer

R. Vijayalakshmi and Soundharya Ravindran

9.1 INTRODUCTION

For cell proliferation, differentiation, growth and development, zinc is an essential micronutrient. Among other things, it plays a role in homeostasis, DNA synthesis, RNA transcription, cell division and activation. It takes part in cell division, protein synthesis and growth which helps infants, children, adolescents and pregnant women to maintain health. Studies have shown that zinc-responsive stunting and more rapid body weight gain in malnourished children supplemented with zinc (Roohani et al., 2013).

Zinc exists in two forms in the human body as protein-bound zinc, which is involved in the stabilisation and functioning of proteins, and free zinc which is labile and chelatable. The signalling process is triggered by the changes in the level of free zinc at both extracellular and intracellular sites. This type of zinc signalling can be categorised into fast and slow. Fast signalling occurs quickly in a few seconds to minutes and the latter takes place over a longer period. In these processes, zinc functions as an intracellular second messenger. In the instance of the late signalling pathway, extracellular stimuli cause the transcriptional regulation of zinc-related proteins such as zinc importers (ZIPs) and zinc transporters (ZNTs) to occur.

9.2 MECHANISM OF ZINC IN CANCER

Any dysregulation of and/or mutations in the ZIP and ZNT genes have been linked to a variety of functional diseases, including diabetes and cancer. There is a connection between the expression of zinc transporters and their dysregulation or dysfunction in several cancer types, suggesting that any change in the intracellular concentration of zinc and its homeostasis could affect the malignancy. ZNT expression levels/activities were altered in a variety of malignancies, however, no specific mutations or variations of ZNT or ZIP were

linked to a specific type of cancer. Depending on the subtype of the transporter, ZIPs and ZNTs can be tissue-specific or globally expressed in bodily tissues (Michalczyk and Cymbaluk-Ploska, 2020).

Cellular ZIPs have been extensively studied and found increased in patients with various tumour types. This could explain why zinc levels are higher in almost all tumour types. For example, in prostate cancer malignant cells – ZIP1, ZIP2 and ZIP3 have been found to be down-regulated while in individuals with pancreatic, prostate, lung and ovarian cancers, over-expression of ZIP4 was connected to

DOI: 10.1201/9781003412472-9

enhanced cell proliferation. The expression of ZIP6 and ZIP10 has been associated with lymph node metastases in breast cancer patients. In tamoxifen-resistant breast cancer, higher levels of ZIP7 were linked to enhanced tumour development and invasion. Many enzymes and transcription factors including nuclear factor (NF)-B, are regulated by changes in zinc homeostasis (Bafaro, 2017).

Many dietary substances were found to be potential cancer-prevention agents. Zinc, for example, is recognised as playing a critical part in the host's defence mechanisms against the genesis and progression of a variety of cancers (Dhawan and Chadha, 2010). Zinc levels in serum and malignant tissues of patients with various types of cancers are known to be aberrant, indicating zinc's role in the development of cancer. Serum zinc levels have been found to be lower in patients with cancer of breast, gall bladder, lung, colon, head and neck as well as bronchus (Skrajnowska and Korczak, 2019). Even while serum zinc levels are lower in most cancer contexts, tumour tissues in breast and lung cancer have higher zinc levels when compared to normal tissues. The degree of expression of ZNTs in human tumours is related to their malignancy, implying that changes in intracellular zinc homeostasis can contribute to cancer severity. In most cancers, certain ZIPs are increased, allowing tumour cells to avoid apoptosis and promote cell survival through autophagic mechanisms.

RING E3 ubiquitin ligases are involved in a wide range of biological functions, and abnormalities in them are linked to cancer formation as well as found to be over-expressed in malignancies. The RING family of E3 ubiquitin ligases is the most common form and it can bind two zinc ions in a cross-brace arrangement. Both the structural integrity of the RING finger domain and the stabilisation of the E2-E3 complex during the ubiquitin pathway are influenced by zinc binding. The identification and development of new anticancer drugs target certain RING E3 ligases.

9.3 ZINC DEFICIENCY AND CANCER

Several studies revealed the association between cancer of cervix, bladder, prostate, etc., and zinc levels which showed decreased expression of the latter. Its deficiency has been linked to an increased risk of cancer in several epidemiological studies that include larger tumours, later stages of cancer and head and neck cancer. Clinical studies have suggested that Zn deficiency could impair cellular mechanisms that respond to and repair DNA damage resulting in the accumulation of DNA mutations and increased cancer risk (Ho, 2004). There is a connection between the expression of ZNTs and their dysregulation/dysfunction in many cancer types, suggesting that any change in the intracellular concentration of zinc and its homeostasis could affect the condition (Wang et al., 2020). Zinc insufficiency affects the oxidant defence system, causing it to malfunction. Oxidative stress is a major contributor of chronic diseases including cancer. NADPH oxidase, a family of plasma-membrane-associated enzymes, catalyses the formation of superoxide radicals from oxygen by using NADPH as an electron source. This enzyme is known to be inhibited by zinc.

Zinc is known to stimulate the formation of metallothionein (MT), which is a hydroxyl radical scavenger. The generation of hydroxyl radicals is catalysed by iron and copper ions while zinc prevents the two molecules from attaching to the cell membrane. Many antioxidant defence-system proteins, such as glutathione peroxidase, MTs and Cu/Zn superoxide dismutase, need zinc to function. Multiple proteins involved in DNA damage and repair responses require zinc. Its deficiency has been shown to produce oxidative DNA damage and affect the latter's repair process both *in vitro* and *in vivo*. Clinical studies have suggested that zinc deficiency could impair cellular mechanisms that respond to and repair DNA damage resulting in the accumulation of DNA mutations and increased cancer risk. (Ho, 2004).

9.4   ZINC IN PROSTATE CANCER

In the year 1954, Mawson and Fisher reported that there is a decrease in zinc levels in patients with prostate cancer (Song and Ho, 2009). Many clinical studies that followed further confirmed the above finding. There was a significant decrease in zinc levels to the tune of 60 to 80% when compared to normal or benign prostate subjects (Costello et al., 2004). The decrease also showed a concomitant decrease in citrate levels as an early incident. The down-regulation of zinc and citrate levels is a characteristic hallmark and clinically relevant markers in differentiating normal and cancer cells.

Of the three zones of the human prostate gland, the peripheral zone plays a major role in accumulating zinc while the central gland consists of a lesser amount of zinc than the former. This is because of the presence of glandular secretory epithelial cells in the peripheral zone which accumulate zinc. There is a 10-fold higher accumulation of zinc in the peripheral zone than in any other soft tissues. Due to the toxicity of zinc, mammalian cells typically have mechanisms that prevent zinc from building up inside the cells, but in secretory cells, the build-up of extra zinc is necessary to suppress m-aconitase activity. The exact mechanism of zinc accumulation is unclear but understood that the ZIP1 zinc transporters are important for this uptake by prostate cells. The increased expression of ZIP1 transporters increases the accumulation of zinc in the cells and their downregulation leads to a decrease.

A number of clinical studies have shown decreased levels of zinc in patients with prostate cancer. A consistent relationship between decreased levels of zinc and citrate is observed in early malignancy irrespective of variation in patient population, stage of cancer, sampling process and assay procedures. It has been proved that downregulation of zinc is a hallmark characteristic of prostatic cancer. It is thought that ZIP1 downregulation is the key mechanism for decreased zinc accumulation. When prostate cancer progresses, the RAS-responsive element binding protein (RREB) is increased. This down regulates ZIP1 transporters resulting in the decrease of zinc levels in prostate cancer. Other zinc importers such as ZIP2, ZIP3 and ZnT4 are also shown to have decreased expression in prostate cancer. It has been observed that testosterone and prolactin help in the regulation of ZIP1 expression, which in turn helps in the uptake of zinc.

It has also been shown that zinc is involved in the apoptosis and metabolism of prostate cells. Zinc suppressed the metastatic potential of prostate cancer by inhibiting NF-Kβ signalling and suppressing the invasive potential of proteolytic enzyme – urokinase-type plasminogen activator, aminopeptidase N and prostate-specific antigen (Song and Ho, 2009). It is known that zinc deficiency disrupts the function of critical zinc-dependent proteins that maintain the DNA integrity in prostate cells, ultimately resulting in increased DNA damage.

It has been observed that there were no zinc-accumulating and citrate-producing malignant cells found in the cases of prostate cancer. Malignant cells are manifested only when there is a metabolic transformation when normal cells have lost the ability to accumulate zinc and citrate oxidising cells (Costello and Franklin, 1998) . This observation opens up the opportunity to utilise zinc as a treatment regimen. Additionally, it has been observed that the cellular uptake of zinc did not take place by a simple diffusion process and there was a need to restore or enhance the uptake transport process. A combination of hormonal regimens by including prolactin or testosterone along with increasing zinc uptake is considered to be the most efficacious approach.

9.5   ZINC IN LUNG CANCER

Lung cancer is the most common form worldwide resulting in death in the population which is divided into Small Cell Lung Carcinoma (SCLC) and Non-Small Cell Lung Cancer (NSCLC). The latter accounts for 85% of lung cancers and a vast majority of epithelial tumours are known to be associated with reduced intra-tumoral and plasma zinc levels. Zinc supplementation has shown growth inhibitory effects against NSCLC cells and might increase docetaxel efficacies (Kocdor et al., 2015). Several meta-analysis studies have revealed the association between serum zinc levels and lung cancer (Issell et al., 1981; Wang et al., 2019). In some studies, the serum copper results were found to be higher in the controls and the zinc levels lower in tumour group. Copper and zinc are closely related trace elements that are involved in cell proliferation, growth, gene expression, apoptosis and other processes.

Zinc microenvironment influences the behaviour of malignant cancer cells, and this contributes to the prevention of lung cancer. Compared to other cancers, the role of zinc in lung cancer was not much investigated. A study conducted by Gomez et al. (2016) showed the association of zinc in preventing lung cancer.

9.6   ZINC IN BREAST CANCER

Breast cancer is a complex condition that is primarily caused by mutations or aberrant activation of genes that regulate cell growth and proliferation (Alam and Kelleher, 2012). Zinc is a cofactor of proteins that regulate responses to DNA damage, intracellular signalling enzymes, and Matrix Metallo Proteinases (MMP), which are involved in the development of breast cancer besides a transcription factor for enzymes involved in the synthesis of DNA and RNA. The changes in zinc concentration may play a significant role in cell dysfunction and proliferation, including the development and progression of the disease.

More than 25 different species of proteases make up the large family of MMPs, which are crucial for healthy tissue remodelling and depend on zinc. The MMPs are divided into five classes which share two structural components *viz.* zinc ion and pro-peptide. The former is located at the catalytic site of the protein and the latter contains a cysteine residue. MMPs are known to be involved in the propagation of neoplasms including breast cancer. Several studies have observed changes in the expression and increase in the proteolytic activity of these enzymes in both invasive and metastatic cancers. MMPs are thought to contribute to the metastasis process by opening a passageway in the extracellular matrix for the cancer cells to use to colonise other tissues. According to Do Nascimento Holanda et al. (2017), deregulation of MMPs is crucial in numerous phases of the growth of breast cancer and in processes that need zinc binding to the catalytic site. Several studies evaluated the concentration of zinc in plasma and its relationship and association it has with breast cancer. It has been observed that the low concentration of zinc in the plasma is positively related to an increase in the plasma concentration of MMP-2 and MMP-9 in patients with breast cancer when compared to the control group.

9.7   ZINC IN OESOPHAGEAL CANCER

Oesophageal cancer (OC) is the sixth leading cause of cancer deaths worldwide. Due to a lack of early clinical symptoms and presentation, its diagnosis occurs at an advanced stage accompanied by a poor prognosis. The two types of OC include Oesophageal Squamous Cell Carcinoma (OSCC) and Oesophageal Adeno-Carcinoma (OAC). Epidemiological and clinical studies revealed that OAC arises mainly due to risk factors such as gastroesophageal reflux disease and Barret's oesophagus while OSCC is closely associated with smoking, alcoholism and deficiency of vitamins and micronutrients including zinc. The exact mechanism of zinc's cancer-prevention role in OC is not understood clearly but evidence from clinical and animal studies has demonstrated that zinc is important for healthy oesophageal epithelium and its deficiency results in abnormal cell proliferation promoting tumour development. As a cofactor for more than 300 enzymes and contributor to the stabilisation of protein 3-D structure, zinc plays a multifunctional role in maintaining normal and healthy oesophageal epithelium.

Zinc deficiency is a common phenomenon in high incidence of OC and in the initiation and progression of the disease. Several animal and clinical studies have explained the molecular mechanism and correlation between zinc deficiency and OC. The latter induced single and double-stranded DNA breaks elevating oxidative stress and compromising immunological function were implicated as mechanisms and showed an inverse relationship between them. A study by Islami et al. (2009) showed that the blood zinc levels in spring and fall in the Linxian province of China (a region with a high risk of developing OC) were only 72% and 62% respectively of the recommended dietary requirement. Another study suggested that zinc deficiency contributed to the increase in the risk of oesophageal cancer and its increased dietary supplementation could decrease such exigency in Chinese populations. It has been discovered that high-risk countries have a lower mean zinc supply than low-risk nations.

Several *in vitro* and animal studies have that zinc deficiency affected the development of OC by regulating the miRNA expression. Cell line study with KYSE170 and ECA109 by He et al. (2016) showed that the relative levels of miR-21, miR-31 and miR-93 were significantly decreased while the expression of miR-375 was significantly increased in oesophageal cells cultured with zinc. There was no difference in the cell status indicating the correlation between zinc levels and miRNA expression.

9.8   ZINC IN OVARIAN CANCER

Ovarian cancer is one of the most lethal gynaecological malignancy due to pelvic and abdominal metastasis. Several combination approaches are used for ovarian cancer therapy including surgery, chemotherapy and targeted therapy but the overall survival rate of patients with ovarian cancer is still very low. 75% of ovarian tumours are found to be metastatic as the tumour cells detach from the ovaries and spread into the peritoneal organs including liver, spleen and intestines. It has been suggested that the epithelial-to-mesenchymal transition (EMT) is associated with ovarian tumour metastasis. There hasn't been enough research done on zinc's potential link to ovarian cancer.

However, a study demonstrated that zinc activates the metal response element-binding transcription factor-1 (MTF-1), a cellular zinc sensor that directly binds to six highly conserved zinc finger domains to maintain zinc homeostasis. By encouraging EMT, the MTF-1 has been shown to act as an oncogene in ovarian cancer. It has been established that zinc promotes EMT in ovarian cancer cells by inducing MTF-1 expression and activating the downstream ERK1/2 and AKT pathways. This suggests zinc's role in ovarian tumour metastasis. Additionally, zinc depletion in ovarian cancer cells prevented EMT by reducing MTF-1 expression and weakening the ERK1/2 and AKT pathways. According to this study, zinc deficiency has the potential to treat ovarian cancer by suppressing EMT. Since MTF-1 expression in ovarian cancer cells is strongly connected with zinc concentrations, the loss of MTF-1 resulted in a considerable decrease in intracellular zinc levels (Zhang et al., 2020).

ZIP13 is a member of the SLC39A/ZIP family, found in the Golgi apparatus and internalised compartments of various cells. Recent research has identified ZIP13 as a zinc transporter in the dermis and linked it to various dermal diseases. A thorough assessment of the ZIP family's relationships with ovarian cancer was conducted by Cheng et al. (2021), which revealed that ZIP13 is an independent predictive marker in ovarian cancer patients that could increase cell proliferation, invasion, adhesion and metastasis. It was discovered that patients with ovarian cancer had less overall survival if ZIP13 expression was high. In ovarian cancer cells, deletion of ZIP13 inhibited cell growth resulting in its less migration and invasion besides decline in peritoneal metastasis. Further, ZIP13 deletion appeared to enhance the levels of vesicular zinc. ZIP13 transporter also regulates the intracellular distribution of zinc influencing the expression of genes involved in the extracellular matrix and mediating the metastasis of ovarian cancer.

9.9   CONCLUSION

Zinc has a potential role in the activation and stabilisation of a large number of enzymes and transcription factors besides possessing antioxidant response, causing apoptosis and in carcinogenesis. The homeostasis of zinc is crucial for the maintenance and regulation of different functions and its dysregulation or imbalance has adverse effects in human health including malnutrition, stunted growth and development of cancer. Recent research has shown that zinc and its compounds, as well as nanoparticles, have importance in cancer biology. The exact role and molecular mechanism with which zinc exhibits its effects is still not clear. The element has been shown to have contradictory effects depending on the type, stage and other clinical manifestations of the patients. However, zinc supplementation has been indicated to have positive effects on prostate cancer and promote EMT-mediated metastasis in patients with ovarian cancer. Zinc could be used as a potent biomarker or as a complimentary chemo-preventive agent depending upon the patient's clinical status. More details are required to understand the molecular mechanisms and the biological relevance of the element as well as to use it in cancer on a larger scale.

REFERENCES

Alam S, Kelleher SL. 2012. Cellular mechanisms of zinc dysregulation: A perspective on zinc homeostasis as an etiological factor in the development and progression of breast cancer. Nutrients 4, 875–903. 10.3390/nu4080875.

Bafaro E, Liu Y, Xu Y, Dempski RE. 2017. The emerging role of zinc transporters in cellular homeostasis and cancer. Signal Transduct. Target. Ther. 2, 1–12. 10.1038/sigtrans. 2017.29.

Cheng X, Wang J, Liu C, Jiang T, Yang N, Liu D, Zhao H, Xu Z. 2021. Zinc transporter SLC39A13/ZIP13 facilitates the metastasis of human ovarian cancer cells via activating Src/FAK signaling pathway. J. Exp. Clin. Cancer Res. 40, Art. 199. 10.1186/s13046-021-01999-3.

Costello LC, Feng P, Milon B, Tan M, Franklin RB. 2004. Role of zinc in the pathogenesis and treatment of prostate cancer: Critical issues to resolve. Prostate Cancer Prostatic Dis. 7, 111–117. 10.1038/sj.pcan.4500712.

Costello LC, Franklin RB. 1998. Novel role of zinc in the regulation of prostate citrate metabolism and its implications in prostate cancer. Prostate 35, 285–296.

Dhawan DK, Chadha VD. 2010. Zinc: A promising agent in dietary chemoprevention of cancer. Indian J. Med. Res. 132, 676–682.

Do Nascimento Holanda AO, De Oliveira ARS, Cruz KJC, Severo JS, Morais, JBS, Da Silva BB, Do Nascimento Marreiro D. 2017. Zinc and metalloproteinases 2 and 9: What is their relation with breast cancer? Rev. Assoc. Med. Bras. 63, 78–84. 10.1590/1806-9282.63.01.78.

Gomez NN, Biaggio VS, Ciminari ME, Chaca MVP, Alwarez SM. 2016. Zinc: What is its role in lung cancer?, in: Erkekogulu P, Gumusel BK (Eds.), Nutritional Deficiency. IntechOpen. 10.5772/63209.

He Y, Jin J, Wang L, Hu Y, Liang D, Yang H, Liu Y, Shan B. 2016. Evaluation of miR-21 and miR-375 as prognostic biomarkers in oesophageal cancer in high-risk areas in China. Clin. Exp. Metastasis. 34, 73–84. 10.1007/s10585-016-9828-4

Ho E. 2004. Zinc deficiency, DNA damage and cancer risk. J. Nutr. Biochem. 15, 572–578. 10.1016/j.jnutbio.2004.07.005.

Islami F, Ren JS, Taylor PR, Kamangar F. 2009. Pickled vegetables and the risk of oesophageal cancer: A meta-analysis. Br. J. Cancer 101, 1641–1647.

Issell BF, Macfadyen BV, Gum ET, Valdivieso M, Dudrick SJ, Bodey GP. 1981. Serum zinc levels in lung cancer patients. Cancer 47, 1845–1848.

Kocdor H, Ates H, Aydin S, Cehreli R, Soyarat F, Kemanli P, Harmanci D, Cengiz H, Kocdor MA. 2015. Zinc supplementation induces apoptosis and enhances antitumor efficacy of docetaxel in non-small-cell lung cancer. Drug Des. Devel. Ther. 9, 3899–3909. 10.2147/DDDT.S87662.

Michalczyk K, Cymbaluk-Płoska A. 2020. The role of zinc and copper in gynecological malignancies. Nutrients 12, 1–21. 10.3390/nu12123732.

Roohani N, Hurrell R, Kelishadi R, Schulin R. 2013. Zinc and its importance for human health: An integrative review. J. Res. Med. Sci. 18, 144–157.

Skrajnowska D, Bobrowska-Korczak B. 2019. Role of zinc in immune system and anti-cancer defense mechanisms. Nutrients 11. 10.3390/nu11102273.

Song Y, Ho E. 2009. Zinc and prostatic cancer. Curr. Opin. Clin. Nutr. Metab. Care 12, 640–645. 10.1097/MCO.0b013e32833106ee.Zinc.

Wang J, Zhao H, Xu Z, Cheng X, 2020. Zinc dysregulation in cancers and its potential as a therapeutic target. Cancer Biol. Med. 17, 612–625. 10.20892/j.issn.2095-3941.2020.0106.

Wang Y, Sun Z, Li A, Zhang Y. 2019. Association between serum zinc levels and lung cancer: A meta-analysis of observational studies. World J. Surg. Oncol. 17, 1–8. 10.1186/s12957-019-1617-5.

Zhang R, Zhao G, Shi H, Zhao X, Wang B, Dong P, Watari H, Pfeffer LM, Yue J. 2020. Zinc regulates primary ovarian tumor growth and metastasis through the epithelial to mesenchymal transition. Free Radic. Biol. Med. 160, 775–783. 10.1016/j.freeradbiomed.2020.09.010.

10 Zinc-Based Therapeutic Approach for Diabetes Mellitus

Tao Ming Sim and Dinesh Kumar Srinivasan

10.1 INTRODUCTION

Diabetes mellitus (DM) is a disease of abnormal carbohydrate metabolism which is characterized by hyperglycemia (elevated blood glucose levels). It is due to either a relative or absolute impairment in the secretion of insulin by the pancreas and may be associated with varying degrees of peripheral resistance to the action of the hormone insulin. The two major kinds of DM are type 1 DM and type 2 DM, with the latter being far more common (>90%).

- Type 1 DM: The pathogenesis of type I DM involves autoimmune destruction of pancreatic β cells, resulting in an absolute insulin deficiency. Type I DM accounts for about 5-10% of all cases of diabetes (Mobasseri et al., 2020).
- Type 2 DM: The most common form of diabetes and the focus of this chapter is a heterogenous disorder with differing prevalence across different ethnicities. On the basis of current trends, the incidence of type 2 DM is rising due to aging and the soaring rates of obesity with its prevalence expected to double by the year 2030 (Wild et al., 2004).

The causes of T2DM are multi-factorial and involve both environmental and genetic factors that affect β cell function and the sensitivity of tissues to insulin. It is characterized by hyperglycemia due to a progressive decline in insulin secretion by pancreatic β cells superimposed on a backdrop of peripheral tissue insulin resistance, resulting in a metabolic state of relative insulin deficiency. Most patients with T2DM are asymptomatic at presentation and are detected as hyperglycemic in screening tests, prompting further confirmatory laboratory tests. The classic symptoms of T2DM include polyuria, polydipsia, polyphagia and weight loss (Turner et al., 2010). While the symptoms seem insidious, the complications of T2DM are far more severe and it has been largely agreed that a major complication is a reduced life expectancy. Different numerical values of years of life lost have been found by the studies on populations in different geographical areas, ranging from 3.1 to over 10 years, but all point towards the same conclusion that T2DM

DOI: 10.1201/9781003412472-10

increases mortality (Leung et al., 2015; Wright et al., 2017; Kang et al., 2017). Other complications of T2DM are a consequence of the chronic hyperglycemia, due to which, there is an accumulation of advanced glycation end products, giving rise to diabetic angiopathy, nephropathy and neuropathy.

Diabetic patients are also more predisposed to infections as T2DM impairs the immune system, with well-documented mechanisms including disruptions to the complement system, secretion of cytokines and chemotaxis and phagocytosis of monocytes (Geerlings and Hoepelman, 1999). Furthermore, T2DM is commonly associated with a state of chronic low-level inflammation that is typically linked to obesity (Wellen and Hotamisligil, 2005).

Currently, T2DM is managed by a combination of lifestyle modifications and pharmacological interventions. These interventions aim to prevent or delay the onset of diabetic complications through regulated control of fasting and post-prandial glucose levels. Oral hypoglycemic agents (OHAs) are the most commonly prescribed medications in the long-term management of T2DM but are associated with adverse effects including hypoglycemia, weight gain, lactic acidosis and fluid retention (Manandhar Shrestha et al. 2017). To date, an ideal antidiabetic agent, marked by good efficacy, high therapeutic window, low cost and high clinical utility, has remained rather elusive and is still the focus of much research (Lebovitz, 1999).

10.2   ZINC AND TYPE 2 DM

Zinc (Zn) is a metallic trace element found in biological systems and serves an instrumental role as a co-factor of more than 300 enzymes in the human body involved in diverse processes such as mitosis and apoptosis (McCall et al., 2000). A peculiar relationship between Zn and T2DM has been proposed following the ground-breaking discovery that Zn is a component of insulin crystals (Scott and Fisher, 1935). The amount of Zn in biological systems is regulated in enzyme function by transporters and Zn-binding proteins such as metallothionein (MT) which can tightly bind Zn and release it depending on its redox status (Maret and Krężel, 2007).

A study was conducted on patients with T2DM to investigate the levels of Zn in their serum and the effects of factors such as age, gender, glycemic control and duration of diabetes as compared to healthy controls (Masood et al., 2009). It was found that its levels in serum were significantly lower in patients with T2DM as compared to healthy subjects. Further, no association was found between factors such as age, gender, glycemic status, duration of DM and the serum concentration of Zn. It is therefore highly suggestive that the decreased concentration of Zn in diabetic patients is due to the disease process itself rather than possible confounding factors such as age, gender and duration of the disease.

The altered metabolism of Zn in diabetes has been the subject of numerous studies and Zn supplementation has been demonstrated to have beneficial effects in diabetic animal models (Kumar et al., 2012) and humans, shedding light on future therapeutics focused on Zn (Simon and Taylor, 2001; Anderson et al., 2001). Research elucidating the roles of Zn transporter (ZnT) - 8 and MT in the pathogenesis of T2DM further concretizes the possible underpinning of altered Zn physiology in the development of T2DM.

10.3   MECHANISMS OF ACTION OF Zn AND ITS COMPLEXES AGAINST DM

Despite the seminal discovery that Zn chloride ($ZnCl_2$) had insulin-mimetic activity and stimulated lipogenesis in rat adipocytes (Coulston and Dandona, 1980), Zn-containing compounds received particularly less attention as potential anti-diabetic agents than other metal complexes (Azam et al., 2018). Nevertheless, several mechanisms have been published to explain the anti-diabetic effects of Zn and its complexes.

10.3.1   ANTIOXIDANT PROPERTY OF ZINC

Persistent low levels of chronic inflammation found in patients with type 1 and 2 DM are associated with oxidative damage from free radicals (Oberley, 1988). These free radicals come from varied sources such as activated leukocytes, increased transition metal bioavailability and auto-oxidation of glucose in plasma (Wolff et al., 1991). A study has shown that the radical-scavenging antioxidant capacity in patients with type 1 and 2DM is lower than in age-matched controls which suggests that oxidative stress and the inability to cope with this pathological stimulus in DM patients contribute to the pathogenesis of the disease (Maxwell et al., 1997). Zn levels are decreased in diabetic patients and correction of this deficiency has been found to decrease lipid peroxidation (Faure et al., 1995) and results in the improvement of glucose homeostasis (Faure et al., 1993).

Two main mechanisms have been proposed to explain the antioxidant properties of Zn in biological systems (Bray and Bettger, 1990). The first mechanism involves the protection of sulfhydryl groups against oxidation which has been reported in studies on the enzyme δ-aminolevulinate dehydrase which contains 4 reactive sulfhydryl groups. Zn has been found to stabilize δ-aminolevulinate dehydrase by reducing the reactivity of its sulfhydryl groups through direct binding, chelating in close proximity to the sulfhydryl group which increases steric hindrance and causing a conformational change to the enzyme subunit (Gibbs et al., 1985). These protective mechanisms of Zn reduce the reactivity of sulfhydryl groups which are closely associated with the activity of the enzyme, therefore protecting the enzymes against oxidative stress. These mechanisms can be extrapolated and applied to many biological peptides in which binding of Zn protects a sulfhydryl group.

The second mechanism involves the prevention of the production of oxygen free radicals such as OH· and O2·⁻. Zn has been reported to act as a co-factor for the enzyme superoxide dismutase (SOD), regulates glutathione metabolism and MT expression besides inhibiting nicotinamide adenine dinucleotide phosphate-oxidase (NADPH-oxidase) (Foster and Samman, 2010). SOD is an essential antioxidant enzyme that regulates detoxification of reactive oxygen species (ROS) and catalyses of dismutation of superoxide anion into oxygen and hydrogen peroxide.

10.3.2   ZN AND ITS INFLUENCE ON INSULIN SIGNALLING

Zn has a potent effect on glucose homeostasis by promoting insulin sensitivity and pancreatic β cell function (Foster and Samman, 2010). It is also known to promote activation of phosphatidylinositol protein 3-kinase and protein kinase B (Akt), the two important steps in the intracellular signalling of insulin, which ultimately lead to increased transport of glucose uptake into cells (Barthel et al., 2007). In the insulin signalling pathway, activated insulin receptor kinase (IRK) which is a receptor tyrosine kinase (RTK) phosphorylates tyrosine residues on insulin receptor substrate-1 (IRS1) and insulin receptor substrate-2 (IRS2), culminating in a phosphorylation cascade to transmit the signal from binding of insulin to insulin receptor (Hançer et al., 2014).

The insulin-sensitizing effect of Zn has been linked to the inhibition of the tyrosine-phosphatase activity of the intracellular protein tyrosine phosphatase-1B (PTP1B) which in its active state, is a negative regulator of the insulin signalling pathway. (Figure 10.1).

The downstream effect of signalling via the insulin pathway is the translocation of glucose transporter type 4 (GLUT4) to the plasma membrane of

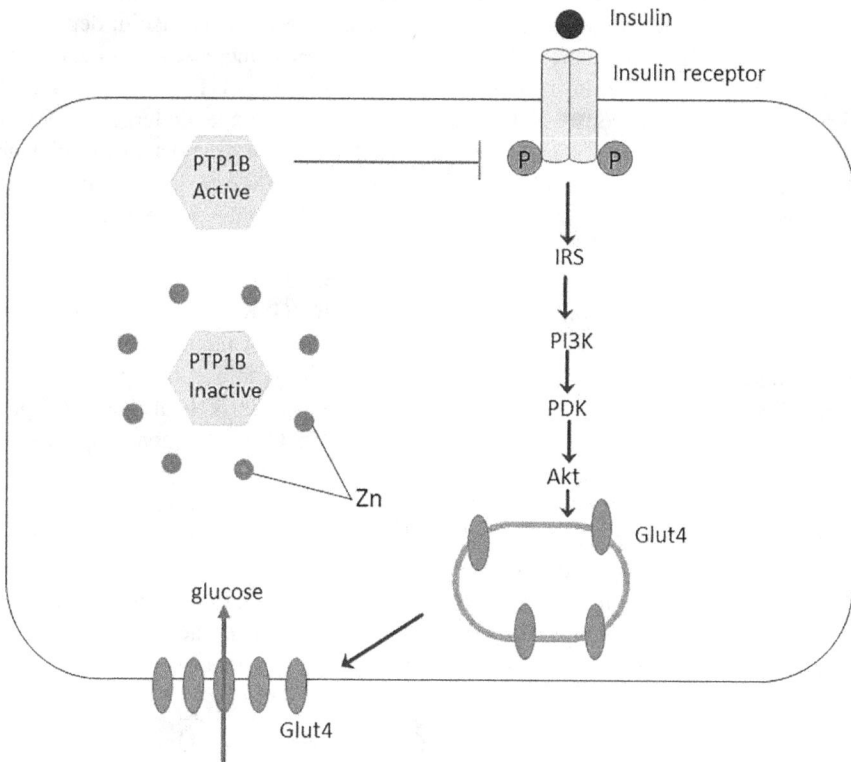

FIGURE 10.1 Insulin signalling pathway and insulin mimicking the function of Zn ions. Zn might inhibit the activity of PTP1B, which activates the insulin-signalling pathway. PI3K, phosphatidylinositol-3-kinase; IRS, insulin receptor substrate; PKD, protein kinase D; Glut4, glucose transporter type 4.

Source: Fukunaka and Fujitani, 2018.

adipocytes and muscle cells. Skeletal muscles are the major reservoirs of Zn in the body and hold approximately 60% of it. The Zn transporter SLC39A7 (ZIP7) has been found to be associated with insulin-regulated glycemic control within skeletal muscles. A knockdown model of ZIP7 revealed that glucose metabolism in skeletal muscles is enhanced by ZIP7 transporter via Akt phosphorylation and that ZIP7-mediated Zn activates insulin receptor signalling through binding with PTP1B (Myers et al., 2013).

Zn has been found to have cytoprotective effects on the islets of Langerhans (pancreatic islet cells). In rat models, supplementation of Zn prior to treatment with diabetogenic drugs dithizone and alloxan was found to prevent islet destruction and hyperglycemia (Mikhail and Awadallah, 1977).

In addition to Zn-mediated Akt phosphorylation, Zn is also shown to affect insulin-responsive aminopeptidase (IRAP) which is a Zn-dependent molecule that is expressed in adipocytes and muscle cells and is required for the maintenance of normal GLUT4 levels required for glucose uptake into these cells. IRAP is usually stored in intracellular membrane compartments and its translocation to the plasma membrane is mediated by insulin (Garvey et al., 1998). This insulin-dependent translocation of IRAP is found to be impaired in DM. The increase in Zn-dependent IRAP activity can cause an increase in translocation of GLUT4 to the surface of adipocytes and muscle cells which could be a potential therapeutic target in DM. It was reported that Zn was able to induce translocation of GLUT4 to the plasma membranes of peripheral tissues, with the resultant uptake of glucose into cells causing a consequent lowering in blood glucose levels, conferring better glycemic control (Ezaki, 1989).

Another target of Zn that has been identified is GSK-3b which is part of the insulin signalling pathway. Glycogen synthase kinase-3β (GSK-3β) inhibits IRS and therefore raised levels of GSK-3β can interfere with insulin signalling. GSK-3β has been found to be elevated in the muscles of patients with T2DM, giving rise to impaired glycogen synthase activity and insulin resistance in skeletal muscles (Ilouz et al., 2002). It was also found that in the presence of Zn, GSK-3β is noncompetitively inhibited which leads to increased glucose uptake.

Altogether, Zn affects the insulin signalling pathway at multiple levels by upregulating Akt phosphorylation and by decreasing GSK-3β levels possibly as a direct result of Akt phosphorylation. The regulation of the levels of these intracellular signalling molecules confers Zn its insulinomimetic properties. As Zn affects the molecular signalling pathway of insulin, it is therefore a probable therapeutic target in the treatment of DM and insulin resistance.

10.3.3 ZINC AND ITS EFFECTS ON THE CYTOKINE-MEDIATED DESTRUCTION OF β CELLS

Type 1 DM is caused by the destruction of β cells in the islets of Langerhans in the pancreas. Though the major immunological process involved in the destruction of β cells is T-cell mediated cytotoxicity, cytokine-induced cell death and the presence of auto-antibodies have also been documented as alternate pathways in the destruction of β cells (Donath et al., 2003; Cnop et al., 2005). The same cytokine pathways have also been indicated to have a role in the pathogenesis of type 2 DM. In addition to the insulin-like effects of Zn as previously alluded, Zn has also been found to have cytoprotective effects on β cells, especially in protecting β cells against destruction through proinflammatory cytokine pathways.

The effects of Zn on cytokine production are controversial and might be dependent on the dose or the time of exposure to Zn. *In vitro* administration of Zn up to concentrations of 500 μM has been found to stimulate cytokine production. A combination of Zn and the ionophore pyrithione decreases proinflammatory cytokine production, indicating that high intracellular Zn exerts an inhibitory effect on cytokine production while its low intracellular content promotes the latter (Wellinghausen et al., 1996)]. It has been suggested that Zn inhibits cyclic nucleotide phosphodiesterase (PDE), causing an increase in intracellular cyclic guanosine monophosphate (cGMP) which mediates the inhibitory effect of Zn on release of proinflammatory cytokines TNF-α and IL-1β by suppressing LPS-induced activation of inhibitory kappa B (IκB) kinase β (IKKβ) and NFκB (von Bülow et al., 2005). Another cellular mechanism proposed to explain the inhibitory actions of Zn on the secretion of proinflammatory cytokines is the induction of mRNA of A20, a Zn-finger protein inhibiting NFκB activation (Prasad et al., 2004).

10.4   ZINC AS AN ANTI-DIABETIC AGENT

Zn has been shown to display insulinogenic properties which may, in part, explain the mechanism of action of Zn as an anti-diabetic agent. Among various Zn compounds, mononuclear Zn coordination complexes have been shown to exhibit potent anti-diabetic effect (López-Viseras et al., 2014; Philip et al., 2017). Coulston and Dandona (1980) reported that Zn exerts a potent stimulatory lipogenic effect similar to but independent of endogenous insulin, and therefore has an additive effect with insulin when both are incubated *in vitro*.

There has also been interest in the use of Zn nanoparticles in anti-diabetic treatment. El-Gharbawy et al. (2016) investigated the use of Zinc oxide (ZnO) nanoparticles as an anti-diabetic agent and compared its efficacy against Vildagliptin, a DPP4 inhibitor and a standard diabetic medication. It was found that the ZnO nanoparticles improved many indices of diabetic dysfunction including glucose tolerance, insulin levels, frustosamine levels and weight loss. A treatment regimen of both Vildagliptin and ZnO nanoparticles further improved these indices. It was concluded that ZnO nanoparticles exhibited significant anti-diabetic effects while the addition of Vildagliptin exerted a synergistic action.

Novel Zn compounds have also been experimented in the development of potent anti-diabetic drugs with minimal biological toxicity. Vijayaraghavan et al. (2012) developed a novel Zn-3 hydroxy flavone that showed a significant anti-hyperglycemic activity comparable to that of gliclazide, a standard antidiabetic drug in animals. Another study involved the synthesis and evaluation of Zn complexes in a murine model. The study concluded that the compounds bis(3-hydroxy-2-methyl-4(H)-pyran-4-thiono) Zn and bis(3-hydroxy-2-methyl-4(H)-pyran-4-seleno) Zn had potent glucose-lowering properties and improved hemoglobin A1c (HbA1c) profile (Nishiguchi et al., 2017). Another team studied a new Zn-mixed ligand complex (metformin-3-hydroxyflavone) and found it to have significant anti-diabetic properties (Koothappan et al., 2018).

10.5 ZINC IN DIABETES-RELATED COMPLICATIONS

10.5.1 Diabetic Retinopathy

In the past decades, several human studies on the efficacy of Zn supplementation have been conducted. Faure et al. (1995) tested the efficacy of oral Zn gluconate in insulin-dependent diabetic patients with early retinal degenerative lesions and found that it reduced lipid peroxidation, a process having an important role in degenerative ocular diseases (Njie-Mbye et al., 2013).

10.5.2 Diabetic Neuropathy

A double-blind randomized study conducted to test oral Zn therapy in diabetic neuropathy concluded that a six-week therapy offered significant improvement in blood glucose profiles and improved motor nerve conduction velocity (Gupta et al., 1998).

10.5.3 Oxidative Damage

The effects of Zn supplementation in altering plasma thiobarbituric acid reactive substances (TBARS), a marker of oxidative damage was studied by Anderson et al. (2001) and concluded that Zn supplementation decreased oxidative damage. A further study confirmed the results of the earlier one and established the potent antioxidant properties of Zn therapy (Roussel et al., 2003).

It is well known that Superoxide dismutase (SOD) is a key antioxidant enzyme that breaks down potentially harmful intracellular oxygen species. A clinical trial examined the effects of Zn on SOD activity and gene expression in diabetic patients (Nazem et al., 2019) and showed that daily intake of Zn (50 mg) increased both gene expression and enzyme activity of SOD and also increased levels of insulin. There was also a significant reduction in HbA1c and fasting blood glucose levels.

10.5.4 Lipid and Glycemic Control

A study explored the effects of a combination of Zn and vitamin A on lipid and glycemic profiles in type 1 DM patients (Shidfar et al., 2010). Similar to earlier investigations on vitamins with Zn, this one also yielded positive results in controlling lipid and glycemic index in DM patients.

10.5.5 Inflammation and Diabetic Foot

Another study found Zn supplementation had beneficial effects on ulcer size and metabolic parameters in DM patients with diabetic foot issues (Heravi et al., 2017).

The study also found a significant decrease in C-reactive protein (CRP), a biomarker of inflammation.

10.5.6 VASCULAR COMPLICATIONS

A double-blind, randomized, placebo-controlled human trial investigated the antioxidant properties of Zn in reducing microvascular complications of DM by observing the levels of microalbuminuria, a marker of glomerulopathy. In the three-month trial, it was found that Zn intake significantly reduced urinary albumin excretion (Parham et al., 2008). Another study on Zn therapy using biomarkers of oxidative damage and vascular function in DM patients showed that improving the Zn status minimized the impact of oxidative damage and vascular function (Seet et al., 2011).

10.5.7 CARDIO VASCULAR COMPLICATIONS

The effect of Zn administration on diabetic cardiovascular complications was investigated by Heidarian et al. (2009). The three-month, randomized, double-blind, controlled, crossover study involving T2DM patients found that Zn supplementation reduced serum homocysteine and increased folate and vitamin B12 levels a indicating potential cardioprotective effect.

10.5.8 OVERALL PROTECTIVE EFFECT

Dietary consumption of Zn on the development and risk of DM was studied in Australian middle-aged women involving 8921 participants with a follow-up period of six years. 333 cases of DM were recorded during the period and after correcting the dietary and non-dietary factors. It was found that the highest quintile of dietary Zn intake had almost half the odds of developing type II DM. It was also found that a high Zn/iron ratio has a similar protective effect against the development of type II DM (Vashum et al., 2013).

A clinical trial on the effects of genetic variation in SLC30A8 observed a three-way interaction between this genotype, body mass index, Zn to iron ratio and inferred that a high Zn:iron ratio conferred a protective effect on T2DM risk and this was more pronounced among T-allele carriers of the *SLC30A8* gene (Drake et al., 2017).

The human studies which evaluated the effects of supplementation of Zn in patients with DM are summarized in Table 10.1.

10.6 CONCLUSION

In the past few decades, there has been much interest in zinc as a therapeutic option for DM. This trace element has been found to have significant positive effects on various metabolic pathways implicated in the pathogenesis of DM and human trials have found that supplementation offered a preventive effect on both the development and progression of the disease. Clinical studies have revealed vast benefits of

TABLE 10.1

Zinc Supplementation-based Therapeutics for Patients with DM

Study Details	Country	Findings and Conclusions	Reference
T1DM, 2 groups (n = 18) 3 months, Case-control	France	Zn deficiency in insulin-dependent diabetic patients corrected by Zn supplementation. Also, supplementation decreased lipid peroxidation.	Faure et al., 1995
T2DM, 3 groups (n = 50), 6 weeks, Randomized, double-blind placebo-controlled	India	Oral Zn supplementation was found to improve glycemic control and the severity of diabetic peripheral neuropathy.	Gupta et al., 1998
T2DM, 4 groups (n = 110), 6 months, Randomized, double-blind placebo- controlled	Tunisia	Potential antioxidant effect of individual and combined supplementation of Zn and Chromium in patients.	Anderson et al., 2001
T2DM, 2 groups (n = 56), 6 months, Randomized, double-blind placebo-controlled	Tunisia	Zn supplementation offered beneficial antioxidant effects in diabetic patients.	Roussel et al., 2003
T2DM, 4 groups (n = 69), 3 months, Randomized, double-blind placebo-controlled	Iran	Combination of vitamins and minerals such as Zn decreased blood pressure.	Farvid et al., 2004
T2DM, 4 groups (n = 69), 3 months, Randomized, double-blind placebo-controlled	Iran	Combination of vitamins and minerals such as Zn improved glomerular function.	Farvid et al., 2005
T2DM 2 groups (n = 39), 6 months, Randomized, double-blind placebo-controlled	Iran	Supplementation of Zn reduced microalbuminemia, a manifestation of renal microvascular complication of DM.	Parham et al., 2008
T2DM, 2 groups (n = 50), 3 months, Randomized, double-blind placebo-controlled crossover	Iran	Zn supplementation decreased homocysteine level which is an independent risk factor for cardiovascular complications of T2DM.	Heidarian et al., 2009
T1DM, 2 groups (n = 48), 12 weeks,	Iran	Combination of vitamin A and Zn improved serum lipid profile.	Shidfar et al., 2010

TABLE 10.1 *(Continued)*
Zinc Supplementation-based Therapeutics for Patients with DM

Study Details	Country	Findings and Conclusions	Reference
Randomized, double-blind placebo-controlled T2DM, 2 groups (n = 40), 3 months, Randomized, single-blind placebo-controlled	Singapore	No beneficial effects on vascular function and oxidative damage.	Seet et al., 2011
T2DM, 2 groups (n = 8921), 6 years, Cohort	Australia	Higher total dietary Zn intake and a high zinc/iron ratio are associated with a lower risk of developing T2DM.	Vashum et al., 2013
T2DM patients with grade 3 diabetic foot ulcer, 2 groups (n = 60), 12 weeks, Randomized, double-blind placebo-controlled	Iran	Zn supplementation caused significant reduction in severity of diabetic foot ulcers and beneficial effects on glucose metabolism, cholesterol profile and in decreasing biomarkers of inflammation.	Hervari et al., 2017
T2DM, 2 groups (n = 26132), 19 years, Cohort	Sweden	Dietary Zn supplementation and high zinc/iron ratio found to lower the risk of developing T2DM. Protective effect of Zn therapy influenced by other factors such as obesity and *SLC30A8* genotype.	Drake et al., 2017
T2DM, 2 groups (n = 70), 8 weeks, Randomized, double-blind placebo- controlled	Iran	Daily supplementation of Zn gluconate (50 mg) was useful in the management of oxidative damage associated with T2DM by reducing superoxide dismutase activity. Also, it controlled lipid and glycemic markers.	Nazem et al., 2019
T2DM, 3 groups (n = 98), 3 months, Interventional open- labelled randomized	Egypt	Combination therapy of high-dose vitamins A, E and Zn improved glycemic control, insulin secretion and beta cell function.	Said et al., 2021

Zn as an anti-diabetic agent with strong antioxidant, anti-inflammatory and insulin-secreting capabilities. It also altered the sequelae of DM such as microvascular complications (microalbuminuria and diabetic retinopathy) and macrovascular complications (cardiovascular disease). Despite many exciting findings and data from trials, it is important to investigate the detailed effects of various dosages and durations of Zn therapy across a wide variety of populations. Given the known association between DM, metabolic syndrome and Zn metabolism, the studies could examine the differences in response to consumption of Zn-fortified products and the development progression of DM as well as associated risk factors.

REFERENCES

Anderson RA, Roussel AM, Zouari N, Mahjoub S, Matheau JM, Kerkeni A. 2001. Potential antioxidant effects of zinc and chromium supplementation in people with type 2 diabetes mellitus. J. Amer. Coll. Nutr. 20, 212–218. doi: 10.1080/07315724.2001. 10719034.

Azam A, Raza MA, Sumrra, SH. 2018. Therapeutic application of zinc and vanadium complexes against diabetes mellitus a coronary disease: a review. Open Chem. 16 (1), 1153–1165. 10.1515/chem-2018-0118.

Barthel A, Ostrakhovitch EA, Walter PL, Kampkötter A, Klotz LO. 2007. Stimulation of phosphoinositide 3-kinase/Akt signaling by copper and zinc ions: mechanisms and consequences. Arch. Biochem. Biophys. 463, 175–182. doi: 10.1016/j.abb.2007. 04.015. Epub 2007 May 4. PMID: 17509519.

Bray TM, Bettger WJ. 1990. The physiological role of zinc as an antioxidant. Free Rad. Biol. Med. 8, 281–291. doi: 10.1016/0891-5849(90)90076-U.

Cnop M, Welsh N, Jonas J-C, Jörns A, Lenzen S, Eizirik DL. 2005. Mechanisms of pancreatic beta-cell death in type 1 and type 2 diabetes: many differences, few similarities. Diabetes. 54, S97–S107. doi: 10.2337/diabetes.54.suppl_2.s97.

Coulston L, Dandona P. 1980. Insulin-like Effect of Zinc on Adipocytes. Diabetes. 29, 665–667. doi: 10.2337/diab.29.8.665.

Donath MY, Størling J, Maedler K, Mandrup-Poulsen T. 2003. Inflammatory mediators and islet β-cell failure: a link between type 1 and type 2 diabetes. J. Mol. Med. 81, 455–470. doi: 10.1007/s00109-003-0450-y.

Drake I, Hindy G, Ericson U, Orho-Melander M. 2017. A prospective study of dietary and supplemental zinc intake and risk of type 2 diabetes depending on genetic variation in SLC30A8. Genes Nutr. 12, 30. doi: 10.1186/s12263-017-0586-y.

El-Gharbawy RM, Emara AM, Abu-Risha SE. 2016. Zinc oxide nanoparticles and a standard antidiabetic drug restore the function and structure of beta cells in Type-2 diabetes. Biomed. Pharmacother.. 84, 810–820. doi: 10.1016/j.biopha.2016.09.068.

Ezaki O. 1989. IIb group metal ions (Zn2+, Cd2+, Hg2+) stimulate glucose transport activity by post-insulin receptor kinase mechanism in rat adipocytes. J Biol Chem. 1989 Sep 25;264(27), 16118–16122. PMID: 2550432.

Farvid MS, Jalali M, Siassi F, Hosseini M. 2005. Comparison of the effects of vitamins and/ or mineral supplementation on glomerular and tubular dysfunction in type 2 diabetes. Diabetes Care. 28, 2458–2464.

Farvid MS, Jalali M, Siassi F, Saadat N, Hosseini M. 2004. The Impact of Vitamins and/or Mineral Supplementation on Blood Pressure in Type 2 Diabetes. J Am Coll Nutr. 23(3), 272–279. doi: 10.1080/07315724.2004.10719370.

Faure P, Benhamou PY, Perard A, Halimi S, Roussel AM. 1995. Lipid peroxidation in insulin-dependent diabetic patients with early retina degenerative lesions: effects of an oral zinc supplementation. Eur J Clin Nutr. Apr 49, 282–288. PMID: 7796786.

Faure P, Corticelli P, Richard MJ, Arnaud J, Coudray C, Halimi S, Favier A, Roussel AM 1993. Lipid peroxidation and trace element status in diabetic ketotic patients: influence of insulin therapy. Clin Chem. May 39, 789–793. PMID: 8485869.

Foster M, Samman S. 2010. Zinc and redox signaling: perturbations associated with cardiovascular disease and diabetes mellitus. Antioxid Redox Signal. 13(10), 1549–1573. doi: 10.1089/ars. 2010.3111.

Fukunaka A, Fujitani Y. 2018. Role of Zinc Homeostasis in the Pathogenesis of Diabetes and Obesity. Int. J. Mol. Sci.. 19, 476. doi: 10.3390/ijms19020476.

Garvey WT, Maianu L, Zhu JH, Brechtel-Hook G, Wallace P, Baron AD. 1998. Evidence for defects in the trafficking and translocation of GLUT4 glucose transporters in skeletal muscle as a cause of human insulin resistance. J Clin Invest. 101, 2377–2386. doi: 10.1172/JCI1557.

Geerlings SE, Hoepelman AI. 1999. Immune dysfunction in patients with diabetes mellitus (DM). FEMS Immunol Med Microbiol. 26, 259–265. doi: 10.1111/j.1574-695X.1999.tb01397.x.

Gibbs PNB, Gore MG, Jordan PM. 1985. Investigation of the effect of metal ions on the reactivity of thiol groups in human 5-aminolaevulinate dehydratase. Biochem J 1 February 1985. 225, 573–580. doi: 10.1042/bj2250573.

Gupta R, Garg VK, Mathur DK, Goyal RK. 1998. Oral zinc therapy in diabetic neuropathy. J Assoc Physicians India. Nov;46, 939–942. PMID: 11229219.

Hançer NJ, Qiu W, Cherella C, Li Y, Copps KD, White MF. 2014. Insulin and metabolic stress stimulate multisite serine/threonine phosphorylation of insulin receptor substrate 1 and inhibit tyrosine phosphorylation. J Biol Chem. 2014 May 2;289, 12467–12484. doi: 10.1074/jbc.M114.554162.

Heidarian E, Amini M, Parham M, Aminorroaya A. 2009. Effect of zinc supplementation on serum homocysteine in type 2 diabetic patients with microalbuminuria. Rev Diabet Stud. 2009 Spring, 64–70. doi: 10.1900/RDS.2009.6.64. Epub 2009 May 10. PMID: 19557297; PMCID: PMC2712914.

Heravi MM, Barahimi E, Razzaghi R, Bahmani F, Gilasi HR, Asemi Z. 2017. The effects of zinc supplementation on wound healing and metabolic status in patients with diabetic foot ulcer: a randomized, doubleblind, placebocontrolled trial. Wound Rep and Reg. 25, 512–520. doi: 10.1111/wrr.12537.

Ilouz R, Kaidanovich O, Gurwitz D, Eldar-Finkelman H. 2002. Inhibition of glycogen synthase kinase-3β by bivalent zinc ions: insight into the insulin-mimetic action of zinc. Biochem Biophys Res Commun. 295, 102–106.

Kang, Yu Mi, Cho, Yun Kyung, Lee, Seung Eun, Park, Joong-Yeol, Lee, Woo Je, Kim, Ye-Jee, & Jung, Chang Hee (2017). Cardiovascular Diseases and Life Expectancy in Adults With Type 2 Diabetes: A Korean National Sample Cohort Study. The Journal of Clinical Endocrinology & Metabolism, 102(9), 3443–3451, doi: 10.1210/jc.2017-00643.

Koothappan M, Vellai RD, Subramanian IP, Subramanian SP. 2018. Synthesis of a new zinc-mixed ligand complex and evaluation of its antidiabetic properties in high fat diet: low dose streptozotocin induced diabetic rats. Diabetes Metab J. 2018 Jun;42, 244–248. doi: 10.4093/dmj.2018.0002. Epub 2018 Apr 24. PMID: 29885106; PMCID: PMC6015968.

Kumar SD, Vijaya M, Samy RP, Dheen ST, Ren M, Watt F, Kang YJ, Bay BH, Tay SSW. 2012. Zinc supplementation prevents cardiomyocyte apoptosis and congenital heart defects in embryos of diabetic mice. Free Radic. Biol. Med. 53, 1595–1606. doi: 10.1016/j.freeradbiomed.2012.07.008.

Lebovitz HE. 1999. Type 2 diabetes: an overview. Clin Chem. 1999 Aug;45(8 Pt 2), 1339–1345. PMID: 10430816.

Leung, Man-Yee Mallory, Pollack, Lisa M., Colditz, Graham A., & Chang, Su-Hsin (2015). Life Years Lost and Lifetime Health Care Expenditures Associated With Diabetes in the U.S., National Health Interview Survey, 1997–2000. Diabetes Care, 38(3), 460–468. doi: 10.2337/dc14-1453.

López-Viseras ME, Fernández B, Hilfiker S, González CS, González JL, Calahorro AJ, Colacio E, Rodríguez-Diéguez A. 2014. In vivo potential antidiabetic activity of a novel zinc coordination compound based on 3-carboxy-pyrazole. J InorgBiochem. 2014 Feb;131, 64–67. doi: 10.1016/j.jinorgbio.2013.10.019. PMID: 24252384.

Manandhar Shrestha J, Shrestha H, Prajapati M, Karkeen A, Maharjan A. 2017. Adverse effects of oral hypoglycemic agents and adherence to them among patients with type 2 diabetes mellitus in Nepal. J. Lumbini Med. Coll. 2017. 5, 34–40. doi: 10.22502/jlmc.v5i1.126

Maret W, Krężel A. 2007. Cellular zinc and redox buffering capacity of metallothionein/thionein in health and disease. Mol Med. 13, 371–375. doi: 10.2119/2007-00036. Maret.

Masood N, Baloch GH, Ghori RA, Memon IA, Memon MA, Memon MS. 2009. Serum zinc and magnesium in type-2 diabetic patients. J Coll Physicians Surg Pak. 2009 Aug;19, 483–486. PMID: 19651009.

Maxwell SR, Thomason H, Sandler D, Leguen C, Baxter MA, Thorpe GH, Jones AF, Barnett AH. 1997. Antioxidant status in patients with uncomplicated insulin-dependent and non-insulin-dependent diabetes mellitus. Eur J Clin Invest. 1997 Jun;27, 484–490. doi: 10.1046/j.1365-2362.1997.1390687.x. PMID: 9229228.

McCall KA, Huang CC, Fierke CA. 2000. Function and mechanism of zinc metalloenzymes. J. Nutr. 130, 1437S–1446S. doi: 10.1093/jn/130.5.1437S.

Mikhail TH, Awadallah R. 1977. The effect of ATP and certain trace elements on the induction of experimental diabetes. Z Ernährungswiss. 16, 176–183. doi: 10.1007/BF02024790.

Mobasseri, Majid, Shirmohammadi, Masoud, Amiri, Tarlan, Vahed, Nafiseh, Hosseini Fard, Hossein, & Ghojazadeh, Morteza (2020). Prevalence and incidence of type 1 diabetes in the world: a systematic review and meta-analysis. Health Promotion Perspectives, Mar 30; 10(2), 98–115. doi: 10.34172/hpp.2020.18.

Myers SA, Nield A, Chew GS, Myers MA. 2013. The zinc transporter, Slc39a7 (Zip7) is implicated in glycaemic control in skeletal muscle cells. PLOS ONE. 2013, 8, e79316. doi: 10.1371/journal.pone.0079316.

Nazem MR, Asadi M, Jabbari N, Allameh A. 2019. Effects of zinc supplementation on superoxide dismutase activity and gene expression, and metabolic parameters in overweight type 2 diabetes patients: A randomized, double-blind, controlled trial. Clin Biochem. 2019 Jul;69, 15–20. doi: 10.1016/j.clinbiochem.2019.05.008. Epub 2019 May 23. PMID: 31129183.

Nishiguchi T, Yoshikawa Y, Yasui H. 2017. Anti-diabetic effect of organo-chalcogen (Sulfur and Selenium) zinc complexes with hydroxy-pyrone derivatives on leptin-deficient type 2 diabetes model ob/ob mice. Int. J. Mol. Sci. 18, 2647. doi: 10.3390/ijms18122647.

Njie-Mbye YF, Kulkarni-Chitnis M, Opere CA, Barrett A, Ohia SE. 2013. Lipid peroxidation: pathophysiological and pharmacological implications in the eye. Front. Physiol. 4, 366. doi: 10.3389/fphys.2013.00366.

Oberley, Larry W. (1988). Free radicals and diabetes. Free Radical Biology and Medicine, 5(2), 113–124. doi: 10.1016/0891-5849(88)90036-6.

Parham M, Amini M, Aminorroaya A, Heidarian E. 2008. Effect of zinc supplementation on microalbuminuria in patients with type 2 diabetes: a double blind, randomized, placebo-controlled, cross-over trial. Rev Diabet Stud. Summer; 5, 102–109. doi: 10.1900/RDS.2008.5.102. Epub 2008 Aug 10. PMID: 18795212; PMCID: PMC2556442.

Philip JE, Shahid M, Prathapachandra-Kurup MR, Velayudhan MP. 2017. Metal based biologically active compounds: design, synthesis, DNA binding and antidiabetic activity of 6-methyl-3-formyl chromone derived hydrazones and their metal (II) complexes. J. Photochem. Photobiol. B: Biol. 175, 178–191. doi: 10.1016/j.jphotobiol.2017.09.003.

Prasad AS, Bao B, Beck FWJ, Kucuk O, Sarkar FH. 2004. Antioxidant effect of zinc in humans. Free Radic. Biol. Med.. 37, 1182–1190. doi: 10.1016/j.freeradbiomed.2004. 07.007.

Roussel AM, Kerkeni A, Zouari N, Mahjoub S, Matheau JM, Anderson RA. 2003. Antioxidant effects of Zinc supplementation in Tunisians with Type 2 diabetes Mellitus. J Am Coll Nutr. 22(4), 316–321, doi: 10.1080/07315724.2003.10719310.

Said E, Mousa S, Fawzi M, Sabry NA, Farid S. 2021. Combined effect of high-dose vitamin A, vitamin E supplementation, and zinc on adult patients with diabetes: a randomized trial. J. Adv. Res. 28, 27–33. doi: 10.1016/j.jare.2020.06.013.

Scott DA, Fisher AM. 1935 May. Crystalline insulin. Biochem J. 29, 1048–1054. doi: 10. 1042/bj0291048. PMID: 16745762; PMCID: PMC1266596.

Seet RCS, Lee CYJ, Lim ECH, Quek AML, Huang H, Huang SH, Looi WF, Long LH, Halliwell B. 2011. Oral zinc supplementation does not improve oxidative stress or vascular function in patients with type 2 diabetes with normal zinc levels. Atherosclerosis. 219, 231–239. doi: 10.1016/j.atherosclerosis.2011.07.097.

Shidfar F, Aghasi M, Vafa M, Heydari I, Hosseini S, Shidfar S. 2010. Effects of combination of zinc and vitamin A supplementation on serum fasting blood sugar, insulin, apoprotein B and apoprotein A-I in patients with type I diabetes. Int J Food Sci Nutr. 61(2), 182–191, doi: 10.3109/09637480903334171.

Simon SF, Taylor CG. 2001. Dietary zinc supplementation attenuates hyperglycemia in db/db Mice. Exp. Biol. Med. 226, 43–51. doi: 10.1177/153537020122600107.

Turner B, Williams S, Taichman D, Vijan S. 2010. Type 2 diabetes. Ann Intern Med. 152, ITC3-1. doi: 10.7326/0003-4819-152-5-201003020-01003.

Vashum KP, McEvoy M, Shi Z, Milton AH, Islam MR, Sibbritt D, Patterson A, Byles J, Loxton D, Attia J. 2013. Is dietary zinc protective for type 2 diabetes? Results from the Australian longitudinal study on women's health. BMC EndocrDisord. 13, 40. doi: 10.1186/1472-6823-13-40.

Vijayaraghavan K, Pillai SI, Subramanian SP. 2012. Design, synthesis and characterization of zinc-3 hydroxy flavone, a novel zinc metallo complex for the treatment of experimental diabetes in rats. Eur. J. Pharmacol.. 680, 122–129. doi: 10.1016/j.ejphar.2012.01.022.

Von Bülow V, Rink L, Haase H. 2005. Zinc-mediated inhibition of cyclic nucleotide phosphodiesterase activity and expression suppresses TNF-alpha and IL-1 beta production in monocytes by elevation of guanosine 3,5′-cyclic monophosphate. J Immunol 175:4697–4705pmid:16177117.

Wellen KE, Hotamisligil GS. 2005. Inflammation, stress, and diabetes. J Clin Invest. 115, 1111–1119. doi: 10.1172/JCI25102.

Wellinghausen N, Driessen C, Rink L. 1996. Stimulation of human peripheral blood mononuclear cells by zinc and related cations. Cytokine. 1996 Oct 8, 767–771. doi: 10.1006/cyto.1996.0102. PMID: 8980878.

Wild, Sarah, Roglic, Gojka, Green, Anders, Sicree, Richard, & King, Hilary (2004). Global prevalence of diabetes: Estimates for the year 2000 and projections for 2030. Diabetes Care, 27(5), 1047–1053. PMID: 15111519. doi: 10.2337/diacare.27.5.1047.

Wolff SP, Jiang ZY, Hunt JV. 1991. Protein glycation and oxidative stress in diabetes mellitus and ageing. Radic. Biol. Med. 10, 339–352. doi: 10.1016/0891-5849(91) 90040-A.

Wright, Alison K., Kontopantelis, Evangelos, Emsley, Richard, Buchan, Iain, Sattar, Naveed, Rutter, Martin K., & Ashcroft, Darren M. (2017). Life Expectancy and Cause-Specific Mortality in Type 2 Diabetes: A Population-Based Cohort Study Quantifying Relationships in Ethnic Subgroups. Diabetes Care, 40(3), 338–345. doi: 10.2337/dc1 6-1616.

11 Zinc in Indian Traditional Medicines

Ethirajan Sukumar and Dasarathan Dharani

11.1 INTRODUCTION

Traditional medicines form part of a country's heritage and civilisation as they were founded several hundred years ago by indigenous experts. With their constant observation, inquisitiveness and long years of experience, they accumulated a wealth of information and passed it on to their successive generations. These systems grew with the culture and traditions of the race and faced onslaughts of several internal forces and external invasions. But most of them still serve a considerable section the society in many parts of the world. Balinese Medicine, Chinese Medicine, Egyptian Medicine, Greek Medicine, Korean Medicine, Mesopotamian Medicine, Persian Medicine, Zulu Medicine besides Indian systems of medicine (that consist of Ayurveda, Siddha and Unani) are the best examples of traditional systems of the world today.

Some of these medicines are recognised by the World Health Organization (WHO) and included in the health care programs of national and regional governments of many countries. In fact, in some countries, traditional medicines are the only choice available to people who reside in terrains away from cities and towns. The National Center for Complementary and Alternative Medicines (NCCAM), an arm of National Institutes of Health, Bethesda, USA has taken up research on some of the drugs of traditional systems on modern lines to substantiate the claims made as well as to bring out new drug entities for many life-threatening diseases. During COVID-19 outbreak in recent times, many traditional medical systems were put into action world over, mainly to safeguard people from instant death as no definite medicines or vaccines were available in the early stages of the pandemic.

11.2 TRADITIONAL MEDICINES OF INDIA

The history of medicine in India can be traced to the remote past. Evidences for the well-organised system of medicine in India could be obtained from the archaeological remains of Harappa and Mohenjo-Daro. In the Indus Valley civilisation, it was reported that a system of medicine had prevailed in which drugs from natural sources were used. The progress of medicine in India could be inferred through different stages from the works of classical authors, compilers and the accounts of travellers from abroad. Archaeological evidence is available to confirm the medical

DOI: 10.1201/9781003412472-11

practices of ancient civilisations over different periods of time (Sharma, 1992). During the period of its glory, many nations of the civilised world of those days were eager to obtain information on the healing art from India and thus the influence of Indian medicine permeated far and wide into several countries such as China, Egypt, Greece, Rome and Arabia which moulded their medical systems. India was considered by them as a luminous centre of knowledge with immense potential that attracted philosophers and sages of antiquity to study the science of life. There is evidence in Greek and Roman medicines of the influence of Indian medicine.

11.3   METALLIC DRUGS OF TRADITIONAL MEDICINE

Traditional medicines use naturally occurring materials such as plants, metals, minerals and animal products as drugs to treat various diseases. Metals such as gold, silver, mercury, tin, copper, zinc, iron and lead or their ores are employed in drug preparations. Some of these materials are toxic *per se* and hence subjected to detoxification by the process of *Sodhana* or *Putam* in which heating to high temperature, drying under the sun, soaking in herbal juices and repeated triturations are adopted. These processes free the metals from toxicity and render them therapeutically useful entities. Further, the final products are prescribed in minute quantities for short periods of time or with breaks in the duration.

In recent years, many metallic drugs of traditional systems have been subjected to research on modern lines incorporating chemistry, nanotechnology, pharmacology, bioinformatics, etc. These studies not only prove the validity of metallic drugs but also bring out the science behind them.

11.4   AYURVEDIC MEDICINE

The term 'Ayurveda' means the 'Science of Life' (*Ayur* means Life and *Veda* means Science). In a strict sense, it is a science which tells how to lead a disease-free healthy life. Most of the approaches of Ayurveda deal with the prevention of diseases during the entire lifetime in various phases. Besides, the science also explains the principles of maintenance of health and it has also developed a wide range of therapeutic measures to cure illnesses. These principles of health-promotive aspects relate to the physical, mental and spiritual welfare of human beings. Thus, Ayurveda has become one of the oldest systems of medicine dealing with both preventive and curative aspects of life in a most comprehensive way.

The most important and massive ancient compilation of the school of medicine is known as '*Charaka Samhita*' which contains several chapters dealing at length with therapeutics. About 600 drugs of herbal, mineral and animal origin are described in it. In addition, this compendium also deals with other branches of Ayurveda such as anatomy, physiology, aetiology, diagnosis, pathology, treatment and medicine. An equally exhaustive ancient compilation '*Sushruta Samhita*' relates to the school of surgery. It deals primarily with fundamental principles and theory of surgery in addition to other topics such as anatomy, embryology, toxicology and therapeutics besides a mention of about 650 drugs. *Sushruta* was the first Indian author to make a classification of drugs and remedies and construct scientific terminology.

TABLE 11.1

Therapeutic uses of Zinc and its Compounds in Ayurvedic Medicine

S. No.	Metal/Compound	Use
1	Zinc *(Yasada)*	Wound healing, asthma, diabetes, eye diseases
2	Zinc oxide *(Pushpanjana)*	Irregular fever, promotion of eyesight, conjunctivitis
3	Zinc carbonate *(Rasaka)*	Fever, skin diseases, Diabetes, eye diseases
4	Brass *(Pittala)*	Diseases of liver and spleen, anemia, parasite infections, skin diseases

11.5 ZINC AND ZINC-CONTAINING FORMULATIONS OF AYURVEDA

In Ayurveda, metallic zinc is referred to as *Yasada* and used in treating eye diseases, diabetes, colds, cough, bronchial asthma, wound healing, etc. Also the ore (*Rasaka*) and compounds of zinc such as zinc carbonate (*Kharpara*), zinc oxide (*Puspanjana*) and brass (*Pittala*—an alloy of zinc and copper) are used in the preparation of drug formulations (Panda and Rout, 2006). The therapeutic uses of zinc and its compounds/ alloy in Ayurvedic medicine are given in Table 11.1. Acharya Nagarjuna, a Buddhist philosopher is considered to be the 'Father of Metallic drugs in India'. A special branch, *Rasa Sastra* (Alchemy), was developed in the subsequent year that deals with metallic and herbo-mineral drugs. *Kakshyaputa Tantra, Rasaratna Sammuchaya, Rasa Tarangini, Rasayoga Sagar, Rasendrasara Samgraha* and *Bhaisaya Ratnabali* are some of the treatises of Ayurveda dealing with metallic and herbo-mineral drugs. Zinc-containing formulations were used in Ayurveda for topical use in 2nd century CE and oral consumption from 14th century CE onwards (Anonymous, 1998).

A total of 84 zinc-based drugs are mentioned in Ayurvedic texts (Sharma, 1992; Sen, 1996). However, thirty formulations containing zinc are in use now. Of them, zinc ore/zinc carbonate (*Rasaka/Kharpara*) is added in twenty formulations, while metallic zinc (*Yasada*) in five, brass (*Pittala*) in four and zinc oxide (*Puspanjana*) in only one formulation. The list of zinc-containing Ayurvedic formulations and their indications are presented in Table 11.2. The ores calamine and zincovite are the natural sources of zinc carbonate. About 40% of zinc-based drugs are prescribed for different types of fevers whereas brass-containing formulations are used mainly to treat tumours and vascular ailments.

11.6 MODERN STUDIES ON ZINC-CONTAINING DRUGS OF AYURVEDA

Among the zinc-based drugs of Ayurveda, *Yasada bhasma* is one of the most studied formulations on modern lines starting with spectroscopic analysis to infer the presence of the metal (Thakur et al., 1986). Pharmacological studies have confirmed the safety of the drug and recommended for long-term use to treat diseases (Mukherjee and Shaw, 1993). The drug was found to possess significant

TABLE 11.2

Zinc-Containing Ayurvedic Formulations and Indications

S. No.	Formulation	Indications
A. Formulations containing Zinc metal (*Yasada*)		
1	*Mehahara Rasa*	Diabetes
2	*Mrutasanjivani Rasa*	Edema, Fever, Anemia
3	*Silajawadi Vati*	Dysuria, Diabetes
4	*Trivanga Bhasma*	Diabetes, Infertility
5	*Yasadamruta Lepa* (External use)	Fistula, Wounds, Skin diseases
B. Formulation containing Zinc oxide (*Puspanjana*)		
1	*Churnanhana*	Eye diseases, Cataract
C. Formulations containing Zinc ore/Zinc arbonate (*Kharpara*)		
1	*Chintamani Rasa*	Fever, Tuberculosis, Spleen and Liver diseases
2	*Baidyanath Rasa*	Anorexia, Intermittent Fever
3	*Kharpara Bhasma*	Epilepsy, Chronic Fever
4	*Kasturi Bhairava Rasa*	Fever, Oligispermia, Tuberculosis, Diabetes
5	*Kshyakeshri Rasa*	Anemia, Tuberculosis, Fever
6	*Laxmivilasa Rasa*	Jaundice, Anemia, Bronchitis
7	*Mahajvarankusa Rasa*	Intermittent Fever, Malaria
8	*Navagraha Rasa*	Gastero-intestinal diseases, Piles, Fistula
9	*Pradarantaka Lauha*	Leucorrhoea
10	*Pratapatapana Rasa*	Irregular Fever
11	*Ratnaparava Vati*	Body pain, Leucorrhoea
12	*Sannipatantaka Rasa*	Chronic Fever
13	*Sarvarogahara Rasa*	Arthritis, Edema
14	*Somanath Rasa*	Diabetes
15	*Sutikahara Rasa*	Edema, Anemia
16	*Suvasantamalatee Rasa*	Piles, Intermittent Fever
17	*Trishnadi Gutika*	Bronchitis, Debility, Fever
18	*Tyhikari Rasa*	Diseases of spleen, Intermittent Fever
19	*Vangastaka Rasa*	Fevers, Urological Diseases, Diabetes
20	*Visveswara Rasa*	Fevers
D. Formulations containing Brass (*Pittala*)		
1	*Akadasayasa Rasa*	Tonic, Vericocele, Hydrocele
2	*Pitala Rasayana*	General health, Vitality
3	*Sadanana Rasa*	Gastero-intestinal disorders, Dysentery
4	*Swarna Sindura Rasa*	Nervine tonic, Anorexia, Impotency

hypoglycemic activity in rats (Prasad and Sharma, 1989). It was reported that the drug prepared by the traditional three-step process *Shodhana, Bhavana (Jarana) and Marana* is free from toxicity with improved therapeutic efficacy and safe (Saritha and Sridurga, 2019). A later study analysed the elemental composition and particle size to establish quality control of the finished product to help maintaining safety, efficacy and batch-wise uniformity (Chandran et al., 2017).

Pareek and Bhatnagar (2020) who analysed the drug reported that inclusion of plant extracts during the calcination process enhanced its medicinal efficacy and reduced it to a nano size. They also mentioned that modern techniques such as X-ray Diffraction (XRD), Dynamic Light Scattering (DLS), Scanning Electron Microscopy (SEM) and Energy Dispersive X-ray Analysis (EDAX) revealed a significant transformation of original zinc metal to *Yashada bhasma* in the process. XRD further revealed the conversion of zinc metal into crystallite hexagonal zinc oxide phase after proper incineration process, which was in line with the elemental analysis done through EDAX analysis. DLS and SEM revealed a drastic reduction in particle size of zinc metal which makes it suitable for therapeutic use. These techniques also could help to correlate toxicity issues and be considered as a step towards establishing the scientific reasons behind the safety and efficacy of ancient medicines.

11.7   SIDDHA MEDICINE

The Tamils inhabiting the Southern peninsula of India had a hoary past and they undertook a systematic study of nature and its elements and developed a highly systematised medical science known as 'Siddha'(Shanmugavelan, 1992). The founders of Siddha system known as '*Siddhars*', were experts in the preparation of drugs from metals, minerals, plants and animal products. Among the traditional medicines, Siddha system makes use of metals and minerals in the drug preparation to a large extent. Even metals with high toxicity such as mercury, gold and silver as well as non-metals such as arsenic and cyanides were used by Siddhars after detoxification by several processes. They chose suitable metals for drug preparations based on characters such as shape, colour, lustre, fracture, hardness, taste, odour, etc. (Sornamariammal, 2016). The present-day researchers of modern science are eager to look into these metallic preparations to understand the basics of detoxification processes and the nanotechnology involved in it. While administering these metallic drugs to the patients, the physicians followed certain guidelines such as dosage and duration. The dosages used to be very little and to administer them, vehicles or carriers known as *anupanams* such as ghee, honey, hot water, milk, etc., were used, depending upon the nature of the disease. Thus, the same preparation could be used to treat various types of diseases by using different *anupanams*.

11.8   ZINC AND ZINC-CONTAINING FORMULATIONS OF SIDDHA

In Siddha medicine, metallic drugs occupy a special place and are prescribed in the treatment of chronic and life-threatening diseases. They are prepared from metal ores by the process of calcination (heating to high temperatures) with the addition of several herbal ingredients to remove the toxicity and improve the therapeutic efficacy of the final products. Though they are prepared from toxic metallic ores, the manufacturing process makes the final product virtually devoid of toxicity and prescribed in very small quantities for short periods to have better therapeutic effects.

In Siddha medicine, zinc is known as *Naagam*, meaning cobra as it makes a hissing noise like the snake upon melting. The synonymous names of zinc are *Seeral, Porumal, Pongal, Iraichal, Aimbugai, Sorum, Vasuki, Vennaagam, Thambirathin Vidhai* and *Vaathathirku Uyir*. Two types of zinc are mentioned in Siddha literature - *Sirukan Naagam* and *Perukan Naagam*, probably indicating the size of the particles obtained after the purification process from the ore and named as small (*Siru*) and big (*Peru*) (Sornamariammal, 2016). The metal is present in nature as a mixture with a few other metals and minerals. After extraction, it becomes a shining solid with a whitish-blue colour. Zinc has an astringent taste, coagulant and tonic actions besides controlling haemorrhage. It is employed in treating venereal diseases, diarrhoea, chronic ulcers, mental disorders, leucorrhoea, etc. (Thiagarajan, 2008). Though there are several formulations of zinc mentioned in Siddha literature, mostly *Naaga Parpam, Naaga Chendooram* and brass are widely used.

The Siddha system considers zinc as a superior metal that has the property to cure genito-urinary (*megham*) diseases, sexually transmitted diseases (*vettai*), dysentery (*krani*), haemorrhoidal ulcers and fistula (*mulaippun*), hyperthermia (*azhal*) and psychiatric problems (*paithyam*). The selective indications of the metal zinc could be purely interpreted for its astringent, styptic, antiulcer or wound healing, demulcent and anti-viral actions in general. The quote '*Mella Thuraththum Vyathikalai*' means that its actions are best achieved through long-term use and '*Uranudaimai Undakkum Undavarai*' which literally means that the metal is capable of improving one's vitality, strength and immunity (tonic and immuno-modulator) (Mudaliar and Uthamarayan, 2006; Thiagarajan, 2013).

11.9   PURIFICATION OF ZINC IN SIDDHA

Purification of metals or their ores is an important step in traditional practices as most of them exhibit toxicity. Zinc is purified by mixing the metal powder with the paste of rosary-nut (*Elaeocarpus Sphaericus*) and the pollen honey of the plant Alexandrian laurel (*Calophyllum Inophyllum*) and insolated (dried under hot sun) for a month. This process detoxifies zinc that could be stored for long years and is suitable for preparing *parpam* and *chendooram*. There are also two other methods to purify zinc that could be used depending on the availability of materials and cost-effectiveness (Thiagarajan, 2008).

11.10   IMPORTANT ZINC-BASED FORMULATIONS OF SIDDHA

There are three important drug formulations containing zinc *viz. Naaga Parpam, Naaga Chendooram, Pithalai Parpam* and frequently prescribed by practitioners. Occasionally *Naaga Sangu Parpam* is also used by some (Thiagarajan, 2008; Pillai, 2007 & 2020; Saheb, 2002 & 2014). Table 11.3 gives a select list of zinc-based drugs of Siddha and indications.

TABLE 11.3

Zinc-Containing Siddha Formulations and Indications

S. No.	Formulation	Ingredients	Indications
1	*Naaga Parpam*	Zinc powder	Diseases of the skeletal system, dyspepsia, tuberculosis, gastric ulcer, vitiligo, anaemia, skin diseases, leprosy, epilepsy, leucorrhoea eye diseases, hemiplegia, menorrhagia, gonorrhoea, etc.
2	*Naaga Chendooram*	Zinc powder	Dropsy, respiratory infections, body pain, cramps, fever associated with bone diseases, diarrhea, psychiatric disorders, tremours, etc.
3	*Pithalai Parpam*	Zinc and Copper	Astringent, expectorant, diuretic, bleeding haemorrhoids, anemia, abdominal pain and asthma.
4	*Thaamira Naaga Parpam*	Zinc metal	Fistula in-ano, urinary diseases, ulcer, sexual disorders, etc.
5	*Naaga Rasa Chendooram*	Zinc metal, sublimed mercury, sulphur and Aloe juice	Piles, fistula in-ano, urinary diseases, skin diseases and early stages of leprosy.
6	*Naaga Rasa Chunnam*	Zinc (small variety), purified tin, mercury and ginger juice	Tuberculosis, bronchitis and cough.
7	*Navaloga Rathnaadhi Maathirai*	Nine metallic powders including zinc and a few plant and animal products.	96 types of eye diseases.
8	*Naaga Sanku Parpam*	Zinc metal, conch shell and leaf juice of *Pergularia Daemia.*	Hemorrhoids, fistula *in ano*, leucorrhoea and venereal disease.

11.11 MODERN STUDIES ON ZINC-CONTAINING DRUGS OF SIDDHA

Many metallic preparations of Siddha medicine were investigated on modern lines utilising the sophisticated instrumentation and technological developments of the day. Among the zinc-based drugs, *Naaga Parpam* has been studied extensively and pioneered by the investigations done by Ilango et al. (2009a & 2009b). This formulation was tested by them for its composition (on X-ray Fluorescence Spectrometer), toxicological profile and anti-hyperlipidemic activity in rat models.

In another study, *Naga Parpam* was found to have a crystallite size of 32–34 nm and had 95% pure zinc when analysed on X-ray Diffraction (XRD) and Energy Dispersive X-ray Analysis (EDAX) techniques. The drug possessed a negative zeta potential in the physiological pH and did not induce blood cell aggregation or protein adsorption. The activation potential towards the complement system or platelets was found to be negligible. However, the drug was highly cytotoxic and platelet-activating which induced red-blood cell lysis also. Due to its crystallite size within 28–34 nm, it has been suggested that the drug could reach the affected site on oral administration via intestinal absorption and possibly release ions in a sustained manner required for therapeutic action (Paul and Sharma, 2011).

In yet another study, EDAX studies on *Naga Parpam* revealed the presence of zinc, oxygen, magnesium, silica and potassium in it. XRD showed characteristic peaks of zinc oxide nano-particles and the secondary phase showed the presence of other elements besides nano-particles size as 37.15 nm. Fourier Transform Infra-red analysis (FTIR) also confirmed the presence of zinc oxide, silica and magnesium. SEM analysis showed the surface morphology and confirmed the particle size to be <100 nm. The photoluminescence study (PL) showed a broad peak at 525 nm due to deep-level emission which is attributed to both oxygen vacancy and disruption of structure due to the presence of Mg. Based on these, *Naga Parpam*, a nano compound, is considered an active drug for the depleted zinc at cell level which is observed in auto-immune disorders and cancer (Ramanathan and Ramasamy, 2019).

The investigation on *Naaga Sangu Parpam* by Kingsley (2018) in rats revealed its safety up to 200 mg/kg body weight dose on repeated administration (28 days) which was further confirmed by histopathological studies. The drug also possessed significant styptic, anti-inflammatory and analgesic activities and was fit for treating hemorrhoids, fistula *in-ano*, leucorrhoea and venereal disease.

11.12 UNANI MEDICINE

The Unani system of medicine owes its origin to Greece. It was enriched by imbibing the contemporary systems of traditional medicine in Egypt, Syria, Iraq, Persia, India, China and other Middle and Far East countries. It has also benefited from the native medical systems in vogue at the time in various parts of the world with different names such as Greco-Arab Medicine, Arab Medicine, Islamic medicine, Traditional medicine, Oriental medicine, etc. (Anonymous, 2002). The system has a long and impressive record in India. It was introduced in the country by the Arabs and Persians sometime around the 11th century. It is so popular on Indian soil that the Government of India has adopted the medicine as one of the Indian Medical Systems. India is one of the leading countries as far as the practice of Unani medicine is concerned and has the largest number of Unani educational, research and healthcare institutions in many parts of the country (Dharani, 2006).

11.13 ZINC AND ZINC-CONTAINING FORMULATIONS OF UNANI

Metals and minerals are integral parts of Unani Medicine which uses seven metals of antiquity such as gold, silver, copper, tin, lead, iron and mercury since prehistoric times. The medicinal use of lead and its various compounds in the formulations is reported in pre-Dynastic Egypt and by the ancient Indian civilisations from 3200 to 2800 CE (Azhar, 2022). In Unani medicine, preparations from metals and minerals are generally known as *Kushta*. It also means the incinerated finely powdered substance prepared from known toxic metals such as mercury (*Seemab*), arsenic

TABLE 11.4

Zinc-Containing Unani Formulations and Indications

S. No.	Formulation	Main Ingredient	Indications
1	*Marham Sendur*	Zinc carbonate	Ulcers
2	*Marhan Qooba*	Zinc oxide	Fungal infection
3	*Marham Saeda Chob Neemwala*	Zinc oxide	Haemorrhoids
4	*Marham Kafoor*	Zinc oxide	Ulcer, Wounds
5	*Marham Kharish Jadeed*	Zinc oxide	Daad, Fungal Infection, Itching, Kharish & Fasd e Dum
6	*Zimad Rahat*	Zinc oxide	Pain, Inflammation
7	*Kushta-e-Justh*	Zinc oxide	PME, Leucorrhoea, Aphrodisiac
8	*Shiyaf Abyaz*	Zinc oxide	Conjunctivitis
9	*Shiyaf Akseer Chashm*	Zinc oxide	Stye
10	*Zimad Bawaseer*	Zinc oxide	Haemorrhoids, Anal Fissure

(*Sam Al-Far*), mercuric chloride (*Sangraf*) and iron rust (*Khubs Al-Hadid*). Medicines made with these metals/minerals, when used with caution and expertise, are effective and may exhibit significant toxic side effects if used bypassing guidelines (https://www.britannica.com/science/family-practice).

A select list of zinc-containing Unani formulations is presented in Table 11.4.

11.14 MODERN STUDIES ON ZINC-CONTAINING DRUGS OF UNANI

When compared to Ayurveda and Siddha systems, very limited scientific studies have been carried out on Unani single drugs and formulations. One among them is *Kushta Jast* which contains zinc (*Jast*) and the plant *Phyla Nodiflora* L. (Synonym: *Lippia Nodiflora* L.) (*Bukan Booti*). In Unani medicne, the formulation *Kushta Jast* is used as a cardiac tonic, brain tonic, antipyretic, aphrodisiac, stomachic, blood purifier and an immunity booster (Al-Din, 1985; Al-Din, 2007).

In a study, Ahmad et al. (2020) has presented literary evidence for the potential benefits of *Kushta Jast* as a prophylactic measure in COVID-19. Although there is no direct mention of antiviral activity in Unani medicine, *Kushta Jast* reduced fever (*humma*) and improved immune function to fight against viral infections. Going by the Unani concept of holism of strengthening the entire body, *Kushta Jast* has been proposed as a prophylactic traditional preparation effective drug against SARS-Cov-2 infection/COVID-19.

A suspension formulation of *Kushta Jast* has been evaluated for a few physical properties and tested for hypoglycemic activity in rabbits. It was found that administration of the drug decreased blood glucose levels in hyperglycemic animals with concomitant increase in plasma zinc levels (Agarwal et al, 1998).

11.15 CONCLUSION

It is amply clear that the metallic drugs of traditional medicines have promise in treating several chronic and baffling diseases. The zinc-based drugs in particular possess curative properties against many prevailing diseases besides being immune-builders that could be put into wider use and included in the Central and State Government healthcare programmes. This immune-boosting property of zinc formulations also helps in paediatric care to reduce infant mortality during the outbreak of diseases such as diarrhea and dysentery. However, further research on zinc formulations that have been in use in traditional medicines for hundreds of years is warranted to bring them into wider use.

ACKNOWLEDGEMENTS

The authors are grateful to Dr. T. Porchelvi MD (Siddha), Chief Medical Officer, Sakthi Siddha and Varma Clinic, Mathur, Chennai, Dr. S. Waseem Aathir BUMS, Asst. Medical Officer (Unani), District Headquarters Hospital, Vellore (Tamil Nadu) and Dr. Afeefa Muskan BUMS, Internee, Govt. Unani Medical College, Chennai for useful discussions and help in providing additional information related to their fields. They also thank Dr. S. Brindha PhD, Research Scholar, Department of Historical Studies, Bharathi Women's College, Chennai for her help in the compilation work.

REFERENCES

Agarwal SP, Prakash A, Kohli K. 1998. Pharmacological studies of a "Kushta Just" suspension formulation. Indian J Pharm Sci. 60, 225–227.

Ahmad T, Zakir M, Fatma SH, Kazmi MH, Javed G, Ali S. 2020. Kushta Jast, a conventional herbo-mineral immunity booster tonic: potential use in COVID-19. Cell Med. 10, 24.1–24.6.

Anonymous. 1998. Sarangadhara Samhita, Purbardha, 1st Edition, Chaukhamba Sanskrit Academy, Varanasi.

Anonymous. 2002. An Overview of Ayurveda, Yoga-Naturopathy, Unani, Siddha and Homoeopathy. Department of Indian Medicine and Homoeopathy, Ministry of Health and Family Welfare, Government of India, New Delhi, p. 25.

Al-Din K. 2007. Ilmul Advia Nafeesi. Aijaz Publishing House, New Delhi, India.

Al-Din R. 1985. Kanzul Advia Mufrada. University Publication Unit, Aligarh, India.

Azhar MM. 2022. Usrub (lead) and its compounds used in Unani Medicine: Overview. Asian J Res Chem. 15, 358–366. doi: 10.52711/0974-4150. 2022. 00064.

Chandran S, Patgiri P, Bedarkar P, Gokarna RA, Shukla VJ. 2017. Particle size estimation and elemental analysis of Yashada bhasma. Int. J. Green Phar. 11, S765–S773.

Dharani D. 2006. Historical Perspectives of the Practice of Indigenous Medicines in Tamil Nadu with special reference to Siddha Medicine, PhD Thesis, University of Madras, Chennai.

Ilango B, Sharief SD, Vinothkumar K, Rajkumar R, Prathiba D, Sukumar E. 2009a. Histopathological studies on the effect of Naga parpam, a zinc-based drug of Siddha medicine, in rats. J. Cell Tissue Res. 9, 1869–1873.

Ilango, B, Sharief SD, Vinothkumar K, Rajkumar R, Viswanathan S, Sukumar E. 2009b. Effect of Naga parpam, a zinc-based siddha medicine on hyperlipidemia in rats. J. Complement. Integ Med. 6, Art. 2, 1–7.

Kingsley J. 2018. Standardization and pharmacological screening of Naaga Sangu Parpam, MD. (Siddha) Dissertation, The Tamil Nadu Dr. MGR Medical University, Chennai.

Mudaliar K, Uthamarayan CS. 2006. Siddha Vaidhya Thirattu, 2nd Edition. Directorate of Indian Medicine and Homoeopathy, Chennai, p. 80.

Mukherjee B, Shaw BP. 1993. Effect of Yasada bhasma on different diseases of G.I.T. with special reference to Gastric ulcer, MD Thesis, Calcutta University.

Panda AK, Rout S. 2006. Zinc in ayurvedic herbo-mineral products. Nat. Prod. Radiance. 5, 284–288.

Pareek A, Bhatnagar N. 2020. Physico-chemical characterization of traditionally prepared Yashada bhasma. J. Ay. Integr. Med. 11, 228–235.

Paul W, Sharma PC. 2011. Characterizations of Rasa Chenduram (Mercury based) and Naga Parpam (zinc based) preparations: Physicochemical and in vitro blood compatibility studies, Trends Biomater. Artif. Organs, Oct. 2011.

Pillai CK. 2020. Kannusami Paramparai Vaithiyam (5th Edition), B. Rathina Naicker & Sons, Chennai, 414.

Pillai CK. 2007. Chikitsa Ratna Deepam. B. Rathina Naicker & Sons, Chennai, pp. 231–249.

Prasad CM, Sharma AV. 1989. Yasada bhasma: an effective hypoglycaemic drug. Ancient Sci. Life. 9, 69–70.

Ramanathan R, Ramasamy R. 2019. Characterization of zinc oxide-based nano siddha medicine naga parpam – potential nutritive supplement in cancer. J. Pharm. Sci. Innov. 8, 100–104.

Saheb PMA. 2002. Anuboha Vaidya Navaneedham (Part 10). Thaamarai Noolagam, Chennai, p. 131.

Saheb PMA. 2014. Anuboha Vaidya Navaneedham (Part 5). Thaamarai Noolagam, Chennai, p. 70.

Saritha, Sridurga Ch. 2019. Pharmaceutical standardization of Yashada bhasma. Int. J. Ay. Pharma Res. 7, 13–20.

Sen G. 1996. Bhaisajya Ratnavali, 12th Edition, Ed: Ambica Dutta Sastri, Chaukhamba Sanskrit Samsthan, Varanasi; Sharma PH. 1983. Rasayogasagara, Vol. 1 & 2, 1st Edition, Krishnadas Academy.

Shanmugavelan A. 1992. Siddhars Science of Longevity and Kalpa Medicine of India. Directorate of Indian Medicine and Homoeopathy, Madras, p. 1.

Sharma PV. 1992. History of Medicine in India. Indian National Science Academy, New Delhi, p. 3.

Sornamariammal I. 2016. Siddha Minerals in Bogar 7000. World Siddha Trust, Chennai, p. 25.

Thakur SN, Srinivas C, Deshpande PJ. 1986. Spectroscopic analysis of Yasada bhasma (Zinc salt). Ancient Sci. Life 5, 24–242.

Thiagarajan R. 2008. Siddha Materia Medica, Mineral and Animal Section. Department of Indian Medicine and Homoeopathy, Chennai, p. 135.

Thiagarajan R. 2013. Gunapadam (Thathu Jeeva Vaguppu), 8th Edition, Parts 2 and 3. Department of Indian Medicine and Homoeopathy, Chennai, pp. 165–179.

12 Role of Zinc in Veterinary Medicine

Neha Singh, Kanisht Batra, Digvijay Singh, S. Mathan Kumar, and Sushila Maan

12.1 INTRODUCTION

For animals and birds, zinc is a crucial trace mineral. The first ever evidence suggesting the role of zinc was obtained from laboratory animals by observing the associations between zinc and carbonic anhydrase, an enzyme containing zinc in its structure (Keilin and Mann, 1940). Experimental zinc deprivation in domesticated species has effected chicks (Dell and Savage, 1957), swine (Stevenson and Earle, 1956), lambs (Ott et al., 1964) and calves (Miller et al., 1967). Zinc deficiency in animals was characterised by loss of appetite, sluggish development, dermal anomalies and unsuccessful pregnancies. Parakeratosis is a condition that causes thickening and hardening of the skin of pigs fed with grain-based diets which are rich in calcium. Zinc supplementation has been shown to prevent this condition (Tucker and Salmon, 1955). In ruminants, which have the ability to break down various anti-nutritional substances like phytic acid in the rumen, zinc-responsive diseases were far more difficult to demonstrate. It is challenging to pinpoint the rate-limiting factor in zinc-responsive diseases because there are more than 300 zinc-dependent enzymes with various structural and functional effects as well as greater amount of zinc proteins that are functional (Coleman, 1992). Measurements of free intracellular zinc at femtomolar concentrations have been attempted to better understand the cause of deficiency of zinc in animals, whereas therapeutic benefits of zinc have been discovered at near-molar quantities. Unfounded claims that a variety of "organic" zinc supplements have higher bioavailability are being used as a marketing ploy to capitalise on consumer fear about zinc deficiency. As a result, many functions as well as the repercussions of zinc depletion in animals have been detailed in this chapter Figure 12.1.

12.2 GENE EXPRESSION

All the signalling, as well as metabolic activities, need around 2000 transcription factors that depend on zinc for both structural and functional integrity (Cousins et al., 2006). It is crucial for foetal development and has an impact on DNA synthesis, protein and nucleic acid metabolism and foetal growth. According to Cousins and coworkers (2003) zinc regulated genes involved in energy use, growth, reduction-oxidation processes, signal transduction and responses to stress, as well as receptors

DOI: 10.1201/9781003412472-12

FIGURE 12.1 Role of zinc in different body systems.

for the cytokine T-cells. When cells are quickly dividing or synthesising, zinc deprivation's principal effects on gene expression can cause teratogenic abnormalities (Chesters, 1992). According to various studies, acute zinc deficiency caused rats to prefer gnawing on food instead of eating a complete meal, which had an impact on both the pattern and rate of food consumption. Deprivation on diets with a tiny amount of additional zinc, i.e., 3 to 6 mg per kg of dry matter (DM)—avoids such issues, though the zinc was insufficient for healthy development and growth. Although cholecystokinin gene expression is increased by zinc depletion (Cousins et al., 2003), zinc's participation in this process is likely multifactorial. Moreover, there was a decrease in the expression of pyruvate kinase, which is strictly regulated by insulin, and an increase in the expression of leptin (Kwun et al., 2007).

12.2.1 ANTIOXIDANT DEFENCE

Zinc is a component of the Zinc-finger domain, which is comprised of tetrahedral coordination of zinc to cysteine and histidine residues (Berg, 1990). This domain has scavenging characteristics which can help to battle intracellular oxidative stress. According to various *in vitro* studies, zinc may protect cells against peroxidation of lipids brought on by iron via blocking iron binding sites on the surface of cells. Additionally, the superoxide dismutase, Cu-ZnSOD works to shield cells from the superoxide radical. Zinc deficiency makes endothelium cells more vulnerable to oxidative stress (Beattie and Kwun, 2004).

12.2.2 IMMUNE SYSTEM

The immune system protects host cells against pathogens by discriminating between self and non-self. Several subpopulations of immune competent cells, such as monocytes, natural killer (NK) cells, antigen-presenting cells T-cells and B-cells as well as cytokines and complement proteins, are included in the immune system's constituent parts. Many vitamins, selenium, copper, iron, magnesium, zinc, and copper are among the micronutrients that have a big impact on the immune system and health. The enzymes containing zinc are responsible for the continuous proliferation of B-cells and T-cells in the immune system, which would be impossible without zinc.

12.2.3 VISION

It is presumed that zinc is necessary for the retinal cells' metabolism as well as for changing the plasma membranes of photoreceptors, controlling the light-rhodopsin reaction, and influencing the transmission of synapses. In mice, MT and the zinc transporters ZnT3 and ZnT7, which are located in various retinal coatings, support zinc homoeostasis and take part in ocular activities. Furthermore, taurine levels in the retina must be maintained for zinc to work on TAUT, a transporter that moves taurine across tissues (Marquez et al., 2017). Electro retinograms of cats with reduced taurine levels showed pronounced anomalies in timing and amplitude of the b-wave in rod and cone, in addition to fundal lesions. The average concentration of taurine in aqueous humour in healthy dogs is 50 mM, whereas in dogs with primary glaucoma, a drop in taurine levels results in photoreceptor impairment and cell death (Jacobson et al., 1987). Dogs are known to have eyeshine, which is a phenomenon caused by the tapetum lucidum reflecting light. In healthy dogs, zinc is present in the tapetum; although, zinc levels were reduced or nonexistent in the irregularly formed cells of tapetum from Beagles with a hereditary autosomal recessive condition. Zinc chelators, such as ethambutol, have also been blamed for the decline in tapetal zinc concentration since they cause ultrastructural disorders and the loss of tapetum colour (Pereira et al., 2021).

12.2.4 REPRODUCTION

Zinc ions perform the structural and regulatory functions of the prostate-specific arginine esterase enzymes, which keep the prostate and spermatozoa functioning normally in a dog's seminal plasma. Zn is located in the spermatozoa's outer thick filament of the tail of sperm, and forms Zn-mercaptide complexes with the cysteine's sulfhydryl (-SH) groups. It prevents the outer, thick fibres from oxidising too soon (Henkel et al., 2001). It also affects spermatozoa's ability to move by regulating how the system of ATP utilises energy, which is implicated in the energy control of phospholipids and contraction (Hidiroglou and Knipfel, 1984). Since lipid is the main source of energy needed for spermatozoa to travel, the presence of Zn in the centre of sperm and its relationship to the lipoprotein portion indicates that Zn is involved in lipid degradation. Zn is also essential for maintaining nuclear chromatin throughout the male genome transfer's decondensation stage and is essential for conception (Kvist, 1982).

12.2.5 ZINC HOMEOSTASIS AND METABOLISM

Zinc is typically absorbed from the stomach, and the majority of it is absorbed from the intestine's brush border membrane with the aid of carrier-mediated proteins (Rucker et al., 1994). Zinc is more excreted from ruminants in their rumen and reticulum than is absorbed in their stomachs. Various low molecular weights binding ligands such as EDTA, glucose, citrate and amino acids (histidine and glutamate), and picolinate promote the uptake of zinc in monogastric animals by preventing the formation of the zinc phytate complex in their diets (Hambidge et al., 1986). In contrast, feeding zinc-methionine or zinc chloride coupled with EDTA can lower the amount of zinc that rats' duodenal loops absorb (Hempe and Cousins, 1989). Due to their hydrophilic nature, these ions cannot enter the cell through passive diffusion. The zinc may enter the cells through the help of cation channels resulting in facilitated diffusion which is mediated by amino acids or with the help of specific receptors (Simkin, 1997). Given that a significant amount of zinc is still coupled to transferrin, the studies suggest that the transferrin receptor (CD71) plays a crucial role in the absorption of Zn (Cunningham-Rundless et al., 1980). According to recent literature, the transfer of ZnT1-ZnT8, ZnT10 and other molecules mediates zinc uptake (Hill et al., 2009). There are four zinc transporter proteins such as ZnT-1, ZnT-2, ZnT-3 and ZnT-4, which allow zinc to enter the cells and perform its function there. In terms of the distribution of zinc in cellular compartments, the nucleus holds 30–40% of the total amount, the cytoplasm holds around 50%, and the cell membrane holds the remaining zinc. Instead of being present in its free form, the cellular zinc is bound, either as zinc proteins, zinc enzymes, or zinc nucleotides (Smeyers Verbeke et al., 1977). The enzyme tyrosine kinase of various proteins and cyclic AMP as well as cyclic GMP routes, both depend on zinc, which also plays a significant role in cell signalling (Wellinghausen et al., 1996).

12.3 CLINICAL CONSEQUENCES OF ZINC DEFICIENCY

12.3.1 ABNORMALITIES INVOLVING THE SKIN AND LIMBS

All species exhibit skin thickening, hardness and fissures (parakeratosis) as a late symptom of zinc deficiency. Ogawa et al. (2018) claim that zinc deficiency affects the rear limbs and teats of dairy cows, the hind feet and feathers of chickens, the young pigs' extremities, the muzzle, neck, ears and back, and the region of the lamb's scrotum, hoof and area surrounding its eyes. Parakeratosis, a condition in small animals, i.e., sheep, goats and piglets and young cattle, affecting the surface of skin, is one specific disorder brought on by Zn deficiency. There was a distinctive shedding of coat in calves which affects the limbs, neck, head and area around the eyes ("glasses"). According to Miller et al. (1979), zinc deficiency slows the rate of healing in experimentally produced skin wounds, and wounds brought on by ectoparasites or other skin infections are likely to worsen the consequences of parakeratosis. The usual ring structure of new horn growth in horned lambs vanishes, and the horns eventually fall off, causing spongy, squishy outgrowths which bleeds continuously (Miller et al., 1967); alterations to the shape of hoofs is another possibility. The entire fleece may be shed if the wool fibres become thin, loose, and lose their crimp. Lambs lacking in zinc may develop posthitis and vulvitis, which are marked by expansion of the sebaceous glands (Demertzis, P. N. 1972).

12.3.2 SKELETAL DISORDERS

In pigs deficient in zinc, the femur becomes smaller and less powerful, but in comparisons with controls that were fed in pairs show that these alterations are due to a lower intake of feed (Brugger et al., 2019). Long bone thickening and shortening have also been observed in chicks. Zinc deficiency during embryonic development drastically impairs skeletal growth, and in chick embryos generated from severely deficient hen eggs, severe deformities of the head, appendages and vertebrae have been discovered (Huang et al., 2019). Zinc-deficient calves experience joint stiffness, hock swelling and bowing of the hind limbs. Zinc is a crucial mineral needed for healthy skeletal development and maintenance of equilibrium in bones. Additionally, it appears that zinc can promote bone regrowth. In maintaining bone homeostasis and regeneration, zinc is beneficial because it can influence the activity of chondrocytes as well as osteoblast while repressing the osteoclast response. The development of treatment strategies utilising zinc to enhance bone regeneration is based on how zinc affects the composition of skeletal cells and how it aids in development of skeleton O'Connor et al., 2020).

12.3.3 REPRODUCTIVE DISORDERS

In early studies with pigs, zinc deficiency reduced litter size and impaired hatchability in marginally deficient hens (Hoekstra et al., 1967). Zinc-deficient bull calves and ram lambs exhibit hypogonadism (Miller et al., 1968). Some data suggests that Zn may be required for the hypothalamo-pituitary-gonad axis to function naturally. According to Kumar (2003), supplementing with Zn raises the level of testosterone in the blood of buffaloes and crossbred bulls. Sexually mature bucks revealed a lack of libido and shrinkage of the testicles when zinc was severely depleted (Neathery et al., 1973). Moderate zinc insufficiency in lambing sheep reduces the number of kiddings and their birth weights but does not cause congenital defects. A very low-zinc diet during pregnancy decreases the newborn lamb's chance of survival, and anorexia in the ewe can have a secondary effect of pregnancy toxaemia.

12.3.4 WOUND HEALING

A wound's complex healing process is governed by a number of cytokines that influence the host's immune system. The zinc applied locally on the surface can have an injury-repairing action. Greater exogenous expression of the gene for insulin growth factor-I was seen in pigs throughout the wound healing process. Zinc's immunopotentiating ability to survive reactions of inflammation due to sepsis in the model of pig has led to speculation that it may be used as a preventative measure before any surgical operation. Pigs given iatrogenic endotoxaemia responded better to zinc supplementation with 30 ppm of zinc, but growth performance was unaffected. Different stages of wound healing were aided by zinc's capacity to control immune homeostasis, a number of wound healing processes, including granulation and re-epithelization, platelet cell haemostasis, ECM remodelling via matrix metalloproteinases (MMPs) regulation, inflammation and host defence response and membrane repair via its association with tripartite motif (TRIM) family proteins.

Zinc plays a crucial part throughout the repair mechanism of injury present in many cells. After injury, there is clot formation and coagulation in the wound tissue for establishing haemostasis. Immune infiltration and inflammation immediately follow this occurrence to eliminate the damaged tissue from the wound and microorganisms, causing granulation and avoiding infection. The proliferation and migration of different cells, such as endothelial cells, fibroblasts, epithelial cells and keratinocytes, into wound deposits ECM and repopulates the injured site, facilitating wound healing. Later, there is deposition of the matrix and removal of control scar development. Polymorphonuclear leucocytes (PMN) and NETosis use a unique kind of programmed neutrophil death that results in the production of NETs and extracellular matrix (ECM). During these processes of wound healing, Zinc increases platelet activity and aggregation by tyrosine phosphorylating platelet proteins through protein kinase C (PKC) (Watson et al., 2016). Moreover, PKC inhibitors can be used to prevent the tyrosine phosphorylation that exogenous zinc promotes on several high molecular weight proteins. Alpha-granules, which are found in platelets, contain a variety of proteins and factors, including CXCL1, GRO- (growth regulated protein), CXCL4, CXCL5, ENA-78, epithelial-derived neutrophil-activating protein 78, PPBP (pro-platelet basic protein), -TG (beta-thromboglobulin), CCL3 (macrophage inflammatory protein 1-) and CCL5 (RANTES, regulated upon activation normal T cell expressed and secreted). These factors are responsible for recruiting innate immune cells to the location of the wound for activation. Zinc induces release of alpha-granules (Taylor et al., 2016), which are responsible for pathogen recognition function of immune cells and result in inflammatory response via cytokines and chemokines (Morrell et al., 2014).

The topical zinc sulphate ($ZnSO_4$) has been widely used at a concentration of 3% due to its antioxidant effect during wound healing (Rostan et al., 2002). The largely insoluble zinc oxide (ZnO) or 1% zinc chloride ($ZnCl_2$) has been used during wound healing. ZnO prolongs the supply of zinc to wounds and improves their healing ability. Furthermore, ZnO promotes the degradation of collagen in necrotic wounds. The application of Topical zinc increases the mRNA expression of metallothionein protein having anti-UV photoprotective effect (Pinnell, 2003).

Moreover, due to effective cell penetration of zinc, recent drug delivery systems have utilised zinc oxide nanoparticle (ZnO-NPs) for the treatment of wounds. However, extensive pharmacodynamic and toxicology studies are still required before widespread application in ligand-based drug delivery.

12.4   ZINC DEFICIENCY IN FARM ANIMALS

Variable and often less conclusive results have been obtained about zinc's growth-promoting effects on farm animals. The most common form of dietary zinc supplementation is organic (zinc-methionine/zinc-lysine) or inorganic (zinc oxide, zinc sulphate, etc.). Pregnant cattle and sheep had higher apparent zinc absorption than non-pregnant cattle and sheep. It has been determined that gestational status has an impact on zinc retention, which is higher in the third trimester of pregnancy (Vierboom et al., 2003). Although severe zinc deficiency is known to reduce immunity supplementing with more zinc than necessary does not always boost an animal's immunity (Spears et al., 2002). Zinc deficiency is common in grazing cattle, and supplementation is a common farming practise. Stress results in decreased feed intake as well as decreased zinc retention, which necessitates supplemental zinc intake in addition to diets.

12.4.1   ZINC DEFICIENCY IN MONOGASTRIC ANIMALS

Zinc's importance in pig nutrition has been documented since 1955 due to its role in treating skin disorders such as parakeratosis. However, chelated copper and zinc did not improve growth performance when compared to their inorganic salts. Dietary zinc levels, rather than the source, have a significant influence on serum zinc concentration (Lee et al., 2001). The role of zinc oxide as a growth promoter and anti-diarrhoeal agent has been extensively evaluated (Sun et al., 2022). Zinc supplementation, at 3 g of dietary zinc, was found to have a beneficial effect on growth and intensity of diarrhoea in enterotoxigenic *E. coli*-challenged pigs. Adequate zinc level in the body is a must for maintaining normal health, moderate antibody response with no detectable immune impairment. However, in pigs fed with 150 ppm of zinc supplementation in addition to the 30 ppm of basal diet, it has no effect on feed intake, feed efficiency, or cellular immunity.

12.4.2   ZINC DEFICIENCY IN POULTRY

The source of dietary zinc has been found to influence serum zinc concentration in birds. Minerals that are organically bound, such as chelated and complexed minerals, have a higher bioavailability in poultry birds (Lee et al., 2001). Zinc deficiency in broilers may cause lymphocyte depletion, thymus degeneration and a reduction in lymphoid follicles in the Fabricius bursa (Burns, 1983). Zinc fortification in poultry birds results in a greater magnitude of antibody response (Pimentel et al., 1991). In zinc-deficient chicks, antibody titres against Newcastle disease and Marek's disease viruses were reduced, due to a decreased response of B cells to lipopolysaccharides and the T cells to mitogens such as Concanavalin-A

(Zhang et al., 1999). The weight and growth index of the bursa, spleen, and thymus were significantly reduced in ducklings fed 22.9 mg of zinc/kg diet. The progeny chicks of hens fed with a zinc-fortified diet showed an enhanced antibody response, showing the compound's immunomodulatory characteristics (Stahl et al., 1989). The mononuclear phagocytic system in turkey was strengthened by dietary zinc supplementation, which led to the early clearance of *E. coli* from circulation (Kidd et al., 1994). Recently, Lim and his colleagues found that feeding birds zinc-met alone, or in combination with Mn-met, did not significantly raise serum IgG levels. Similar to that, adding more zinc has no positive effects on egg quality or laying efficiency (Lim and Paik, 2003).

12.4.3 Zinc Deficiency in Pets

Zinc was designated as a possibly essential mineral for dogs in 1962, with a dry food requirement of 1 mg/1000 kcal of metabolisable energy (ME). Zinc intake for dogs was advised to be 15 mg/1000 kcal ME for adults, 25 mg/1000 kcal ME for puppies, and 17 mg/1000 kcal ME for females during peak lactation and gestation in the most recent NRC article (NRC, 2006). A well-known condition known as canine zinc-responsive dermatosis comprises two separate syndromes. Syndrome I has been connected to a genetic abnormality that affects intestinal zinc absorption, increasing the need for zinc. In turn, low zinc levels and/or bioavailability in diets have been connected to Syndrome II (Colombini, 1999). Regardless of breed, it is the most prevalent ailment in pups and may be brought on by a lack of zinc, which is necessary for healthy growth and development. Low blood and hair zinc levels, mild generalised seborrhea sicca, ceruminous otitis externa, superficial lymphadenomegaly accompanied by a dry hair coat, and other lesions with zinc-responsive dermatitis were also present (Broek et al., 1986). A chemical step in the biosynthesis of linoleic acid into arachidonic acid is catalysed by the enzyme delta-6 desaturase, which needs zinc as a cofactor. Keratinocytes subsequently employ arachidonic acid to build a functioning epidermis (Reed et al., 2014). In healthy dogs, linoleic acid plus zinc supplementation had a synergistic impact that improved coat gloss and coat scale while reducing transepidermal water loss, according to Marsh and coworkers, 2016. With delayed hair growth, scaly skin and ulcerations on the buccal edges, kittens fed with a diet high in zinc have poor coats. The zinc needs of cats are most likely between 15 ppm and 50 ppm (Kane et al., 1981).

12.5   CONCLUSION

Zinc is a micronutrient that is essential for animal health. Zinc has an important role in many important processes of the body, and minor alterations may result in several diseases. Zinc has a crucial role in the regulation of every stage of wound healing, including membrane repair, oxidative stress, inflammation, coagulation, tissue re-epithelialization, immune defence, angiogenesis and fibrosis/scar formation. The precise mechanisms behind these functions of zinc remain unexplored, though it is crucial. The current search needs improved bioavailability of zinc along with sources that can meet the requirements of all animals. It will be much easier to treat wounds that are difficult to heal if more research is done into the mechanisms of zinc and the various proteins for which it acts as a cofactor.

REFERENCES

Beattie, John H., & Kwun, In-Sook (2004). Is zinc deficiency a risk factor for athero-sclerosis? British Journal of Nutrition, 91(2), 177–181. 10.1079/bjn20031072.

Berg JM. 1990. Zinc fingers and other metal-binding domains. Receptor 29, 31.

Broek AVD, Thoday KL. 1986. Skin disease in dogs associated with zinc deficiency: a report of five cases. Journal of Small Animal Practice 27, 313–323.

Brugger D, Windisch WM. 2019. Zn metabolism of monogastric species and consequences for the definition of feeding requirements and the estimation of feed Zn bioavailability. Journal of Zhejiang University-SCIENCE B 20, 617–627.

Burns RB. 1983. Antibody production suppressed in the domestic fowl (Gallus domesticus) by zinc deficiency. Avian Pathology 12, 141–146.

Chesters JK. 1992. Trace element-gene interactions. Nutrition Reviews 50, 217–223.

Coleman JE. 1992. Zinc proteins: enzymes, storage proteins, transcription factors, and replication proteins. Annual Review of Biochemistry 61, 897–946.

Colombini S. 1999. Canine zinc-responsive dermatosis. Veterinary Clinics of North America: Small Animal Practice 29, 1373–1383.

Cousins RJ, Blanchard RK, Moore JB, Cui L, Green CL, Liuzzi JP, Cao J, Bobo JA. 2003. Regulation of zinc metabolism and genomic outcomes. The Journal of Nutrition 133, 1521S–1526S.

Cousins RJ, Liuzzi JP, Lichten LA. 2006. Mammalian zinc transport, trafficking, and signals. Journal of Biological Chemistry 281, 24085–24089.

Cunningham-Rundles S, Bockman RS, Lin A, Giardina PV, Hilgartner MW, Caldwell-Brown D, Carter DM. 1980. Physiological and pharmacological effects of zinc on immune response. Annals of the New York Academy of Sciences 587, 113–122.

Demertzis, P. N. (1972). EFFECT OF ZINC ON SKIN AND HAIR. The Lancet, 300(7789), 1261–1262.

Dell BL, Savage JE. 1957. Symptoms of zinc deficiency in the chick. In Federation Proceedings 16, 394-394.

Hambidge KM, Casey CE, Krebs NF. 1986. Zinc, In: Mertz W (Ed.), Trace Elements in Human and Animal Nutrition, Vo.2. Academic Press, San Diego. pp. 1–37.

Hempe JM, Cousins RJ. 1989. Effects of EDTA and zinc methionin complex on zinc absorption by rat intestine. The Journal of Nutrition 119, 1179–1187.

Henkel R, Baldauf C, Bittner J, Weidner W, Miska W. 2001. Elimination of zinc from the flagella of spermatozoa during epididymal transit is important for motility. Reproduction Technology 10, 280–285.

Hidiroglou M, Knipfel JE. 1984. Zinc in mammalian sperm: a Review. Journal of Dairy Science 67, 1147–1156.

Hill GM, Link JE. 2009. Transporters in the absorption and utilization of zinc and copper. Journal of Animal Science 87, E85–E89.

Hoekstra WG, Faltin EC, Lin CW, Roberts HF, Grummer RH. 1967. Zinc deficiency in reproducing gilts fed a diet high in calcium and its effect on tissue zinc and blood serum alkaline phosphatase. Journal of Animal Science 26, 1348–1357.

Huang L, Li X, Wang W, Yang L, Zhu Y. 2019. The role of zinc in poultry breeder and hen nutrition: an update. Biological Trace Element Research 192, 308–318.

Jacobson SG, Kemp CM, Borruat FX, Chaitin MH, Faulkner DJ. 1987. Rhodopsin topography and rod-mediated function in cats with the retinal degeneration of taurine deficiency. Experimental Eye Research 45, 481–490.

Kane E, Morris JG, Rogers QR, Ihrke PJ, Cupps PT. 1981. Zinc deficiency in the cat. The Journal of Nutrition 111(3), 488–495.

Keilin D, Mann T. 1940. Carbonic anhydrase. Purification and Nature of the Enzyme. Biochemical Journal 34, 1163.

Kidd MT, Qureshi MA, Ferket PR, Thomas LN. 1994. Dietary zinc-methionine enhances mononuclear-phagocytic function in young turkeys. Biological Trace Element Research 42, 217–229.

Kumar N. 2003. Effect of zinc supplementation on seminal attributes and serum testosterone level with special reference to in vitro fertility tests in cross-bred bulls. M.V.Sc. Thesis, IVRI, Izatnagar, UP, India.

Kvist U. 1982. Spermatozoal thiol disulphite interaction, a possible event underlying physiological sperm nuclear chromatin decondensation. Acta Physiology Scandinavian 115, 503–505.

Kwun IS, Cho YE, Lomeda RAR, Kwon ST, Kim Y, Beattie JH. 2007. Marginal zinc deficiency in rats decreases leptin expression independently of food intake and corticotrophin-releasing hormone in relation to food intake. British Journal of Nutrition 98, 485–489.

Lee SH, Choi SC, Chae BJ, Acda SP, Han YK. 2001. Effects of feeding different chelated copper and zinc sources on growth performance and fecal excretions of weanling pigs. Asian-Australasian Journal of Animal Sciences 14, 1616–1620.

Lim HS, Paik IK. 2003. Effects of supplementing minerals methionine chelates (Zn, Cu, Mn) on the performance and eggshell quality of laying hens. Asian-Australian journal of Animal Science 16, 1804–1808.

Marquez A, Urbina M, Quintal M, Obregón F, Salazar V, Lim L. 2017. Extracellular zinc chelator in vivo on system of taurine in retina: Transport, concentrations and localization of transporter. Journal of Clinical and Experimental Ophthalmology 8.

Miller WJ, Martin YG, Gentry RP, Blackmon DM. 1968. 65Zn and stable zinc absorption, excretion and tissue concentrations as affected by type of diet and level of zinc in normal calves. Journal of Nutrition 94, 391–401.

Miller WJ, Blackmon DM, Gentry RP, Pitts WJ, Powell GW. 1967. Absorption, excretion and retention of orally administered zinc-65 in various tissues of zinc deficient and normal goats and calves. Journal of Nutrition 92, 71–78.

Miller WJ, Morton JD, Pitts WJ, Clifton CM. 1979. The effect of zinc deficiency and restricted feeding on wound healing in the bovine. Proceedings of the Society for Experimental Biology and Medicine 118, 427–431.

Morrell CN, Aggrey AA, Chapman LM, Modjeski KL. 2014. Emerging roles for platelets as immune and inflammatory cells. Blood, The Journal of the American Society of Hematology 123, 2759–2767.

National Research Council. 2006. Nutrient Requirements of Dogs and Cats. National Academies Press.

Neathery MW, Miller WJ, Blackmon DM, Pate FM, Gentry RP. 1973. Effects of long-term zinc deficiency on feed utilisation, reproductive characteristics and hair growth in the sexually mature male goat. Journal of Dairy Science 56, 98–105.

O'Connor JP, Kanjilal D, Teitelbaum M, Lin SS, Cottrell JA. 2020. Zinc as a therapeutic agent in bone regeneration. Materials 13, 2211.

Ogawa Y, Kinoshita M, Shimada S, Kawamura T. 2018. Zinc and skin disorders. Nutrients 10, 199.

Ott EA, Smith WH, Stob M, Beeson WM (1964). Zinc deficiency syndrome in the young lamb. Journal of Nutrition, 82, 41–50.

Pereira, Ana Margarida, Maia, Margarida R. G., Fonseca, António José Mira, & Cabrita, Ana Rita Jordão (2021). Zinc in Dog Nutrition, Health and Disease: A Review. Animals (Basel), 11(4), 978. 10.3390/ani11040978.

Pimentel JL, Cook ME, Greger JL. 1991. Immune response of chicks fed various levels of zinc. Poultry Science 70, 947–954.

Pinnell SR. 2003. Cutaneous photodamage, oxidative stress, and topical antioxidant protection. Journal of the American Academy of Dermatology 48, 1–22.

Reed S, Qin X, Ran-Ressler R, Brenna JT, Glahn RP, Tako E. 2014. Dietary zinc deficiency affects blood linoleic acid: Dihomo-γ-linolenic acid (LA: DGLA) ratio; a sensitive physiological marker of zinc status in vivo (Gallus gallus). Nutrients 6, 1164–1180.

Rostan EF, DeBuys HV, Madey DL, Pinnell SR. 2002. Evidence supporting zinc as an important antioxidant for skin. International Journal of Dermatology 41, 606–611.

Rucker RB, Lonnerdal B, Keen JL. 1994. Intestinal absorption of nutritionally important trace minerals, in: Johnson LR (Ed.), Physiology of the Intestinal Tract, 3rd Ed. Raven Press, New York. pp. 2183–2202.

Simkin PAJ. 1997. Zinc, again. Rheumatology 24, 626–628.

Smeyers-Verbeke J, May C, Drochmans P, Massart DL. 1977. The determination of Cu, Zn. and Mn in subcellular rat liver fractions. Analytical Biochemistry 83, 746–753.

Spears JW, Kegley EB. 2002. Effect of zinc source (zinc oxide vs zinc proteinate) and level on performance, carcass characteristics, and immune response of growing and finishing steers. Journal of Animal Science 80, 2747–2752.

Stahl JL, Cook ME, Sunde ML, Gregor JL. 1989. Enhanced humoral immunity in progeny chicks from hens fed practical diets supplemented with zinc. Applied Agricultural Research (USA) 4, 86–89.

Stevenson JW, Earle IP. 1956 Studies on parakeratosis in swine. Journal of Animal Science 15, 1036–1045.

Sun, Yiwei, Ma, Ning, Qi, Zengkai, Han, Meng, & Ma, Xi (2022). Coated Zinc Oxide Improves Growth Performance of Weaned Piglets via Gut Microbiota. Frontiers in Nutrition, 9, 819722. 10.3389/fnut.2022.819722.

Taylor KA, Pugh N. 2016. The contribution of zinc to platelet behaviour during haemostasis and thrombosis. Metallomics 8, 144–155.

Tucker HF, Salmon WD. 1955. Parakeratosis or zinc deficiency disease in the pig. Proceedings of the Society for Experimental Biology and Medicine 88, 613–616.

Vierboom MM, Engle TE, Kimberling CV. 2003. Effects of gestational status on apparent absorption and retention of copper and zinc in mature Angus cows and Suffolk ewes. Asian-Australasian Journal of Animal Sciences 16, 515–518.

Watson BR, White NA, Taylor KA, Howes JM, Malcor JDM, Bihan D, Sage SO, Farndale RW, Pugh N. 2016. Zinc is a transmembrane agonist that induces platelet activation in a tyrosine phosphorylation-dependent manner. Metallomics 8, 91–100.

Wellinghausen N, Schromm AB, Seydel U, Brandenburg K, Luhm J, Kirchner H, Rink. L. 1996. Zinc enhances lipopolysaccharides induced monokines secretion by a fluidity change of lipopolysaccharides. Journal of Immunology 157, 3139–3145.

Zhang RJ, Zhou YP, Huang Y. 1999. The modulation effects of zinc on immune organs development and function in broilers. Acta Veterinaria et Zootechnica Sinica 30, 504–512.

13 Zinc in Industry

Liju Raju, G. Deviga, M. Mariappan, and E. Rajkumar

13.1 INTRODUCTION

Zinc, the fourth most common metal globally and the third most common nonferrous metal, plays a crucial role in various industries. Its ability to combine with other metals makes it valuable for creating materials with different properties. One significant application of zinc is its role in galvanizing steel to prevent corrosion. Additionally, zinc is utilized in die-casting complex components, making it indispensable in industrial and domestic settings. An important advantage of zinc is its cost-effective recyclability without compromising its physical qualities, enabling easy collection and recovery. This chapter explores the extensive use of zinc and its derivatives across diverse industries, including steel, alloys, batteries, rubber, dietary supplements, pharmaceuticals, pigments, and cosmetics. In industrial applications (Porter, 1991), zinc often replaces lead, and its use in roofing sheets dates back to 1810. Over time, various techniques such as casting and coating technologies have been introduced. The production of pure zinc through electrolytic processes and thermal manufacturing methods further expanded its applications. Zinc has made a significant impact in automotive, battery, steel, construction, and household industries, in addition to corrosion resistance applications (Ilzsg, 2020).

Ancient civilizations, such as the Dacian people and Ancient Romans, utilized zinc alloys, including brass (a combination of zinc and copper), for coins and ornaments (Boni et al., 2003). Despite its historical use, the isolation of pure zinc remained challenging until the 12th century when evidence of zinc smelting was found in India. Overall, zinc's versatility and historical significance have established its importance in various industries, making it a crucial metal for modern society.

13.2 ZINC PRODUCTION AND UTILIZATION AROUND THE WORLD

The demand for zinc has been steadily rising, leading to increased production year after year. The report (Ilzsg, 2022) published by International Lead and Zinc Study Group (ILZSG) for the worldwide zinc production and usage data for the period 2018–2022, highlights the growing importance and demand for zinc in various industries and applications.

                                        DOI: 10.1201/9781003412472-13

In the history of zinc, India is the only country which made pure zinc in the medieval age and had a high volume of zinc-brass alloy. The technique of zinc distillation, a process of vaporizing the zinc from its ore and condensing it, was followed in ancient India as early as in the 16th century (Ray, 1909). Zinc was largely found in the brass industries for the making of idols of deities of Hindus, Buddhism and Jainism (Bhatia, 2007). At Taxila (presently in Pakistan) a big trade center with long history, archeologists found brass objects used in 4th century BC. *Rasaratnakara*of Nagarjuna is ancient literary evidence of the usage of zinc by ancient Indians. Zawar mine (Udaipur) is one of the important mines to know about zinc utilization at medieval age, the Zawar zinc distillation technology was pronounced as a finest technology at the age of pure metal extraction and ore production (Craddock, 1987). Today, India is the third largest producer of zinc globally, with an estimated share of 5.3%. India has a self-sufficient reserve of zinc mines. The largest zinc producers of India include Hindustan Zinc Limited (HZL), Mewat Zinc Limited, Sunrise Zinc Limited, Ambuja Zinc Limited, Rose Zinc Limited, Vedanta Zinc Limited, etc. (Sabnavis et al., 2018). In India, zinc is mainly used in the steel and automobile industries mainly to control the corrosion.

13.3   ZINC IN STEEL INDUSTRY

Steel is an alloy of iron and carbon and is an important material in infrastructural development. Zinc is an essential ingredient in the steel industry; it contributes greatly towards quality and durability of steel (Marder, 2000). The major role of zinc is to protect iron in the steel from atmospheric corrosion. The widely adopted process of introducing zinc into the iron is known as galvanization and the method commonly used is called as Hot Dip Galvanization (HDG) (Galvamozeit, 2022). In HDG, the fabricated steel or iron is brought into a bath of molten zinc, where iron in the steel combines with the zinc to form a tightly bonded alloy coating (Figure 13.1).

This coating acts as a cathodic protection and a barrier between the iron and the corrosive atmosphere (Królikowska & Komorowski, 2015). There are also other methods commercially used in steel industry for zinc coating, such as metallization, zinc-rich painting, electroplating and mechanical plating, with each method having unique characteristics. These procedures are employed depending upon the end use of steel as galvanization affects the physical properties of the final product significantly due to the method of zinc coating.

13.4   METALLIZATION

Metallization or Zinc spraying involves melting of zinc powder/wire in a hot gun and spraying onto the steel components (abraded surfaces) by means of combustion gas or compressed air (Knudsen et al., 2019). Metallization coatings are rough and porous with a density of coating more than the HDG method. Metallization is used when the steel components are too large to undergo HDG and this method can be more advantageous in coating already erected steel structures.

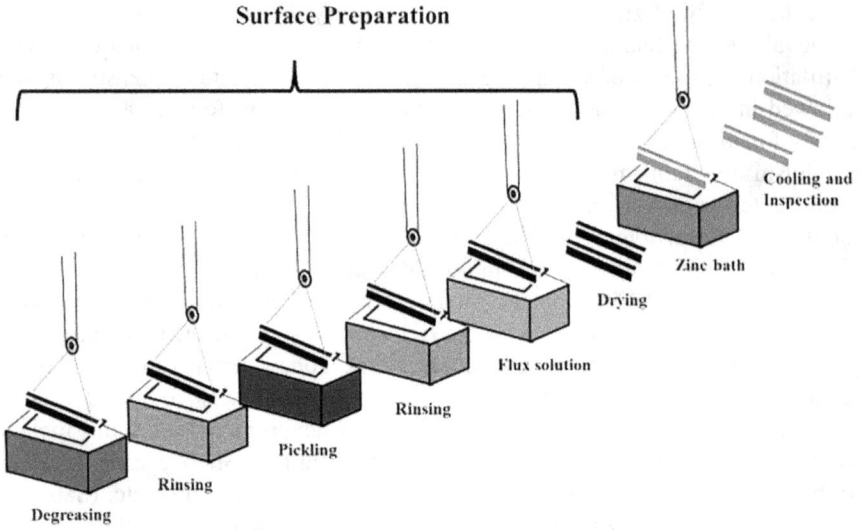

FIGURE 13.1 Representation of hot dip galvanization process.

13.5 ZINC PAINTING

The Zinc paint method, also called cold galvanization, is the application of Zinc coating by means of a brush or spray (Colica, 2021). Zinc paint is a blend of Zinc metal (92–95%) and organic or inorganic binders. Zinc paints are either organic consisting of halogenated hydrocarbons, epoxies, and other organo polymers; or inorganic compounds based on, phosphates, organo alkyl silicates, etc. (Kowalczyk et al., 2019). Prior to painting, the steel surfaces are sandblasted to near white, the Zinc paint mixture is maintained under constant agitation to ensure homogeneity and adhesion of the mix. Though application of Zinc paint on complex structures is tough, it can still be used to coat steel of any size and shape. Zinc paint methods are even used as a primer coat for HDG. Depending on the binder type, the performance of the zinc paint varies, the inorganic zinc paints are resistant to solvents and can withstand a temperature up to 600° K, whereas the organic zinc paints show a solvent-dependent performance, the thermal resistance are only up to 400° K. Organic zinc paints could be degraded easily by exposure to ultraviolet rays and are not as effective as inorganic ones against corrosion of steel. The main drawbacks of the zinc paint approach are the high life cycle cost, the lack of homogeneous coating and the challenge of cleaning the pre-coated surface.

13.6 ELECTROPLATING

Zinc electroplating or electro-galvanization (Figure 13.2) is the process of coating zinc onto steel by electrodeposition (Lindsay and O'Keefe, 1994). In the electrolysis lead/ silver or any other insoluble anodes are used, zinc sulfate salts are employed as the electrolyte. The zinc coating develops on the cathodic steel

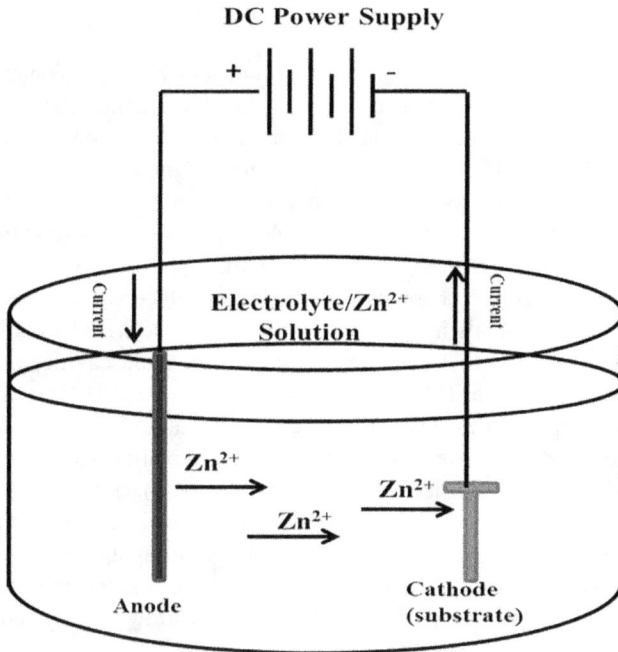

FIGURE 13.2   Pictorial representation of electro galvanization process.

plate or rod by electro-reduction of zinc ions on the surface of the steel. Many additives can be brought into the electrochemical system to refine the coating. The coating of zinc by means of electro-deposition would be highly ductile, which does not allow pealing of the layer on steel deformation. The electroplated coating of zinc would be thinner than that of galvanization and there would be no alloy layers (Nakano and Fukushima, 2002).

13.7   MECHANICAL PLATING

Mechanical plating or mechanical galvanization involves the use of mechanical force to coat small steel or metallic parts with zinc. This method is often employed for coating of small parts which are in the size of 8–10 inches and weighing up to 0.5 kg. The process is carried out in tumbling drums, which are loaded with the metal parts to be coated, along with zinc, glass beads and other proprietary chemicals. The glass beads peen zinc powder onto the component during tumbling. Once finished, the parts are post-treated with a passivation film, dried, and packaged. Unlike HDG process, mechanical plating is carried out at ambient temperature thus allowing the use of metal parts with plastic or other temperature-sensitive assemblies on them. Mechanical plating, like electroplating, does not employ any aqueous electrolyte; there is no effluent produced, and the risk of hydrogen embrittlement is eliminated (Chatterjee, 2005).

13.8   ZINC IN ALLOYS

Zinc is in general known for its galvanization on steel and corrosion control, apart from application in coatings, or as an alloying metal in aluminum, bronzes, brasses and magnesium alloys. Recently zinc alloys have been given considerable attention as a biodegradable metal, an alternative to iron and magnesium (Heiden et al., 2015; Levy et al., 2017). On a global scale about 15% of zinc has been used as a base metal for the alloy industries, they are available as wrought or forged zinc and in the form of sheets, rolls and wires (Pola et al., 2020).

Zinc-based alloys offer characteristic properties which are advantageous with respect to the parent metals in the alloy. They are characterized by a low melting point, with high fluidity, that helps in filling intricate mold cavities and very thin sections, typically as low as 0.1 mm (Sadegh and Worek, 2018). Zinc alloys exhibit good mechanical properties compared to conventional Cu alloys, and they show better finishing and can be easily plated on other metal surfaces owing to their corrosion and wear resistance (Gervais et al., 1985). However, they suffer performance loss above 70^0 C or on continuous exposure to heat and aging. As a result, zinc alloys are generally used in non-structural applications such as toys, electronic and electrical equipment, automobiles, sports articles, etc. (Porter 1991). There are almost twenty zinc alloys which are used worldwide in foundries (Rollez et al., 2015).

13.9   MANUFACTURING TECHNIQUES OF ZINC ALLOYS

On the basis of the proportion of zinc, the alloys are manufactured by various methods viz Hot Chamber Die-Casting, Cold Chamber Die-Casting, Gravity and Sand Casting, and Spin/Slash Casting.

The die-casting technique involves injecting of molten metal solution into a permanent die under high pressure (Apelian et al., 1981). In the Hot-Chamber method, the injection system is submerged into the crucible containing the molten alloy. On the other hand, in Cold-Chamber Die-Casting process the furnace is kept separately from the casting machine and the liquid metal is transferred into the injection system (Kapranos et al., 2014). Owing to their low melting nature, about 90–95% of zinc alloys are typically manufactured by Hot-Chamber Die-Casting technique (Rollez et al., 2015). As the number of other metals in molten alloy increases, Cold-Chamber Die-Casting method is adopted as there could be a risk of metal reaction with the injection system and dissolving. A metal or refractory materials-based mold is used in die casting, whereas a sand mold is used in sand casting. The Spin Casting method utilizes a centrifugal force to fill a mold usually made up of rubber; this process is mostly suitable for low melting alloys of zinc (Grada, 2010). Lastly, the slash casting technique, which is used to manufacture hollow alloys, involves pouring of the molten metal into the mold and allowing it to solidify as a shell on the wall of the mold. The remaining liquid in the core is then poured out, leaving behind a hollow (Bray, 1990; Pola et al., 2020).

13.10 PROPERTIES OF ZINC ALLOYS

Despite the fact that zinc alloys are used only for decorative or non-structural purposes, understanding their performance parameters is very important. Several studies have been focused on the exploration of their mechanical, technological and electro-chemical properties in comparison with other non-ferrous alloys. The mechanical properties of zinc alloys typically rely on two factors - casting method and alloy composition. In view of casting conditions, there will be significant variations in the alloy at the microstructure level due to characteristic solidification and cooling rates in various casting methods. For instance, the High Pressure Die-Casting (D) method is expected to promote rapid solidification and cooling, giving rise to a fine microstructure, while the sand (S) or gravity(G) casting is characterized by slower cooling rates which results in a coarser microstructure (Campbell, 2015). As a result, it can be concluded that the Die Casting Technique exhibited a higher mechanical property than the Gravity or Sand cast technique. Table 13.1 summarizes the mechanical properties of few zinc alloys.

The constituents of the zinc alloy and their proportions are the other factors that influence its properties. Aluminum is found in most zinc casting alloys as an alloying element, it lowers the melting temperature and offers fluidity making it suitable for a wide range of foundry applications and further enhancing mechanical properties. The role of copper is to increase mechanical strength in terms of hardness, tensile strength, and wear resistance. A small portion of added magnesium would provide resistance to intergranular corrosion.

TABLE 13.1

Mechanical Properties of Few Zinc Alloys (Pola et al., 2020). (Open Access Creative Common CC BY License)

Alloy	Tensile Strength (MPa)	Yield Strength (MPa)	E%
Zamak 2 (D)	358	-	7
Zamak 3 (D)	283	-	10
Zamak 5 (D)	331	-	7
Zamak 7 (D)	283	-	13
ZA-8 (S)	248–276	200	1–2
ZA-8 (G)	221–255	207	1–2
ZA-8 (D)	372	290	6–10
ZA-12 (S)	276–317	207	1–3
ZA-12 (G)	310–345	207	1–3
ZA-12 (D)	400	317	4–7
ZA-27 (S)	400–440	365	3–6
ZA-27 (G)	421–427	365	1
ZA-27 (D)	421	365	1–3
ACuZinc 5 (D)	407	338	0.4
ACuZinc 5 (G)	297	-	5

Aging, the reduction of mechanical properties with time at room temperature is one of the performance indicators of zinc alloys. Zinc alloys experience a low tensile strength to the tune of 10–15% depending on the composition and casting thickness with time. The primary causes of aging are a lack of uniform rate of solidification and cooling processes. Because of their low melting temperature, zinc alloys readily undergo aging through the diffusion process.

In zinc alloys, wear resistance is another significant property, this must be considered depending upon the areas of use of zinc alloys ranging from decorative (belts buckles, chain and, zippers, etc.) to the automotive industry (pulleys, small gears, gearboxes, gear racks, etc.). Wear resistance is not an intrinsic property of the material, as it depends solely on tribological system and the test conditions used. Hence, a comparison of the wear behavior of zinc alloys under various conditions (i.e., different applied loads, sliding distance, with or without lubrication, etc.) is not reliable. Zinc Al and Zinc Cu alloys have been extensively studied for their wear resistance, as the concentrations of Al and Cu greatly influence the wear resistance of the alloy. For instance, the Zn-Al alloys, Zamak 2 and Zamak 3 with low Al content, have been compared with high Al/Cu content alloys of zinc for their wear resistance performance. It was found that Zamak 2 and Zamak 3 alloys exhibited poor wear resistance with respect to high-content Al/Cu alloys. However, in Al-rich zinc alloys, an increase in Al content up to 40 to 50 wt. % is not positive for wear resistance. Copper is considered to be beneficial for wear resistance up to 2% wt, whereas higher content does not result in a significant improvement in material performance. This is due to a hardening effect of α-phase solid-solution of Cu up to 2 wt. %, above this other phase formation will take place, resulting in a loss of material performance. Cavitations erosion is also an important performance indicator of alloys, Al rich Zinc alloys are known to offer greater cavitation erosion resistance. This was due to the role of $\beta + \eta$ eutectic phase of the Al-Zn alloy, which is more resistant than primary Zn phases (Pola et al., 2010).

Alloys undergo structural deformation, under pressure and temperature, known as *creep*. The creeps in Zn alloy family are very common because of its low melting point and structural instability with an increase in temperature. Creep in zinc alloys can be divided into three stages. The first stage, known as primary creep, is driven by strain hardening phenomena and is characterized by a decrease in strain rate with time. The strain rate remains constant in the second stage, i.e., secondary creep, since the strain hardening effects are compensated by the recovery effects. The final stage, known as tertiary creep, causes structural weakness in the alloy due to material matrix softening and an inability to withstand load, the addition of copper is reported to enhance creep resistance (Sharma and Martin, 1974). Copper content of approximately 1 wt. % and aging treatment are constructive for creep resistance since they contribute to obstructing grain boundary sliding during tests. This has been exploited in the production of the Al-Cu-Zinc alloys, which exhibit improved creep strength, and also several alloys near the Zn-Al-Cu ternary eutectic. Coarser microstructures exhibit superior creep resistance due to a lower density of grain boundaries. However, ultra-fine microstructures produced in 0.8 mm thick Zamak 5 die-castings demonstrated enhanced creep resistance(Frank et al., 2019).

Corrosion resistance of zinc alloys, like mechanical properties, is consistently related with alloy system composition, impurity level and microstructures. Impurities induced by elements such as lead, tin and cadmium are known to cause intergranular corrosion in zinc alloy materials. And hence standardization organizations set a stringent permissible limit for these elements. Metallic Magnesium is frequently used in trace amounts to prevent corrosion. Cu and Al are found to be more useful in combating air corrosion in Zn alloys than the other alloying elements (Rollez et al., 2015).

13.11   ZINC IN BATTERY INDUSTRY

Batteries are made by one or more cells that create electron flow in a circuit. They mainly contain three components: cathode, anode, and electrolyte. The two major classifications of batteries are primary batteries (non-rechargeable) and secondary batteries (rechargeable) (Hymel, 2016).

Lead is the metal of choice in the field of energy storage systems. But an alternative was in search due to its cost especially in electric vehicles. Lithium has drawbacks such as poor safety and insufficient energy density (Boddula and Asiri, 2020). When compared to the gravimetric and volumetric densities of zinc (500 Wh kg^{-1}cell and 1400 Wh kg^{-1}cell) and with lithium) (350 Wh kg^{-1}cell and 810 Wh kg^{-1}cell respectively), the former was preferred (Stock et al., 2019). Based on these, researchers and scientists started using zinc is an alternate metal, which is inexpensive and has a higher energy density than lithium.

13.12   ZINC AS A PROMISING TOOL FOR BATTERY APPLICATIONS

Zinc has a negative standard potential (−0.78 V), and the safest metal to be used in batteries. Since 1796, when Volta built the first battery with zinc as an anode, for more than a century, zinc-air batteries have been a well-known replaceable energy source and can be kept for long as dry material, as well as a good source of energy (Harting et al., 2012).

Zinc-air batteries (ZAB) are a hybrid technology that stands between regular batteries and fuel cells (Figure 13.3).

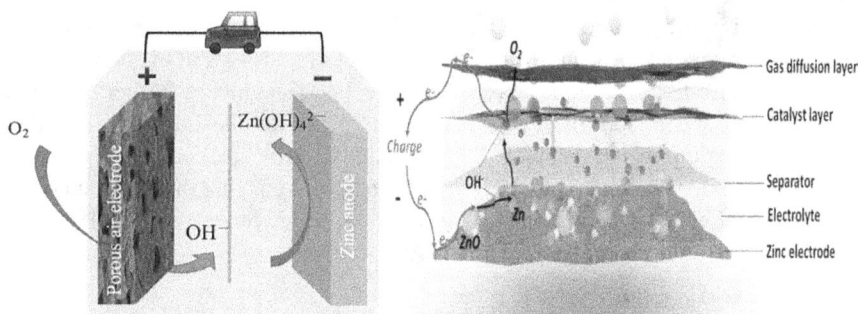

FIGURE 13.3   Schematic representation of ZAB (Fu et al., 2017; Gu et al., 2017). Copyrights obtained from The Royal Society of Chemistry and Wiley.

It works similarly to traditional batteries in that it uses metal as an anode, and it works similarly to fuel cells in that only one reactant, oxygen flows through the cell at all times. Since 1878, the zinc-air battery has been prepared and is still being explored as a viable choice for green energy storage (Harting et al., 2012; Mainar et al., 2018). Zinc electrode was already used in many commercial primary batteries such as Zn/NiOOH, Zn/MnO$_2$, Zn/AgO$_2$ and Zn/O$_2$ (Stock et al., 2019).

13.13   STRUCTURE OF ZAB

The four essential components of a Zinc Air Battery (ZAB) are: (i) Zinc electrode (ii) Separator (iii) Alkaline electrolyte (iv) Catalyst-painted gas diffusion layer

13.13.1   ZINC ANODE

The amount of zinc present in the battery determines the performance of the battery. The cathode material is constructed by using the polymer gel that has been mixed with zinc powder and Teflon (Zhu et al., 2003). Air-breathing cathodes require an active material coating on the carbon Gas Diffusion Layer (GDL). Drop coating or spray coating methods are commonly used (Lee et al., 2014), as it improves the life cycle of ZAB, and it allows the movement of hydroxyl (OH$^-$) ions. Researchers were working on limiting the spontaneous evolution of hydrogen by anode, which is the primary source of corrosion (Fu et al., 2017).

13.13.2   AIR CATHODE

The use of air as an electrode is the main advantage of air batteries. The performance of an air cathode mainly depends on GDL and the electro-catalyst used. Hydrophobic and hydrophilic properties were balanced by the GDL. In an air cathode, both Oxygen Reduction Reaction (ORR) and Oxygen Evolution Reaction (OER) occur through GDL. The electro-catalyst is loaded on the GDL surface, which accelerates the rate of the reaction. The conventional battery had platinum metal along with alloys such as RuO$_2$ or IrO$_2$ and MnO$_2$ as electrocatalyst (Zhang et al., 2019). After the development of metal-based nanomaterials, transition metal oxides, carbon materials, metal nitrides and alloys were also used as catalysts.

ORR and OER reactions are affected by the following factors: corrosion, passivation layer and dendrite formation (Wang et al., 2019). ZnO is deposited on the metal electrode surface to act as a passivation layer and blocks further migration of metal ions and affects the cycle life. A sharp needle-like formation is called dendrite formation, which changes the morphology of the electrode. The dendrites so formed can penetrate into the separators resulting in short circuits and morphological changes (Fu et al., 2017; Gu et al., 2017; Stock et al., 2019).

13.13.3   SEPARATORS

Separators conduct hydroxyl ions from the air cathode, which acts as a physical barrier between two electrodes. The separator material should have high conductivity, electrical insulation and a stable ionic solution. It prevents physical contact between the electrodes (Fu et al., 2017; Gu et al., 2017).

13.13.4 ELECTROLYTES

Electrolytes are the most important factor in determining the life cycle of a battery. Aqueous alkaline electrolytes such as KOH, NaOH, and LiOH are typically widely used in traditional batteries (Zhu et al., 2003). It has drawbacks such as less efficiency, poor rechargeability and formation of dendrite structure. It forms insoluble carbonates which make recharging the battery difficult. After that, alkaline electrolytes are replaced by non-aqueous electrolytes, ionic electrolytes, hybrid electrolytes and solid-state electrolytes (Mainar et al., 2018).

The following examples are commercially used electrolytes for zinc-air batteries and their manufacturers;

- Nera neutral chloride-based electrolyte – EOS Energy Storage
- Sulfonate containing ionic liquid electrolyte – Fleridic Storage
- Zinc particle suspension – Zinc-Nyx

13.13.4.1 Air-Diffusion Layer

Due to the fact that oxygen is poorly soluble at atmospheric pressure, a three-phase reaction setup is required to obtain an adequate amount of oxygen. A cathode consists of a catalyst layer, metallic mesh screen, hydrophobic membrane, and air-diffusion layer (Zhu et al., 2003).

The following are commercial manufacturers of ZAB:

Eevercell, Fluidic energy, Z-power, EOS, Zinc Five, ZNR Batteries, ZAF, Zinum, Zinc Nyx Energy Solutions

13.13.5 APPLICATIONS

- ZAB are used in hearing aids, medicinal devices, navigation devices and railway signal devices
- Hearing aids are one of the real-time applications of ZABs. ZABs replace toxic mercury oxide and also give comparatively better sound quality
- Zinc air cells with alkaline electrolytes for hearing aid batteries were manufactured by Rayovac, Power one, Zein powder, Renata, AER energy sources and metallic power
- Military application-based batteries were manufactured by Electric fuel and Efb power

13.13.6 BIFUNCTIONAL ZAB

Bifunctional ZAB attracted special attention of researchers because of its recharge-able ability. The great advantage of ZAB, where zinc is used as an anode material is safe and cost-effective. An air cathode is also very cheap and abundant, and hence using those kinds of batteries is really cost-effective (Mainar et al., 2018). The retractability and cycle life are the only limitations. A bifunctional air electrode, when discharging, accepts oxygen and releases it while charging. As both ORR and OER have excess potential, constructing a long-term usage battery was quite challenging. Flexible ZAB is another future trust, where flexible devices gaining more attention nowadays. Since 2015, scientists have been paying special attention to developing flexible devices using zinc-air batteries (Fu et al., 2017).

There are a few other batteries that use zinc and they are: (i) Zinc-Bromine flow battery (ii) Nickel-Zinc battery (iii) Zinc-cerium battery (iv) Zn-Mn oxide battery (v) Zn-Silver oxide battery (vi) Hybrid $Zn-O_2$ battery (vii) Hybrid $Zn-Ni/O_2$ battery (viii) Hybrid $Zn-Co/O_2$ battery

Zinc-Bromine flow batteries are efficiently used in remote areas and for telecom applications. Nickel-zinc batteries are good alternatives for lithium-ion batteries. Due to its high cycle life (up to 600 deep cycles), it is used in the medical field. Zn-Mn oxide batteries played a predominant role in storage devices for almost six years in the market (Boddula and Asiri, 2020).

13.14   ZINC IN CEMENT INDUSTRY

Cement is an important building material and the backbone of the country's infrastructure development. Cement manufacturing is a significant source of greenhouse gas emissions, and requires a considerable amount of thermal energy. Zinc components play a crucial role in addressing these challenges. There are several reports available on the effect of zinc components in the cement manufacturing process, focusingOrdinary Portland Cement (OPC), a white-colored product which is used to produce concrete. Cement is primarily composed of clinker (90% clinker + 10% gypsum), a material that consists of calcium silicate, iron and aluminum, the clinker is at a temperature of 1400°C as it leaves the kiln (Osman et al., 2020). The mineralization process involves high temperatures. The use of zinc material reduces the temperature range significantly. During the calcination process, one ton of clinker emits around 850 kg of carbon dioxide and adding a trace quantity of zinc metal, a significant quantity of CO_2evolution is reduced (Flores-Velez and Dominguez, 2002).

Zinc mining companies are globally producing five to seven billion tons of zinc tailing wastewater (Colorado et al., 2016; Šiler et al., 2018). These tailing materials have been shown to be successful in replacing expensive silica fume, which is used as a filler material (Agarwal et al., 2015). One of the notable effluents was EAFD (Electric Arc Furnace Dust) produced during the smelting process of steel and more than 3.7 million tons of EAFD were produced annually. Acid treatment of EAFD results in $ZnFeO_4$ predominance. But the percentage addition of EAFD affects the compressive strength and leachability of cement. A 20% addition of waste material to cement is acceptable, while 10% EADF with OPC shows properties the same as OPC with silica fume filler material. Compressive strength and thermal stability are the same as silica fume. One of the major advantages is adding waste as filler material will make the product cost-effective.

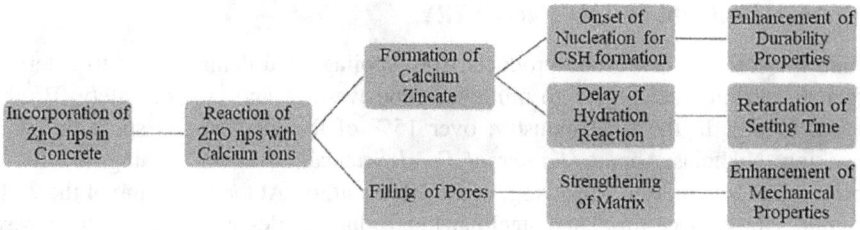

FIGURE 13.4   Addition of ZnO NPs in concrete (Kumar et al., 2021). Copyrights obtained from Elsevier) (Flores-Velez and Dominguez, 2002).

Jarosite is another major industrial effluent that is used in cement manufacturing. During hydrometallurgical operation, a purification process of zinc results in the residual type byproduct called jarosite. It contains heavy metals such as zinc, manganese, strontium, cobalt, barium and copper, which can act as mineralizers. Mineralizer makes mineralization process to the development of clinker mineral phase. The incorporation of jarosite percentage was tested by varying 0.5–1.75%, which resulted in adding 1.5% waste material to cement to get a good compressive strength and complete clinkerisation before 1400°C. Adding more than the above-optimized percentage, the mechanical property of cement would be diminished. However, utilizing byproducts makes one industry's waste as another industry's raw material. ZnO nanoparticles are incorporated with cement to increase its compressive strength, mechanical properties and help to decrease carbon dioxide emission besides reducing energy consumption (Figure 13.4).

13.15   ZINC IN COSMETICS

Zinc is an essential trace element found in the human body, and has a multifunctional role in the biological system (Abendrot and Lis, 2018). It is also a key ingredient in many cosmetic and personal care products, as compounds of zinc are known to exhibit anti-acne, antimicrobial, UV-resistant anti-pigmentation, cleansing or stabilizing activities. It may be noted that the quality of zinc used in cosmetics, food and pharmaceuticals differ from those used in the paint and battery industries. The oxides, citrates, acetate and gluconate of zinc are used in toothpastes; as antibacterial agents and aid in the control of foul breath. Zinc oxide is the most common component employed in the cosmetic sector; in the skin and hair care products (Gupta et al., 2014). It is also used in nano and bulk form, with the former as a recent addition in cosmetics. And also as a primary ingredient in sunscreen products as they act as UV filters and colorant in cosmetics (Smijs and Pavel, 2011). ZnO nanoparticles can effectively block harmful UV rays in the range of 340–400 nm, however in order to cover both UVA-1 (340–400 nm) and UVA-2(320–340 nm) spectrum, zinc oxide is often used in combination with titanium oxide (TiO_2), where ZnO concentration has been properly regulated in sunscreens with a maximum concentration of 25% has been recognized as optimal safe dose (Kim et al., 2017). Studies have explored the use of zinc compounds, including 1% zinc chloride (Rostan et al., 2002) and zinc glycine complexes which offer good protection against sunburns (Masaki et al., 2007).

13.16 ZINC IN FOOD INDUSTRY

Zinc is used to protect food products from spoilage and damages due to external factors, packing techniques, to minimize food wastage and provide safety (Brody et al., 2008). In the food industry, over 15% of the total cost is spent on food packing. Nicholas Appert (*Father of Food Science*) invented canning, a critical process in modern food packaging, in the 19th century. At the beginning of the 20th century, three-piece tin-plated steel cans and glass bottles were making their way into food packaging applications. During World War II, wax-related and petroleum-based products were used to protect ammunition and later, they were used as cereal and biscuit packing materials. Many novel packing techniques were developed during these times that include aluminum foil, polyethylene, polyvinylidene, metal cans, antiseptic packing, and flexible packaging. At end of the 20th century, scientists preferred an active packaging method for effective food protection (Ozdemir and Floros, 2004). The important functions of active food packing materials are as moisture absorbers, oxygen scavengers, ethanol emitters and olefin absorbers (Akbar and Anal, 2014). The demand for ready-to-eat (RTE) foods in the 21st century leads to more food-borne diseases (Espitia et al., 2012). The inclusion of antimicrobial compounds was a widely recommended method to control microbial development in packaged foods. Furthermore, using a material that exhibits both antimicrobial and improved mechanical properties serves a dual purpose in the food packaging industry.

Silver is a proven material for food packing with anti-microbial agent properties. However, in terms of quantity, zinc has been found to be the best alternative. Zinc is present in hundreds of enzymes and protein domains and plays a crucial role in biological functions such as DNA metabolism, neurotransmission, reproduction, vision and taste. Foods like cereals, pulses and meat contain zinc in considerable quantity and can be taken as a supplement to overcome deficiency of zinc (Al-Naamani et al., 2016). As a result, zinc is a good choice for food packaging. There are five zinc compounds that were accepted as "generally regarded as safe (GRAS)" by the United States Food and Drug Administration (USFDA), and zinc oxide was one of them. ZnO possesses good mechanics and is used as an anti-microbial agent in many applications (Espitia et al., 2012).

There are several reports available for the antimicrobial activity of zinc oxide nanoparticles against microbes *Staphylococcus aureus*and *Escherichia coli,* etc. Polyethylene (PE) is the most used polymer for food packing due to its low cost, high impact strength and good chemical strength. The incorporation of ZnO-chitosan nanocomposite on PE gives a great advantage to food packing (Rajamanickam et al., 2012). The nanoparticles in food packing can disperse through the polymer and block oxygen, CO_2 and moisture content and thereby preserve the food materials and maintain freshness (El-Sayed et al., 2020). Chitosan is a deacetylated form of a naturally occurring substance chitin obtained mostly from the shells of crustaceans. According to the USFAD, chitosan is a safe food preservative material with good antimicrobial activity.A bio nanocomposite of chitosan/quargum/zinc oxide was also produced and evaluated for Ras cheese preservation (Ahmed et al., 2019). The addition of clove essential oil (CEO) and ZnO to polymers such as PLA (polylactide) and PCL (polycaprolactone) was tested as CEO gives antimicrobial activity to the packing so that ZnO offers better mechanical property to that coating and restricts the degradation of the polymer. The addition of ZnO is effective in slowing down the food material spoilage, maintaining food color, increasing the mechanical strength of packing and avoiding moisture content.

13.17   ZINC IN THE PRODUCTION OF BULK CHEMICALS AND PHARMACEUTICALS

Many chemical reactions are carried out in the production of fine chemicals, pharmaceuticals and allied products in the industries using metal catalysts. Due to a longer lifetime, more tolerance to polar functional groups and higher turnover, zinc-based catalysts are preferred. Li et al (Li et al., 2015) suggested various zinc catalysts for the interamolecular and intramolecular hydroamination reactions that involve formation of C-N bonds by the direct addition of an organic amine (N-H bond) across the unsaturated C-C bond in alkenes, alkynes and allenes. The application of organo-zinc reagents are preferred in organic synthesis involving cyclopropanation (Simmons-Smith reactions) and transmetallation reactions due to their milder reactivity and excellent chemoselectivity. In contrast to the polar nature of the Grignard reagents (RMgX) and C-Li bonds, carbon-zinc bond (C-Zn) is a highly covalent character, less reactive and allows controlling the functionalized derivatives easily. The representative preparation of an organo zinc catalyst is given below:

$$2i - BuMgBr + ZnCl_2 \rightarrow i - Bu_2Zn + 2MgClBr \qquad 13.1$$

13.17.1   Cyclopropanation Reaction (Simmons-Smith Reaction)

The presence of cyclopropane rings in many natural products revealed several interesting properties, and these compounds are also useful synthons for further synthetic transformations. In the presence of a Zn-Cu couple with diiodomethane, Simmons and Smith synthesized chiral cyclopropane derivatives from alkenes.

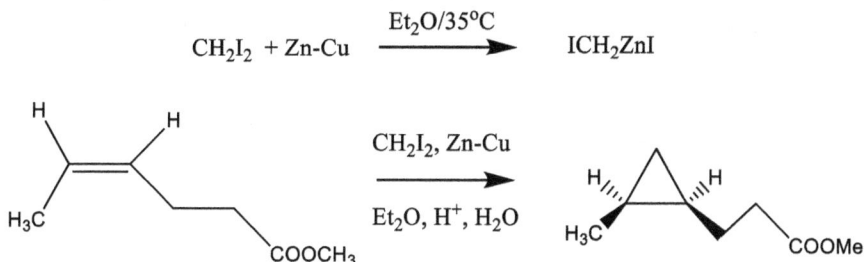

13.17.2   Transmetallation Reaction

Transmetallation reactions can be used to produce organozinc halides (RZnX)

$$\text{RLi/RMgX} \xrightarrow{\text{THF/0−25°C, 1hr}} \text{RZnCl} \tag{13.2}$$

13.17.3 Applications of RZnX

In the presence of N-methyl morpholine, RZnX can react with a variety of electrophilic substrates and its phenyl-substituted catalyst produces a variety of products with high yields.

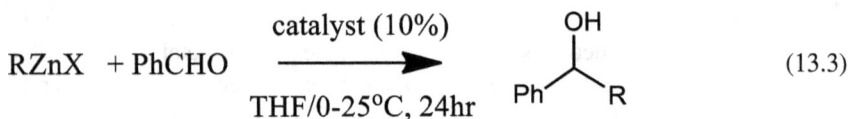

$$\text{RZnX} + \text{PhCHO} \xrightarrow[\text{THF/0-25°C, 24hr}]{\text{catalyst (10\%)}} \underset{\underset{Ph}{}\overset{OH}{\diagdown}R}{} \tag{13.3}$$

(R = Me, Et, X = Br, I):

Catalytic deactivation (by chemical/physical/thermal/mechanical) in the industrial process means the chemical/physical interactions that suppress the catalytic activity or selectivity. Deactivation in Industrial process:

$$4NO + 4NH_3 + O_24N_2 + 6H_2O \tag{13.4}$$

In the biomass combustion process, the catalyst used is V_2O_5-WO_3/TiO_2 and this can selectively reduce NO to N_2 in the presence of NH_3. This reaction takes place by adsorption process and after the end of the process, the Vanadium catalyst can be regenerated by the O_2 present in the flue gas. Here Zinc salt $ZnCl_2$ was used to induce the deactivation of the catalyst by aerosol method (Larsson et al., 2007).

The acylation of phenols, thiophenols, alcohols, and amines can be brought out at room temperature using Zn-dust as a catalyst and solvent-free conditions. Furthermore, used Zn dust can be recycled for future use (Pasha et al., 2010).

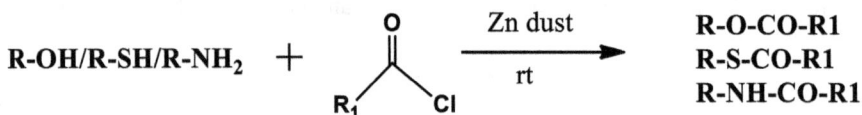

$$\textbf{R-OH/R-SH/R-NH}_2 + \underset{R_1}{\overset{O}{\diagdown}}Cl \xrightarrow[\text{rt}]{\text{Zn dust}} \begin{array}{l} \textbf{R-O-CO-R1} \\ \textbf{R-S-CO-R1} \\ \textbf{R-NH-CO-R1} \end{array}$$

R = R1 = alkyl/aryl

13.18 CONCLUSION

Even though zinc has been employed from ancient civilizations as early as 12th century, its utility as well as that of its compounds and nanoparticles has expanded significantly in recent years. The ever-expanding material science research worldwide utilizes many naturally available and low-cost metals such as zinc and many minerals in the production of instruments, food items, pharmaceuticals, agro products and many more for the benefit of the common man through industries.

Future research and development must focus on the production of nano zinc, its stabilization and preservations.

REFERENCES

Abendrot M, Lis UK. 2018. Zinc containing compounds for personal care applications. International Journal of Cosmetic Science 40, 319–327.

Agarwal SK, Ali MM, Pahuja A, Singh BK, Duggal S. 2015. Mineralising effect of jarosite: a zinc industry by-product in the manufacturing of cement. Advances in Cement Research 27, 248–258.

Ahmed J, Mulla M, Jacob H, Luciano G, Bini T, Almusallam A. 2019. Polylactide/poly (ε-caprolactone)/zinc oxide/clove essential oil composite antimicrobial films for scrambled egg packaging. Food Packaging and Shelf Life 21, 100355.

Akbar A, Anal AK. 2014. Zinc oxide nanoparticles loaded active packaging, a challenge study against Salmonella typhimurium and Staphylococcus aureus in ready-to-eat poultry meat. Food Control 38, 88–95.

Al-Naamani L, Dobretsov S, Dutta J. 2016. Chitosan-zinc oxide nanoparticle composite coating for active food packaging applications. Innovative Food Science & Emerging Technologies 38, 231–237.

Apelian D, Paliwal M, Herrschaft D. 1981. Casting with zinc alloys. Jom 33, 12–20.

Bhatia P. 2007. Mining and metallurgy in ancient India. Indian Historical Review 34, 283–287.

Boddula R, Asiri AM. 2020. Zinc Batteries: Basics, Developments, and Applications. John Wiley & Sons, USA.

Boni M, Gilg HA, Aversa G, Balassone G. 2003. The" calamine" of southwest Sardinia: Geology, mineralogy, and stable isotope geochemistry of supergene Zn mineralization. Economic Geology 98, 731–748.

Bray J. 1990. Properties and selection: nonferrous alloys and special purpose materials. ASM Metals Handbook 92.

Brody AL, Bugusu B, Han JH, Sand CK, McHugh TH. 2008. Innovative food packaging solutions. Journal of Food Science 73, 107–116.

Campbell J. 2015. Complete Casting Handbook: Metal Casting Processes, Metallurgy, Techniques and Design. Butterworth-Heinemann.

Chatterjee B. 2005. Mechanical plating. Galvanotechnik 96, 1589–1599.

Colica, M. (2021). Zinc Spray Galvanizing. International Conference of the European Association on Quality Control of Bridges and Structures; 2021: Springer.

Colorado HA, Garcia E, Buchely MF. 2016. White ordinary Portland cement blended with superfine steel dust with high zinc oxide contents. Construction and Building Materials 112, 816–824.

Craddock P. 1987. The early history of zinc. Endeavour 11, 183–191.

El-Sayed SM, El-Sayed HS, Ibrahim OA, Youssef AM. 2020. Rational design of chitosan/guar gum/zinc oxide bionanocomposites based on Roselle calyx extract for Ras cheese coating. Carbohydrate Polymers 239, 116234.

Espitia PJP, Soares NdFF, Coimbra JSdR, de Andrade NJ, Cruz RS, Medeiros EAA. 2012. Zinc oxide nanoparticles: synthesis, antimicrobial activity and food packaging applications. Food and Bioprocess Technology 5, 1447–1464.

Flores-Velez LM, Dominguez O. 2002. Characterization and properties of Portland cement composites incorporating zinc-iron oxide nanoparticles. Journal of Materials Science 37, 983–988.

Frank, T., Kansy, A., Kallien, L., Leis, W., & Goodwin, F. (2019).Effect of Zinc Alloy Casting Section Thickness on Creep Behavior. Proceedings of the NADCA 2019 Die Casting Congress and Tabletop, Paper T19-102, Cleveland, OH, USA.

Fu J, Cano ZP, Park MG, Yu A, Fowler M, Chen Z. 2017. Electrically rechargeable zinc–air batteries: progress, challenges, and perspectives. Advanced Materials 29, 1604685.

Galvanizeit. 2022. https://galvanizeit.org/hot-dip-galvanizing. (accessed 20 Nov 2022) Processing of steel galvanizing waste using the method of melt immersion. IOP Conference Series: Materials Science and Engineering; 2020: IOP Publishing. UK

Gervais E, Barnhurst R, Loong C. 1985. An analysis of selected properties of ZA alloys. Jom 37, 43–47.

Grada. 2010. http://www.brockmetals.sk/product_acuzinc.html. (accessed 07-05-2022).

Gu P, Zheng M, Zhao Q, Xiao X, Xue H, Pang H. 2017. Rechargeable zinc–air batteries: a promising way to green energy. Journal of Materials Chemistry A 5, 7651–7666.

Gupta M, Mahajan VK, Mehta KS, Chauhan PS. 2014. Zinc therapy in dermatology: a review. Dermatology Research and Practice 2014.

Harting K, Kunz U, Turek T. 2012. Zinc-air batteries: prospects and challenges for future improvement. Zeitschrift für Physikalische Chemie 226, 151–166.

Heiden M, Walker E, Stanciu L. 2015. Magnesium, iron and zinc alloys, the trifecta of bioresorbable orthopaedic and vascular implantation-a review. Journal of Biotechnology & Biomaterials 5, 1.

Hymel S. 2016. https://learn.sparkfun.com/tutorials/what-is-a-battery/all. (accessed 07-05-2022)

Ilzsg. 2020. https://ilzsg.org/pages/1167/document.aspx?page=10&ff_aa_document_type=B&from=1.[accessed 07-05-2022]

Ilzsg. 2022. https://ilzsg.org/pages/1288/document.aspx?page=2&ff_aa_document_type=R&from=5.[accessed 29-03-2022]

Kapranos P, Brabazon D, Midson S, Naher S, Haga T. 2014. Advanced casting methodologies: inert environment vacuum casting and solidification, die casting, compocasting, and roll casting. Comprehensive Materials Processing 5, 3–37.

Katarivas Levy G, Goldman J, Aghion E. 2017. The prospects of zinc as a structural material for biodegradable implants—a review paper. Metals 7, 402.

Kim K-B, Kim YW, Lim SK, Roh TH, Bang DY, Choi SM, Lim DS, Kim YJ, Baek S-H, Kim M-K. 2017. Risk assessment of zinc oxide, a cosmetic ingredient used as a UV filter of sunscreens. Journal of Toxicology and Environmental Health, Part B 20, 155–182.

Knudsen OØ, Matre H, Dørum C, Gagné M. 2019. Experiences with thermal spray zinc duplex coatings on road bridges. Coatings 9, 371.

Kowalczyk K, Przywecka K, Grzmil B. 2019. Influence of novel ammonium-modified zinc-free phosphate nanofillers on anticorrosive features of primer-less polyurethane top-coating compositions. Journal of Coatings Technology and Research 16, 401–414.

Królikowska, A., Komorowski, L. 2015. The corrosion of HDG zinc coatings on the road and urban infrastructures. Solid State Phenomena 227, 217–220.

Kumar, M., Bansal, M., Garg, R. 2021. An overview of beneficiary aspects of zinc oxide nanoparticles on performance of cement composites. Materials Today: Proceedings 43, 892–898.

Larsson A-C, Einvall J, Sanati M. 2007. Deactivation of SCR catalysts by exposure to aerosol particles of potassium and zinc salts. Aerosol Science and Technology. 41, 369–379.

Lee DU, Choi JY, Feng K, Park HW, Chen Z. 2014. Advanced extremely durable 3D bifunctional air electrodes for rechargeable zinc-air batteries. Advanced Energy Materials 4, 1301389.

Li T, Wiecko J, Roesky PW. 2015. Zinc-catalyzed hydroamination reactions. Zinc Catalysis: Applications in Organic Synthesis, 83–118.

Lindsay JH, O'Keefe TJ. 1994. Electrogalvanizing. Modern Aspects of Electrochemistry. Springer, pp. 165–228.

Mainar AR, Iruin E, Colmenares LC, Kvasha A, de Meatza I, Bengoechea M, Leonet O, Boyano I, Zhang Z, Blazquez JA. 2018. An overview of progress in electrolytes for secondary zinc-air batteries and other storage systems based on zinc. Journal of Energy Storage 15, 304–328.

Marder A. 2000. The metallurgy of zinc-coated steel. Progress in Materials Science 45, 191–271.

Masaki H, Ochiai Y, Okano Y, Yagami A, Akamatsu H, Matsunaga K, Sakurai H, Suzuki K. 2007. A zinc (II)–glycine complex is an effective inducer of metallothionein and removes oxidative stress. Journal of Dermatological Science 45, 73–75.

Nakano H, Fukushima H. 2002. Morphology control of zinc deposits of electrogalvanized steel sheets. Tetsu-to-Hagané 88, 236–242.

Osman DAM, Nur O, Mustafa MA. 2020. Reduction of energy consumption in cement industry using zinc oxide nanoparticles. Journal of Materials in Civil Engineering 32, 04020124.

Ozdemir M, Floros JD. 2004. Active food packaging technologies. Critical Reviews in Food Science and Nutrition 44, 185–193.

Pasha MA, Reddy MBM, Manjula K. 2010. Zinc dust: an extremely active and reusable catalyst in acylation of phenols, thiophenol, amines and alcohols in a solvent-free system. European Journal of Chemistry 1, 385–387.

Pola A, Tocci M, Goodwin FE. 2020. Review of microstructures and properties of zinc alloys. Metals 10, 253.

Pola, A Montesano, L Roberti, R. 2010. Nuove Leghe di Zinco per L'industria del Design, In Proceedings of the 30th Convegno Nazionale AIM, Brescia, Italy.

Porter FC. 1991. Zinc Handbook: Properties, Processing, and Use in Design. CRC Press, USA.

Ray PC. 1909. A History of Hindu Chemistry from the Earliest Times to the Middle of the Sixteenth Century, AD: With Sanskrit Texts, Variants, Translation and Illustrations. Vol. 2. Bengal Chemical & Pharmaceutical Works, Limited.

Rollez D, Pola A, Prenger F. 2015. Zinc Alloy Family for Foundry Purposes. World of Metallurgy 68, 358–354.

Rostan EF, DeBuys HV, Madey DL, Pinnell SR. 2002. Evidence supporting zinc as an important antioxidant for skin. International Journal of Dermatology 41, 606–611.

Rajamanickam, U., Mylsamy, P., Viswanathan, S., Muthusamy, P. 2012. Biosynthesis of zinc nanoparticles using actinomycetes for antibacterial food packaging, International conference on nutrition and food sciences IPCBEE. 195–199.

Sabnavis M, Jagasheth UH, Avachat H, Mishra M. 2018. Zinc Industry: the unsung metal of the economy. CARE Ratings: Professional Risk Opinion, Report, 31st October, 2018.

Sadegh AM, Worek WM. 2018. Marks' Standard Handbook for Mechanical Engineers. McGraw-Hill Education.

Sharma R, Martin J. 1974. Creep of dilute zinc-copper alloys. Journal of Materials Science 9, 1139–1144.

Šiler, P., Kolářová, I., Másilko, J., Novotný, R., Opravil, T. 2018. The effect of zinc on the portland cement hydration. Key Engineering Materials 761, 131–134.

Smijs TG, Pavel S. 2011. Titanium dioxide and zinc oxide nanoparticles in sunscreens: focus on their safety and effectiveness. Nanotechnology, Science and Applications 4, 95.

Stock D, Dongmo S, Janek Jr, Schröder D. 2019. Benchmarking anode concepts: the future of electrically rechargeable zinc–air batteries. ACS Energy Letters 4, 1287–1300.

Wang C, Yu Y, Niu J, Liu Y, Bridges D, Liu X, Pooran J, Zhang Y, Hu A. 2019. Recent progress of metal–air batteries—a mini review. Applied Sciences 9, 2787.

Zhang J, Zhou Q, Tang Y, Zhang L, Li Y. 2019. Zinc–air batteries: are they ready for prime time? Chemical Science 10, 8924–8929.

Zhu W, Poole B, Cahela D, Tatarchuk B. 2003. New structures of thin air cathodes for zinc–air batteries. Journal of Applied Electrochemistry, 33, 29–36.

14 Zinc Catalysts in Synthesis and Plastic Waste Management

Vajiravelu Sivamurugan,
Ravikumar Dhanalakshmi, and Gopal Jeya

14.1 INTRODUCTION

Multicomponent reactions (MCRs) are a type of cascade one-pot operations annexing a sequence of organic reactions. The first report on MCR appeared on Strecker's α-amino cyanide synthesis in 1850 was a major breakthrough in organic synthesis. Many MCRs have been developed and some of the landmarks are Hantzsch, Bigenelli, Passerini, Mannich, Kabachnik-Fields and Ugi reactions, etc. MCRs have taken more effort in recent years due to exceptional synthetic performance, selectivity, practical simplicity, higher yield and atom economy (Mahmoud et al., 2018; Neetha et al., 2020).

In view of today's medicinal, material and emerging industrial developments, MCRs play a major role in the construction of complex organic molecules and drugs (Agarwal et al., 2017). Multiple bond constructions of C-C, C-N, C-O, and C-S in a variety of reactions with zinc catalysts revealed its pivotal role (Neetha et al., 2020). A major challenge is the integration of asymmetric approaches in multi-component reactions (Shaabani et al., 2018; Wang et al., 2018; Ramos et al., 2019) such as Passerini or Ugi-type reactions (Riva et al., 2018). For instance, Zinc-proline (ZnP) is used for the synthesis of 1,5-benzodiazepines (Sivamurugan et al., 2004) and 1,4-dihydropyridines (Sivamurugan et al., 2005). ZnP demonstrated to be an eco-friendly, non-volatile, water-soluble catalyst (Sivamurugan et al., 2006), used for imidazole (Agarwal et al., 2017) and pyrazole synthesis (Layek et al., 2018). Several specific transformations such as cycloaddition, cycloisomerization, Sonogashira, C-H functionalization, metathesis and MCRs have also been reported to produce heterocyclics (Khan et al., 2020).

From the standpoint of many reports, zinc contains several coordinating geometries, redox inactive in abundance and forms stable complexes (Poddar, 2017). Many heterogeneous and homogeneous catalysts were used in heterocyclic synthesis with some new zinc-containing catalysts such as ZnO, Zn/Al hydrotalcite, Zn hydroxyapatite, ZnO/Co_3O_4, ZnO nanospheres, $ZnCl_2$/tungstates, Zn-Cu alloy, ZnBr, $Pd(OAc)_2$/ZnBr and $Zn(OTf)_2$ (Mehraban et al., 2018). In this context, we focused on Zn-catalysed heterocyclic reactions. The major applications of zinc

DOI: 10.1201/9781003412472-14

pertain to corrosion and biological preparations. Due to their completely filled electronic configuration, $[Ar]3d^{10}4s^2$, zinc acts as a reducing agent and possesses a high covalent nature (Krishnan et al., 2018).

14.2 CONSTRUCTION OF HETEROCYCLES USING Zn CATALYSTS

14.2.1 INDOLE-BASED SYSTEM

One pot condensation of benzaldehyde, indole and malononitrile in diisopropylethylamine (DIPEA) yields3-substituted indoles in the presence of Zn(salphen) [salphen = N,N-bis(salicylidene)imine-1,2-phenylenediamine]. Comparatively, Zn(salphen) is more efficient than Cu(II) and Au(II)-salphens. The Zn(salphen) complex is an excellent Lewis acidic activator to trigger the catalytic cycle (Neetha et al., 2020).

14.2.2 IMIDAZOLE-BASED SYSTEM

A combination of benzoin, aldehyde and ammonium acetate in water: ethanol mixture is catalysed by ZnP afforded 92% of imidazoles. The addition of ZnP activates aldehyde to form imine through nucleophilic attack and dehydration of ammonia. The second molecule of ammonia forms a diamine intermediate followed by α-keto imine. Intramolecular nucleophilic attack and cyclisation lead to tetra-aryl products (Agarwal et al., 2017).

14.2.3 SYNTHESIS OF PYRAZOLO PHTHALAZINE DERIVATIVES

The MCR of phthalic anhydride, hydrazine monohydrate with aromatic aldehydes and malononitrile catalysed using Zn(II)acetate under solvent-free thermal conditions gives 92% yield of 1H-pyrazolo[1,2-b]phthalazine-5,10-dione derivatives (Figure 14.1).

Primarily, the role of zinc is to initiate the reaction by activating the carbonyl carbon of phthalic anhydride leading to the formation of a phthalazine ring. Secondly, the activation of aldehyde carbon by Zn(II) followed by condensation with malononitrile led to intermediate and aza-Micheal addition to give pyrazolophthalazine derivatives.

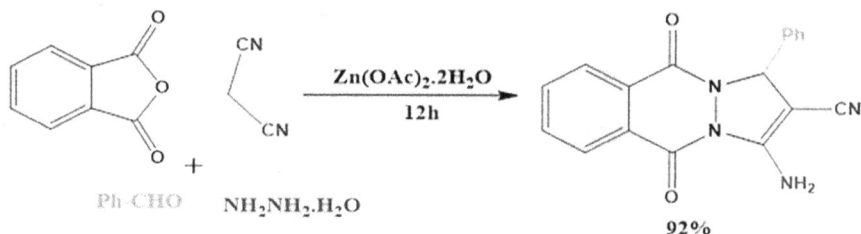

FIGURE 14.1 Synthesis of 1H-pyrazolo[1,2-b]phthalazine-5,10-dione derivatives (Redrawn with permission from Mohamadpour et al., 2018).

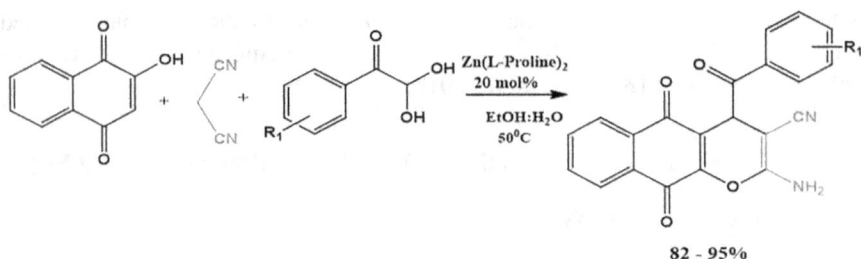

82 - 95%

FIGURE 14.2 Synthesis of 2-amino-4-benzoyl-1,4-napthoquinono[2,3-b]pyran-3-carbonitrile Derivatives (Redrawn with permission from Santos et al., 2020).

14.2.4 Pyran-Based Fused Ring System

Heterocyclic anthraquinone having pyran synthesised using one-pot condensation of 2-hydroxy-1,4-napthaquinone with malononitrile and substituted arylglyoxals promoted by ZnP as catalyst afforded, 2-amino-4-benzoyl-1,4-napthoquinono [2,3-*b*]pyran-3-carbonitriles in 82–95% yield (Santos et al., 2020) (Figure 14.2).

Three-component condensation of 2-naphthol with cyclohexyl isocyanide and substituted propargyl alcohol in the presence of 15 mol% of ZnI_2 and 5 mol% of $FeCl_3$ afforded 3H-benzo[f]chromene-3-carboxamides with 95% yield (Santos et al., 2020).

14.2.5 Azide-Based System

The MCR with the mixture of acyl hydrazide with dialkyl cyanamide in the presence of 10 mol% Zn(II) catalyst in ethanol yielded 3-dialkylamino-1,2,4-triazoles with 76-99%. Comparatively, zinc chloride is a better and economically feasible catalyst than the other zinc(II) catalysts. The catalytic activity of $ZnBr_2$ is lesser than $ZnCl_2$, and comparatively zinc triflate is more expensive than zinc chloride (Yunusova et al., 2018). The reaction is initiated by the activation of acyl hydrazide by Zn(II) followed by nucleophilic addition from nitrile nitrogen of cyanamide as the nucleophile.

14.2.5.1 Synthesis of Propargylamine Derivatives

A^3-coupling reactions are interesting because of their properties and their broader application, such as atom economy, and stoichiometric ratio outlooks (Layek et al., 2018). A^3-coupling reaction involves the association of aromatic aldehyde and secondary amine with aromatic alkyne (slightly excess) in the presence of ZnP yield propargylamine. ZnP coordinates with π-cloud of acetylene to form a zinc-acetylide intermediate and reaction with the iminium ion and the *in-situ* formation of carbonyl compounds and secondary amines afforded the propargyl amines.

14.2.5.2 Synthesis of Tetrazoles

Lipophilic margin and carboxylic acid proxy of tetrazole compounds are considered supreme in pharmaceuticals and coordination chemistry due to their properties and

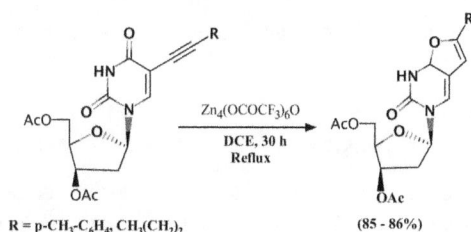

R = p-CH$_3$-C$_6$H$_4$, CH$_3$(CH$_2$)$_2$ (85 - 86%)

FIGURE 14.3 Synthesis of furopyrimidine derivatives via cycloisomerization (Redrawn with permission from Krishnan et al., 2018).

the presence of nitrogen. In the presence of heterogeneous catalysts of ZnO, with the mixture of cyanamide and sodium azideyield aryl aminotetrazoles (Krishnan et al., 2018).

14.2.5.3 Synthesis of Furan Derivatives
2,5-di and 2,3,5-tri-substituted-3-butyn-1-ones give furan-fused pyrimidinone derivatives using ZnCl$_2$-etherate catalyst (Figure 14.3).

The formation of a furan ring was reported through 5-endo-dig cycloisomerization of acetylene with carbonyl oxygen of pyrimidinone at room temperature (Krishnan et al., 2018).

14.2.5.4 Synthesis of Piperidine Derivatives
The ring opening of cyclopropane via Conia-ene cyclization afforded piperidine derivatives. The catalytic promotion of Zn(NTf$_2$)$_2$ proceeds to functionalized piperidines with the reaction of benzyl group attached to propargyl amines with substituted 1,1-cyclopropane methyldiesters (Figure 14.4).

14.2.6 Synthesis of Poly (L-lactide)

Zinc(II) salen complex containing N,N,O,O-tetradendate slain-based ligands with N,O-donors used to produce ring-opening polymerisation of L-lactide with high molecular weight PLLA. The salen ligand and zinc acetate in methanol refluxed and

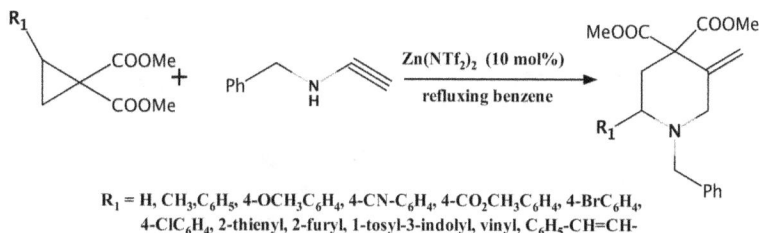

R$_1$ = H, CH$_3$,C$_6$H$_5$, 4-OCH$_3$C$_6$H$_4$, 4-CN-C$_6$H$_4$, 4-CO$_2$CH$_3$C$_6$H$_4$, 4-BrC$_6$H$_4$, 4-ClC$_6$H$_4$, 2-thienyl, 2-furyl, 1-tosyl-3-indolyl, vinyl, C$_6$H$_5$-CH=CH-

FIGURE 14.4 Synthesis of functionalized piperidines using Zn (NTf$_2$)$_2$ catalyst (Redrawn with permission from Krishnan et al., 2018).

stirred at $60°C$ yields a yellow-coloured precipitate with 80% by a coordination insertion mechanism (Rade, 2020).

14.2.7 SYNTHESIS OF ARYL CYANIDE DERIVATIVES

Cyanation of aryl iodide with formamide caused a zinc-catalysed reaction with bisphosphine nixantphos ligand. In this reaction medium, nitrile products were prepared with numerous electron-donating and electron-withdrawing substituted iodides (Zhao et al., 2020). The role of soft base phosphine ligand is to form a binuclear Zn(II) complex.

14.2.8 FUSED RING SYSTEM

14.2.8.1 Synthesis of Pyrazolo Pyridine Fused Coumarin Derivatives

This is a novel way for the synthesis of pyrazolo pyrido coumarins obtained using three-component condensation of easily accessible reactants such as 4-hydroxycoumarin, benzaldehyde and 1-alkyl-5-amino-pyrazole catalysed by 10 mol% zinc chloride catalyst under oil bath afforded fused heterocyclic product in 69–96% yield (Figure 14.5).

Zinc chloride was used to enhance the rate of MCRs with a success of more than 50% and Michael's addition of 1-alkyl-5-aminopyrazoles on the benzylidene-chromene-2,4-diones with a sequence of intramolecular cyclization furnished the pyrazolopyridine coumarin (Shamala et al., 2019).

14.2.8.2 Synthesis of Spiro Isoxazolines

The three-component reaction with oxindole, aldehyde and bromo nitrile oxide in the presence of sodium carbonate base in THF at 45 °C for 10 hours, gives spiro [indoline-3,5'-isoxazol]-2-one in the presence of 10 mol% $ZnCl_2$ catalyst (Figure 14.6).

IR, ^1H, ^{13}C NMR and single-crystal X-ray analysis predicted the structures of products. $ZnCl_2$ catalyst proves their appreciable quality without any loss of region- and diastereoselectivity (Yazdani et al., 2019).

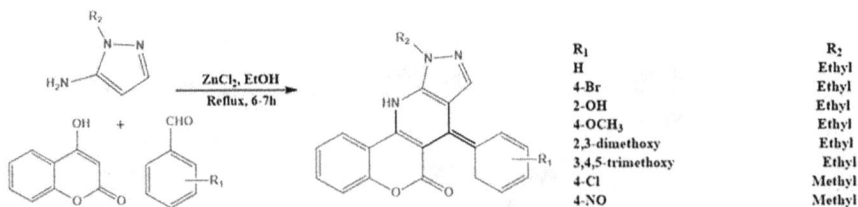

R_1	R_2
H	Ethyl
4-Br	Ethyl
2-OH	Ethyl
4-OCH$_3$	Ethyl
2,3-dimethoxy	Ethyl
3,4,5-trimethoxy	Ethyl
4-Cl	Methyl
4-NO	Methyl

FIGURE 14.5 Synthesis of pyrazolopyrido coumarin derivatives (Redrawn with permission from Shamala et al., 2019).

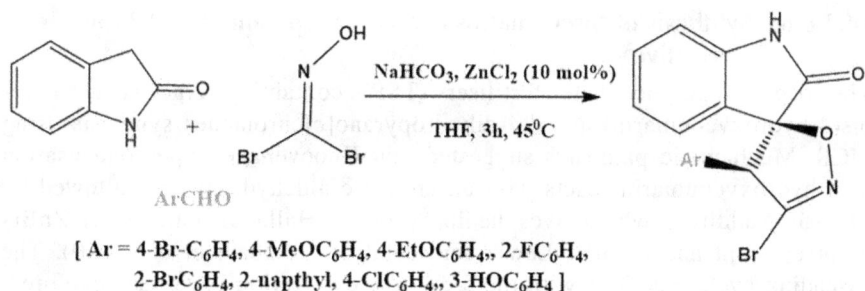

FIGURE 14.6 Synthesis of Spiroindoline-isoxazole-2-one derivatives (Redrawn with permission from Bazgir and Yazdani, 2019).

14.2.8.3 Synthesis of Pyrrolo[2,3-c] and [2,3-b]Carbazoles

One pot condensation of 3-amino carbazoles and propargyl alcohol catalysed by Zn (OTf)$_2$ through heteroannulation. Hydroamination of the triple bond of propargyl alcohol could happen through the formation of π-complex with Zn(II). Thus, zinc triflate and alkyne carbon promoted the migration of nitrogen lone pair to the alkyne carbon and hydrogen transfer generating an aminoketone intermediate and finally on displacement by dehydration afford the preferred product (Krishnan et al., 2018).

14.2.8.4 Synthesis of Bridged Quinolines

One-pot cascade reaction of 1,1-diethyl 2-tert-butyl ethenetricarboxylate with trimethyl silyl-protected 2-ethynylaniline afforded bridged quinolones with 58% yield in the presence of ZnBr$_2$/Zn(OTf)$_2$ catalyst (Figure 14.7).

14.2.8.5 Synthesis of 4-aryl-4H-benzo[g]chromene Derivatives

Three-component condensation of aryl glyoxals, 2-hydroxy-1,4-naphthoquinone and malononitrile with ZnP yields 4-aryl-4H-benzo[g]chromenes. When dehydration of aryl glyoxal by ZnP and Knoevenagel condensation occurs with malononitrile, aryl glyoxal and C-H, activation can be afforded. Michael's addition of naphthoquinone and heterocyclization forms the chromenes (Khalafy et al., 2018).

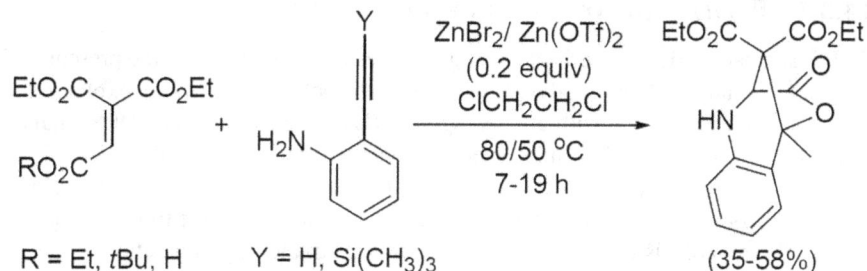

FIGURE 14.7 Synthesis of bridged Quinoline derivatives (Reproduced with permission from Krishnan et al., 2018).

14.2.8.6 Synthesis of Biscoumarins and Dihydropyrano[3,2-c]chromene Derivatives

The triple superphosphate fertiliser (TSP) containing zinc is used for bis(4-hydroxycoumarin) and 3,4-dihydropyrano[*c*]chromenes synthesis using MCR. Mechanistic pathways suggested that Knoevenagel-type condensation of 4-hydroxycoumarin reacts with an aromatic aldehyde group, followed by Michael's addition, which gives the final product (Hallaoui et al., 2020). $ZnBr_2$ is an efficient and popular Lewis acid catalyst (Ghobadi et al., 2020). The cascading cycle reactions with the 5-exo-dig cyclization have been explored, followed by the Friedel – Crafts reaction and ring opening catalysed by $ZnBr_2$led to indeno[1,2-*c*]chromenes (Garkhedkar et al., 2020; Ghobadi et al., 2020). It's an easy and efficient method for the preparation of the polycyclic thiopyranoindol annulated [3,4-*c*] quinolones via domino Knoevenagel-hetero-Diels Alder reactions of indoline-2-thions and novel N-acrylatedanthranil aldehydes (Kiamehr et al., 2017).

14.2.8.7 Synthesis of Furo[3,2-c] Coumarin Derivatives

ZnO/FAP as a heterogeneous catalyst was used to synthesise furan-coumarins in one-pot reaction. FAP ($Ca_{10}(PO_4)_6F_{12}$) evinced hexagonal crystal packing and its calcium and phosphate ions are marshalled over the fluoride ions. ZnO/FAP catalyst was prepared with the mixture of phosphate trisodium dodecahydrate ($Na_3PO_4.12H_2O$) and ZnO. 4-hydroxycoumarin, ethyl-isocyanoacetate/ substituted isocyanide, substituted aldehyde with 2.5% ZnO/FAP catalyst in ethanol yield furo[3,2-c] coumarins with 94-98% yield (Kerru et al., 2020). (Table 14.1).

14.3 ZINC-BASED CATALYSTS IN PLASTICS MANAGEMENT

Eco-friendly synthesis of plastics is an important factor in plastic waste management. Conversion of plastics waste into value-added products by depolymerisation using a Zinc based catalysts could be a viable option. The plastic waste generated is a threat to environmental sustainability and a challenge for the chemical recycling process. However, the polyethylene terephthalate (PET), polyamide (PA) and polycarbonate (PC) can depolymerised into pure monomers using suitable catalysts.

14.3.1 POLYETHYLENE TEREPHTHALATE (PET) WASTE

PET waste can be depolymerised using glycolysis and aminolysis in the presence of zinc-based catalysts. Zinc catalysts in deep eutectic solvents (DES) are explored for depolymerisation of PET with ethylene glycol (Liu et al., 2018). The DES catalyst can convert 100% of PET to monomer, bis(hydroxyethyl) terephthalate (BHET) at 190 °C with selectivity up to 82%. The catalyst showed acid-base synergistic effects between 1, 3-dimethylurea and zinc acetate, and recycled up to five times (Liu et al., 2018). The glycolysis of PET polyester chains is activated by interaction of Zn(II) with ester carbonyl oxygen. The addition of ethylene glycol oxygen interacts with

TABLE 14.1

List of Fused Heterocyclic Systems Synthesised using Zn Catalysts

S.N	Catalyst	Formula	Name of the Product	Ref
1	Zinc chloride	$ZnCl_2$	Pyrazolopyridocoumarin	Shamala et al., 2019
2	Zn(salen)	Zn(salpen)	3-substituted indole	Neetha et al., 2020 & Strianese et al., 2020
3	Zinc Proline	$Zn[(L)Proline]_2$	2-(4'-chlorophenyl)-4,5-diphenyl-1H-imidazole	Agarwal et al., 2017
4	Zinc acetate dihydrate	$Zn(Oac)_2.2H_2O$	1H-pyrazolo[1,2-b]phthalazine-5,10-dione	Mohamadpour et al., 2018
5	Zinc chloride	$ZnCl_2$	Spiroisoxazoline	Bazgir and Yazdani, 2019
6	Zinc Iodide	ZnI_2	N-cyclohexyl 3H-benzo[f]chromene-3-carboxamide	Santos et al., 2020
7	Zinc Proline	$Zn[(L)proline]_2$	2-amino-4-benzoyl-1,4-napthoquinono[2,3-b]pyran-3-carbonitriles	Santos et al., 2020
8	Zinc chloride	$ZnCl_2$	3.Dialkylamino-1,2,4-triazoles	Yunusova et al., 2018
9	Zinc Proline	ZnP	propargylamine	Layek et al., 2018
10	Zinc triflate	$Zn(Otf)_2$	pyrrolo[2,3-c] and [2,3-b]carbazoles	Krishnan et al., 2018
11	Zinc oxide	ZnO	Arylaminotetrazole	Krishnan et al., 2018
12	zinc chloride-etherate	$Zn_4(OCOCF_3)_6O$	2,5-di and 2,3,5-tri-substituted but-3-yn-1-ones	Krishnan et al., 2018
13	Zinc trifimide	Zn(NTf2)2	functionalized piperidines	Krishnan et al., 2018

(Continued)

TABLE 14.1 (Continued)
List of Fused Heterocyclic Systems Synthesised using Zn Catalysts

S.N	Catalyst	Formula	Name of the Product	Ref
14	ZnBr/Zn(OTf)₂	$ZnBr/Zn(OTf)_2$	Bridged Quinolines	Krishnan et al., 2018
15	Zinc Proline	$Zn[(L)Proline]_2$	4-aryl-4H-benzo[g]chromene derivatives	Khalafy et al., 2018
16	Zinc phosphate	$Zn_3(PO_4)_2 \cdot 4H_2O$	bis(4-hydroxycoumarin) and 3,4-dihydropyrano[c] chromenes	Hallaoui et al., 2020
17	Zno/Fap	ZnO/Flourapatite	Furo[3,2-c] coumarin derivatives	Kerru et al., 2020
18	Zinc(II)salen		poly (L-lactide)	Rade, 2020
19	Zn(Oac)₂	Zinc acetate	Aryl iodide	Zhao et al., 2020

the carbocation of ester carbonyl group formed after coordination with Zn(II), which led to the transesterification process.

The glycolysis is improved by adding solvents such as nitrobenzene, N-methyl-2-pyrrolidinone (NMP), aniline or DMSO into the traditional PET glycolysis. The result of PET conversion is 100% and the yield of monomer BHET reaches 82% in the presence of $Zn(OAc)_2$ catalyst in DMSO at 463 K. DMSO was found to be a good solvent for efficient mixing of EG and PET, thus increasing the BHET yield to 82% (comparing to 20% without DMSO) in significantly shorter reaction time at 190 °C (Liu et al., 2018). The organobase in protic ionic salt was applied for PET waste-bottle conversion into BHET. The catalysts are prepared by mixing equal amounts of triazobicyclodecene (TBD) and methanesulfonic acid. The catalyst completely depolymerises PET within two hours, resulting a high yield of BHET up to 91% (Jehanno et al., 2018; Raheem et al., 2019).

The PET waste is depolymerised with ethylene glycol using a porous medium (Al^{3+}, Fe^{3+} and Zn^{2+} supported clay) as a heterogeneous catalyst. This method provides 85–95% of BHET as a glycolysis product. Thus, Lewis acid metal ions such as Al^{3+} and Zn^{2+} loaded on bentonite afforded 90% of BHET at 4 and 5 wt% loading of metal ions. The Al^{3+} and Zn^{2+} -bentonite catalyst offers a higher yield of BHET compared to Fe-bentonite (Jeya et al., 2019). Similarly, the catalytic efficacy of kaolin clay containing Al^{3+}, Fe^{3+} and Zn^{2+} was evaluated for glycolysis. This catalyst gave BHET as a major product in 85% yield (Jeya et al., 2017). The effect of reaction parameters for 100% conversion of PET to yield 80% BHET monomer using zinc acetate as a catalyst was studied (Hu et al., 2020).

14.3.2 Polyvinyl Chloride (PVC)

The polyol derivative of terephthalic acid can be a plasticizer for nitrile-polyvinyl chloride rubber composites. The polyol was obtained by depolymerisation PET wastes with $ZnCl_2$ catalyst and 1-decanol reaction at 190 °C temperature for 4 hours. The depolymerised material was mixed with nitrile-PVC rubber composites and mechanical properties were substantially enhanced (Sirohi et al., 2018). In PVC manufacturing, additives such as thermal/heat stabilisers, plasticizing agents and lubricants are added, resulting in minimised stabiliser compatibility due to the presence of multi-components. For instance, cis-1,2-cyclohexane dicarboxylic acid D-mannitol ester (CAME) complex with Zn alkoxides were used as thermal stabilisers and plasticizers for PVC (Dong et al., 2019).

The pentaerythritol p-hydroxybenzoate ester-based zinc metal alkoxides (PHE-Zn) stabilised PVC and showed higher antibacterial activity to *E. coli* up to 97.92% and *Staphylococcus aureus* up to 91.84% (Zhang et al., 2020). The release of hydrochloric acid during the decomposition of PVC triggers the degradation and is neutralised by forming a Lewis acidic base by Zn Complex (Zhang et al., 2020). Zinc metal-polyol esterbased alkoxide and rhizol-based Zn-containing complex along with zinc urate were used to upgrade the colour and thermal stability of PVC (Dong et al., 2019; Wang et al., 2019; Li et al., 2019). As a distinct type of stabiliser, zinc salt converts the labile chlorine atoms in the PVC chain, leading to better initial whitening, but stimulates a "Zinc burning" event with little without restriction (Wu et al., 2018).

Zinc stearate (ZnSt$_2$) converts the unstable chlorine atom from the PVC chain, leading to better initial whitening. On the other hand, calcium stearate (CaSt$_2$) can absorb and neutralise hydrochloric acid (HCl), which can then improve the long-term stability of polyvinyl chloride (Li et al., 2016; Wang et al., 2017). Zhu and co-researchers prepared a Zn-mannitol complex, which exhibited a low melting point at 166°C and was used as a well-ordered thermal stabiliser for PVC (Zhu et al., 2016). For example, the pentaerythritol stearate ester-based zinc alkoxides (PSE-Zn) were prepared and used as plasticizers (Liu et al., 2017).

14.3.3 POLY LACTIC ACID (PLA)

Chemical degradation of poly lactic acid (PLA) into methyl lactate (ML) by a Zn (II) compound is examined at three different temperatures: 70, 90 and then 110°C, and 94% achieved at a higher temperature (Román-Ramírez et al., 2020). The use of a Zn-acetate comprising ionic liquid 2-[Bmim][OAc]-[Zn(OCOCH$_3$)$_2$], as a catalyst in methanolysis of PLA, could bring about higher depolymerisation of PLA with good yield. The conversion of PLA and the yield of ML were achieved 97% and were reusable five times (Song et al., 2019). Limited degradation of PLA to the formation of alkyl lactates has been achieved by an ethylene diamine Zn (II) complex (Román-Ramírez et al., 2018).

14.3.4 POLYMERISATION OF LACTIDE

Hexamethyldisilazane (HMDS) complex with a binuclear Zn metal centre promoted polymerisation of racemic lactide in THF (Thevenon et al., 2016). Recently a pyrazole-based zinc catalysts was developed for ROP of lactide (LA) (Schafer et al., 2012). Tridentate NNO chelates with different electron-donating substituents in Mg^{2+} and Zn^{2+} complexes containing pyrazoline-β-ketominate ligands showed the highest activity towards the ROP of lactide (Huang et al., 2018). The mono and dimeric Zn(II)-Schiff-base promoted polymerisation of cyclic esters and resulting polylactic acids showed a wide polydispersity index up to 1.83–2.04 (Payne et al., 2020). The recently proven conversion to a similar propylene diamine system, giving better performance, achieves 81% conversion to 50°C within 30 minutes, exemplifying the significance of metal-ligament correlation (Mckeown et al., 2019). The ROP of lactide was catalysed by homoleptic (L$_2$Zn) and heteroleptic (L-Zn-Oar) β—ketoiminate zinc complexes and proved that the heteroleptic complex L-Zn-Oar in a sub-phenoxide or Lewis base coordinated with Zn atom and had the highest catalytic activity (Steiniger et al., 2018).

14.4 CONCLUSIONS AND FUTURE PERSPECTIVES

One-pot Multi-Component Reactions (MCRs) finds extensive applications in materials and pharmaceutical chemistry. The development of novel MCRs using Zn-based catalysts are viable alternative for classical organic synthesis and an efficient method for constructing complex organic molecules. The heterocyclic framework generation offers various applications in drug development in pharmaceutical industries. Most of the MCRs afforded moderate to excellent yield of the product in a single process. Zinc's reaction centre offers an adequate environment to bring synergism with all reactants by making interdependent interactions in the proceeding of the reaction. The usage of different Lewis acidic zinc catalysts and their heterocyclic synthesis discussed in this chapter are neither exhaustive nor limited to the heterocycles. There are plenty of options available for the improvement of zinc complexes, such as stereoselective synthesis and higher product selectivity. Hydrosilylations, reductive coupling reactions, etc., are novel reactions for heterocyclic synthesis.

Polylactic Acid (PLA) a bio-degradable polymer deserves mention in domestic and industrial perspectives for its biocompatibility and to generate manageable wasterecycling and compost preparation. Zinc complexes continue to attract attention as they are inexpensive, non-toxic and could be monitored by NMR spectroscopy because of their magnetic properties. For biomedical utilities, the polymerisation of cyclic esters of nontoxic metal catalysts is essential because of difficulties in completely removing the catalyst residue from the polymer. Zinc is the most preferred metal candidate for such applications, predominantly because of its high performance, low toxicity and low cost.

The role of zinc catalysts in the management of PET polymer wastes has attracted the attention of scientists of late. The chemical process involves a glycolysis reaction of PET polymer into BHET monomer in presence of Zn-based catalysts. From the above observations, it is clear the multi-component reactions involving Zn-based catalysts are valuable techniques in organic synthesis and value-added drug discovery besides obtaining biologically active molecules. Multi-component-reaction (MCR) approaches are also environment-friendly and involve low-cost green synthesis and recyclable heterogeneous catalysts.

REFERENCES

Agarwal S, Kidwai M, Poddar R, Nath M. 2017. A facile and green approach for the one-pot multicomponent synthesis of 2,4,5-Triaryl- and 1,2,4,5-Tetraarylimidazoles by using zinc-proline hybrid material as a catalyst. Chemistry Select 2, 10360–10364.

Bazgir A, Yazdani H. 2019. Lewis acid catalyzed regio- and diastereoselective synthesis of spiroisoxazolines via one-pot sequential knoevenagel condensation/1,3-dipolar cyclo-addition reaction. Synthesis 51, 1669–1679.

Dong T, Li D, Li Y, Han W, Zhang L, Xie G. 2019. Design and synthesis of polyol ester-based zinc metal alkoxides as a bi-functional thermal stabilizer for Poly (vinyl chloride). Polymer Degradation and Stability 159, 125–132.

Garkhedkar AM, Gore BS, Hu WP, Wang JJ. 2020. Lewis acid catalyzed atom-economic synthesis of C_2-Substituted Indoles from o-Amido Alkynols. Organic Letters 22, 3531–3536.

Ghobadi M, Qhazvini PP, Kazemi M. 2020. Catalytic application of zinc (II) bromide (ZnBr2) inorganic synthesis. Synthetic Communications 50, 3717–3738. 10.1080/003 97911.2020.1811873.

Hallaoui AE, Chehab S, Ghailane T, Malek B, Zimou O, Boukhriss S. 2020. Application of phosphate fertilizer modified by zinc as a reusable efficient heterogeneous catalyst for the synthesis of biscoumarins and Dihydropyrano[3,2-c]Chromene-3-Carbonitriles under green conditions. Polycyclic Aromatic Compounds 41, 2083–2102.

Hu Y, Wang Y, Zhang X, Qian J, Xing X, Wang X. 2020. Regenerated cationic dyeable polyester deriving from poly(ethylene terephthalate)waste. Polymer Degradation and Stability 179, 109261. 10.1016/j.polymdegradstab.2020.109261.

Huang Y, Kou X, Duan YL, Ding FF, Yin YF, Wang W. 2018. Magnesium and zinc complexes bearing NNO-tridentate ketiminate ligands: synthesis, structures and catalysis in the ring-opening polymerization of lactides. Dalton Transactions 47, 8121–8133.

Jehanno C, Flores, I, Dove AP, Müller AJ, Ruipérez F, Sardon H. 2018. Organocatalysed depolymerisation of PET in a fully sustainable cycle using thermally stable protic ionic salt. Green Chemistry 20, 1205–1212.

Jeya G, Anbarasu M, Dhanalakshmi R, Vinitha V, Sivamurugan V. 2019. Depolymerization of Poly (ethylene terephthalate) Wastes through Glycolysis using Lewis Acidic Bentonite Catalysts. Asian Journal of Chemistry 32, 187–191.

Jeya G, Ilbeygi H, Radhakrishnan D, Sivamurugan V. 2017. Glycolysis of post-consumer poly (ethyleneterephthalate) wastes using Al, Fe and Zn exchanged kaolin catalysts with lewis acidity. Advanced Porous Materials 5, 1–9.

Kerru N, Gummidi ML, Gangu KK, Maddila S, Jonnalagadda SB. 2020. Synthesis of novel Furo[3,2-c]coumarin derivatives through multicomponent [4+1] cycloaddition reaction using ZnO/Fap as a sustainable catalyst. Chemistry Select 5, 4104–4110.

Khan I, Zaib S, Ibrar A. 2020. New frontiers in the transition-metal-free synthesis of heterocycles from alkynoates: an overview and current status. Organic Chemistry Frontiers 7, 3734–3791.

Khalafy J, Ilkhanizadeh S, Ranjbar M. 2018. A green, organometallic catalyzed synthesis of a series of novel functionalized 4-Aroyl-4H-benzo[g]chromenes through one-pot, three component reaction. Journal of Heterocyclic Chemistry 55, 951–956.

Kiamehr M, Alipour B, Mohammadkhani L, Jafari B, Langer P. 2017. ZnBr₂catalyzed domino knoevenagel-hetero-diels–alder reaction: an efficient route to polycyclic thiopyranoindol annulated [3,4-c]quinolone derivatives. Tetrahedron 73, 3040.

Krishnan KK, Ujwaldev SM, Saranya S, Anilkumar G, Beller M. 2018. Recent advances and perspectives in the synthesis of heterocycles via Zinc catalysis. Advanced Synthesis & Catalysis 361, 382–404.

Lancaster KM. 2018. Reviving up an artificial metalloenzyme. Science 361, 1071–1072.

Layek S, Agrahari B, Kumari S, Anuradha Pathak DD. 2018. [Zn(l-proline)₂] catalyzed one-pot synthesis of propargylamines under solvent-free conditions. Catalysis Letters 148, 2675–2682.

Li M, Zhang J, Xin J, Huang K, Li S, Wang M. 2016. Design of green zinc-based thermal stabilizers derived from tung oil fatty acid and study of thermal stabilization for PVC. Journal of Applied Polymer Science 134, 44679.

Li Y, Li D, Han W, Zhang M, Ai B, Zhang L. 2019. Facile synthesis of di-mannitol adipate ester-based zinc metal alkoxide as a bi-functional additive for poly (Vinyl Chloride). Polymers 11, 813.

Liu H, Li D, Li R, Sun H, Zhang Y, Zhang L, Zhao P. 2017. Synthesis of pentaerythritol stearate ester-based ZincAlkoxide and its synergistic effect with Calcium stearateand Zinc stearate on PVC thermal stability. Journal of Vinyl & Additive Technology 24, 314–323. 10.1002/vnl.21602

Liu B, Lu X, Ju Z, Sun P, Xin J, Yao X. 2018. Ultrafast homogeneous glycolysis of waste polyethylene terephthalate via a dissolution-degradation strategy. Industrial & Engineering Chemistry Research 57, 16239–16245.

Mahmoud NFH, El-Sewedy A. 2018. Multicomponent reactions, solvent-free synthesis of 2-Amino-4-aryl-6-substituted Pyridine-3,5-dicarbonitrile derivatives, and corrosion inhibitors evaluation. Journal of Chemistry 2018, 1–9.

Mckeown P, Román Ramírez LA, Bates S, Wood J, Jones MD. 2019. Zinc complexes for pla formation and chemical recycling: towards a circular economy. ChemSusChem 12, 5233–5238.

Mehraban JA, Azizi K, Jalali MS, Heydari A. 2018. Choline azide: new reagent and ionic liquid in catalyst-free and solvent-free synthesis of 5-Substituted-1H-Tetrazoles: a triple function reagent. Chemistry Select 3, 116–121.

Mohamadpour F, Lashkari M, Heydari R, Hazeri N. 2018. Four-component clean process for the eco-friendly synthesis of 1H-pyrazolo [1,2-b]phthalazine-5,10-dione derivatives using $Zn(Oac)_2.2H_2O$ as an efficient catalyst under solvent-free conditions. Indian Journal of Chemistry 57B, 843–851.

Neetha M, Rohit KR, Saranya S, Anilkumar G. 2020. Zinc-catalysed multi-component reactions: an overview. Chemistry Select 5, 1054–1070.

Payne J, Mckeown P, Mahon MF, Emanuelsson EAC, Jones MD. 2020. Mono- and dimeric zinc (ii) complexes for PLA production and degradation into methyl lactate – a chemical recycling method. Polymer Chemistry 11, 2381–2389.

Poddar R, Jain A, Kidwai M. 2017. Bis[(l)prolinate-N,O]Zn: a water-soluble and recycle catalyst for various organic transformations. Journal of Advanced Research 8, 245–270.

Rade PP, Garnaik B. 2020. Synthesis and characterization of biocompatible poly (L-lactide) using zinc (II) salen complex. International Journal of Polymer Analysis and Characterization 25, 283–299.

Raheem AB, Noor ZZ, Hassan A, Hamid MKA, Samsudin SA, Sabeen AH. 2019. Current developments in chemical recycling of post-consumer polyethylene terephthalate wastes for new materials production: a review. Journal of Cleaner Production 225, 1052–1064.

Ramos ML, Rodrigues OM, Brenno AD. 2019. Neto, mechanistic knowledge and noncovalent interactions as the key features for enantioselective catalysed multicomponent reactions: a critical review. Organic & Biomolecular Chemistry.

Riva R, 2018, Enantioselective four-component Ugi reactions. A chiral organocatalyst condenses four reactants with stereochemical control, sciencemag.org Doi: 10.1126/science.aau7754.

Román-Ramírez LA, Mckeown P, Jones MD, Wood J. 2018. Poly (lactic acid) degradation into methyl lactate catalyzed by a well-defined Zn (II) complex. ACS Catalysis 9, 409–416.

Román-Ramírez LA, Mckeown P, Shah C, Abraham J, Jones MD, Wood J. 2020. Chemical degradation of end-of-life poly(lactic acid) into methyl lactate by a Zn(II) complex. Industrial & Engineering Chemistry Research 59, 11149–11156.

Santos CM Silva AM. 2020. Six-membered ring systems: with O and/or S atoms. Progress in Heterocyclic Chemistry 31, 533–595.

Schäfer PM, McKeown P, Fuchs M, Rittinghaus RD, Hermann, A, Henkel J, Seidel S, Roitzheim C, Ksiazkiewicz AN, Hoffmann A, Pich A, Jones MD, Herres-Pawlis S. 2012. Tuning a robust system: N,O Zinc Guanidine catalysts for the ROPof Lactid. Dalton Transactions 1–3.

Shaabani S, Dömling A. 2018. The catalytic enantioselective ugi four-component reactions. Angewandte Chemie International Edition 57, 16266–16268.

Shamala D, Shivashankar K, Chandra, Mahendra M. 2019. Zinc chloride catalyzed multicomponent synthesis of pyrazolopyridocoumarinscaffolds. Journal of Chemical Sciences 131.

Sirohi S, Dobhal S, Doshi M, Nain R, Dutt K, Pani B. 2018. Eco-friendly synthesis of PET-based polymeric plasticiser and its application in nitrile-PVC rubber blends. Indian Chemical Engineer 61, 206–217.

Sivamurugan V, Deepa K, Palanichamy M, Murugesan V. 2004. [(L)Proline]$_2$Zn catalysed synthesis of 1,5-Benzodiazepine derivatives under solvent-free condition. Synthetic Communications. 34, 3833–3846.

Sivamurugan V, Kumar RS, Palanichamy M, Murugesan V. 2005. Synthesis of hantzsch 1, 4-dihydropyridines under solvent-free condition using Zn[(L)proline]$_2$ as Lewis acid catalyst. Journal of Heterocyclic Chemistry 42, 969–974.

Sivamurugan V, Vinu A, Palanichamy M, Murugesan V, 2006. Rapid and cleaner synthesis of 1, 4-dihydropyridines in aqueous medium. Heteroatom Chemistry: An International Journal of Main Group Elements 17, 267–271.

Song X, Bian Z, Hui Y, Wang H, Liu F, Yu S. 2019. Zn-Acetate-Containing ionic liquid as highly active catalyst for fast and mild methanolysis of Poly(lactic acid). Polymer Degradation and Stability 168, 108937.

Steiniger P, Schäfer PM, Wölper C, Henkel J, Ksiazkiewicz AN, Pich A. 2018. Synthesis, structures, and catalytic activity of homo- and heterolepticketoiminate zinc complexes in lactide polymerization. European Journal of Inorganic Chemistry 2018, 4014–4021.

Strianese M, Pappalardo D, Mazzeo M, Lamberti M, Pellecchia C. 2020. Salen-type luminium and zinc complexes as two-faced Janus compounds: contribution to molecular sensing and polymerization catalysis. Dalton Transactions 49, 16533–16550.

Thevenon A, Romain C, Bennington MS, White AJP, Davidson HJ, Brooker S. 2016. Dizinc lactide polymerization catalysts: hyperactivity by control of ligand conformation and metallic cooperativity. Angewandte Chemie International Edition 55, 8680–8685.

Wang M, Song X, Jiang J, Xia J, Li M. 2017. Binary amide-containing tung-oil-based Ca/Zn stabilizers: effects on thermal stability and plasticization performance of poly(vinyl chloride) and mechanism of thermal stabilization. Polymer Degradation and Stability 143, 106–117.

Wang Q, Wang DX, Wang MX, Zhu J. 2018. Still unconquered: enantioselective passerini and ugi multicomponent reactions. Accounts of Chemical Research 51, 1290–1300.

Wang X, Ma B, Wang Y, Lu S, Ma M, Shi Y. 2019. A new theory of "two-step stabilization mechanism" for triazole-based zinc-containing complex as thermal stabilizer for poly (vinyl chloride). Polymer Degradation and Stability 167, 86–93.

Wu B, Wang Y, Chen S, Wang M, Ma M, Shi Y. 2018. Stability, mechanism and unique "zinc burning" inhibition synergistic effect of zinc dehydroacetate as thermal stabilizer for poly(vinyl chloride). Polymer Degradation and Stability 152, 228–234.

Yunusova SN, Bolotin DS, Suslonov VV, Vovk MA, Tolstoy PM, Kukushkin VY. 2018. 3-Dialkylamino-1,2,4-triazoles via Zn (II)-Catalyzed Acyl Hydrazide–Dialkylcyanamide Coupling. ACS Omega 3, 7224–7234.

Zhang M, Han W, Hu X, Li D, Ma X, Liu H. 2020. Pentaerythritol p-hydroxybenzoate ester-based zinc metal alkoxides as multifunctional antimicrobial thermal stabilizer for PVC. Polymer Degradation and Stability 181, 109340.

Zhao L, Dong Y, Xia Q, Bai J, Li Y. 2020. Zn-CatalyzedCyanation of Aryl Iodides. The Journal of Organic Chemistry 85, 6471–6477.

Zhu L, Wu Y, Shentu B, Weng Z. 2016. Preparation and characterization of zinc-mannitol complexes as PVC thermal stabilizers with high efficiency. Polymer Degradation and Stability 133, 399–403.

15 Unique Complexes of Zinc and Their Applications

Ethirajan Sukumar, Vajiravelu Sivamurugan, and E. Rajkumar

15.1 INTRODUCTION

As is well known, zinc and its compounds play vital roles in medicine, agriculture, drug discovery, nanotechnology, etc. Like other metals, zinc also forms several interesting and useful products that in turn have newer applications. The synthesis of zinc complexes with organic substrates, metals, metal ligands, Schiff bases and drug materials reveals interesting chemistry and yields a plethora of unique products that have widespread applications in many fields. For example, the Metal-Organic Framework (MOF), which consists of new molecules generated using coordination constructions, produces porous three-dimensional structures. These have potential application in the production of high-selectivity catalysts and visible light-absorbing photocatalysts (Öhrström and Noa, 2021). The latter could be used in artificial photo-synthesis and can address pollution issues such as CO_2 reduction, H_2 production and organic pollutant degradation.

Zinc, which occurs in nature in the order of 10^{-6} of the Earth's crust, is a white, lustrous and tarnishable metal. The main source of zinc is sphalerite [(ZnFe)S], which is usually found with galena (PbS). Many processes of extraction are in vogue, where lead and cadmium are also simultaneously obtained with zinc by blast furnace method. The latter is separated out either by distillation or precipitation from their sulphate solutions (Purcell and Kotz, 2010).

15.2 COMPLEXES OF ZINC

In an aqueous solution, zinc reacts with halides to produce complex anions. The presence of the fluoro complex is limited to certain forms, although it is not completely isolated in the solid state. The ZnX_4^{2-} complexes can be isolated as salts of base cations. Zn^{2+} tends to form stronger bonds to fluoride and oxygen. Complex cations with ammonia and amine ligands are obtained as crystalline salts. Dithiocarbamates-based zinc complexes and other sulphur compounds are significant catalysts in the sulphur-induced vulcanization of rubber. Zinc β-diketonates readily form 5-coordinate adducts with water, alcohol and nitrogen bases. The

DOI: 10.1201/9781003412472-15

cation also readily forms complexes in which zinc has the coordination number 5, whereas in organo-compounds it is usually 2 (Krezel and Maret, 2016).

15.3 ORGANO-ZINC COMPOUNDS

These are historically important because they were the first organo-metallic compounds synthesized. Organo-zinc compounds of type R_2Zn and $RZnX$ are known. EtZnI is a polymer with iodide bridges, each iodine atom forming three bonds to zinc two long and one normal. Zinc alkyls and the diaryls are the most conveniently obtained by the reaction of zinc metal with organo-mercury compound or reaction of $ZnCl_2$ with organo-lithium or organo-aluminium or organo-magnesium.

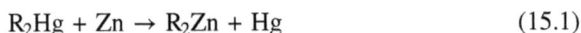

$$R_2Hg + Zn \rightarrow R_2Zn + Hg \qquad (15.1)$$

$$ZnCl_2 + R_2Mg \rightarrow R_2Zn + MgCl_2 \qquad (15.2)$$

The R_2Zn compounds are non-polar liquids or low-melting solids and soluble in most organic liquids. Zinc alkyls react readily with compounds containing active hydrogen such as alcohols.

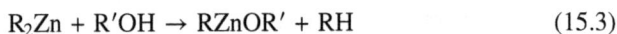

$$R_2Zn + R'OH \rightarrow RZnOR' + RH \qquad (15.3)$$

With lithium alkyls and aryls, zinc forms complexes such as $Li[ZnPh_3]$ and $Li[(Et_2Zn)_2H]$ that contain Zn-H-Zn bridges (Seyferth, 2001).

15.4 ZINC COMPLEXES WITH METALS

Zinc forms alloys with different metals that are used in industries, particularly in die and spin casting for automotive, electrical and hardware applications. The unique properties of these zinc alloys, including their low melting points and viscosities, enable the creation of intricate and delicate shapes. Rapid cooling during casting and assembly is facilitated by the alloys' low working temperatures. Prestel alloy (78% zinc and 22% aluminum) exhibits impressive strength comparable to steel while maintaining the pliability of plastic, owing to its exceptional super-plasticity. The versatility of zinc alloys extends to their moldability using ceramic and cement die casts. The stamping dies are made of 96% zinc and 4% aluminum alloy offer cost-effective alternatives to ferrous metal dies.

Brass is a widely employed zinc alloy, wherein copper is combined with varying percentages of zinc (3–45%, depending on the type of brass). This alloy surpasses copper in ductility, strength and corrosion resistance, making it invaluable in communication equipment, hardware, musical instruments and water valves. Zinc alloys with nickel and silver are used to make typewriter metal, soft aluminum solder and commercial bronze. These alloys find wide applications across diverse industries. Zinc is substituted for lead or tin in modern pipe organs. (http://www.davidrumsey.ch/Technology.htm; Pola et al., 2020).

15.5 ZINC COMPLEXES AS METALLOPHARMACEUTICALS

Metal complexes with unique chemical structures are presently used as alternatives to treat chronic and incurable diseases. In recent times, studies on metal complexes as oral drugs in the treatment of diabetes mellitus were carried out. The development of zinc complexes for anti-metabolic syndrome has also been worked out in the past two decades. Several highly effective anti-diabetic zinc complexes with different coordination structures have been published after studies in experimental animals. In all these unique complexes, zinc has been found to be biologically active by interacting with a few target proteins involved in diabetes mellitus (Yoshikawa and Yasui, 2012).

15.6 ZINC COMPLEXES WITH MIXED LIGANDS

Mixed ligand Zinc (II)—complexes are emerging trends in biological applications and viable candidates for anticancer drugs. Given the different coordination, geometry and structural features, mixed ligand zinc (II)—complexes with oxygen, nitrogen and sulphur appear to offer significant potential. The zinc present in the human body has unique properties such as having variable coordination geometry and capable of undergoing rapid ligand exchange while at the same time maintaining the oxidation state of metal. The charge and steric factors are responsible for the determination of the coordination geometry of the zinc complexes. Several zinc complexes have been reported to possess an anti-cancer property with bio-compatibility and the least toxicity. For bioimaging applications, fluorescent Zn(II)-complexes based on binding units derived from dipicolylamine (DPA), salen and cyclen as well as fluorophore signalling units were used for a variety of targets such as intracellular phosphates, cell membrane, cellular localization, protein labelling, etc. (Kusamoto and Nishihara, 2019; Kumar et al., 2021).

15.6.1 ZINC (II)-COMPLEXES BASED HYDRAZONE-PYRAZOLONE LIGANDS

The reaction between 4-acyl-5-pyrazolones and mono-substituted hydrazone was used to synthesize the proligands H_2L^n (n=1-5). Five zinc (II)-complexes were synthesized based on hydrazone-pyrazolone ligands in a methanol medium. $[Zn(HL^1)_2]$ (**1**) and $[Zn(HL^2)_2]$ (**2**) are mononuclear complexes (Figure 15.1).

Complexes **3** and **5** were isolated as diaqua complexes, $[Zn(HL^n)_2(H_2O)_2]$ (n=3 and 5), whereas complex **4** $[Zn(HL^4)_2]_n$ is a one-dimensional coordination polymer. All the complexes (**1–5**), exhibited enhanced antibacterial activity against *Escherichia coli* and against *Staphylococcus aureus*. The findings from confocal laser scanning microscopy and scanning electron microscopy provided additional evidence for the impairment of membrane permeability in bacterial cells (Marchetti, Pettinari et al., 2022) and highlighted the remarkable efficacy of a Zinc (II) complex containing pyrazolone-based hydrazones against *Trypanosoma brucei*, the causative agent of African sleeping sickness. These complexes were air-stable and *in-vitro* cytotoxic studies revealed that Complex 1 was more active and less toxic against *T. brucei* and Balb/3t3 cells.

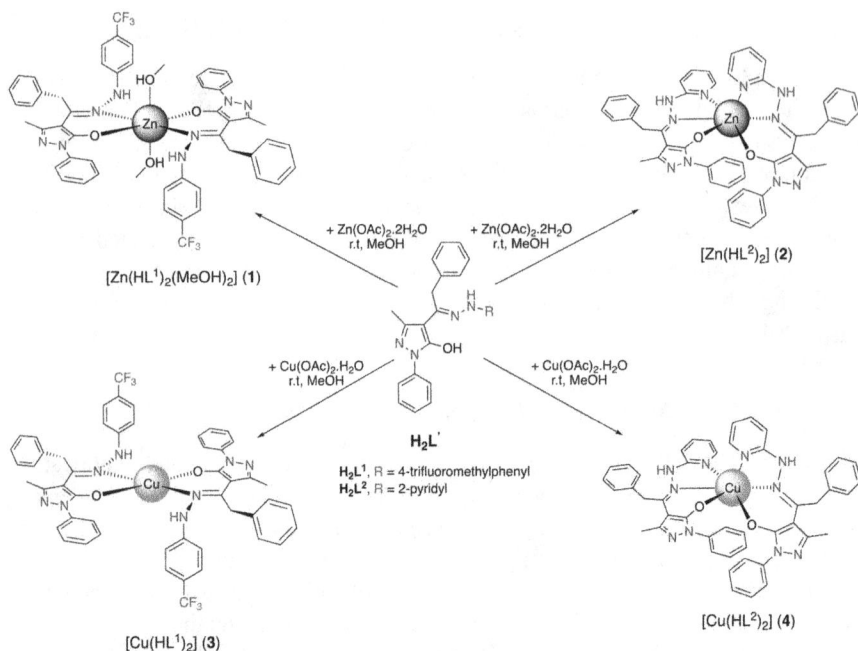

FIGURE 15.1   Zinc (II) and Cu (II) complexes with pyrazolone-based hydrazone ligands (Marchetti, Tombesi et al., 2022.) Copyright granted under the terms of the Creative Commons CC BY license.

15.6.2   ZINC (II)-COMPLEXES BASED N-DONOR LIGANDS

Porchia et al. (2020) have shown that the zinc complexes containing N-donor ligands such as Schiff bases, diimine, terpyridine, pyrazole, etc. had significant cytotoxic activity. In addition, they also revealed that a few complexes adopted different mechanisms of anticancer action when compared to platinum complexes. An extensive study on zinc complexes containing ligands of Schiff base, non-Schiff base and macrocyclic ones revealed that they possessed anti-cancer properties due to (i) rapid ligand-exchange reactions, (ii) capacity to promote Lewis acid activation and (iii) participation in the catalysis of hydrolytic reactions including DNA hydrolysis and cleavage (Pellei et al., 2021).

The octahedral zinc (II) complexes containing terpyridine derivatives altered the DNA conformation by preventing the DNA base packing and exhibited cytotoxic activity against a series of human cell lines such as prostate carcinoma (pc-3), cervix carcinoma (HeLa), liver carcinoma (HepG2) and galactophore carcinoma (MCF-7). The activities were ten times greater than the platinum complex cisplatin (Jiang et al., 2009).

Ermakova et al. (2023) investigated the bioactivity of mixed ligand zinc (II)-complexes with 1H-tetrazole-5-acetic acid and oligopyridine derivatives. They found that the two different coordinations in zinc (II) complexes resulted in the formation of a zig-zag polymer chain arranged along the c-axis of the unit cell (Figure 15.2).

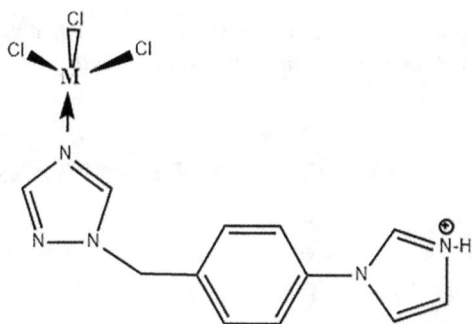

1 : M = Co^{2+}, X = Cl$^-$, 81 %
2 : M = Zn^{2+}, X = NO$_3^-$, 82 %

FIGURE 15.2 The structure of mixed ligand zinc (II)-complexes with 1H-tetrazole-5-acetic acid and oligopyridine derivatives. (Yu et al., 2022). Copyright permission obtained from Elsevier.

In addition, the non-coordinated water molecules exhibited hydrogen intermolecular interactions with N3 of the tetrazole ligand carboxylate ion. The phendione mixed ligand-based Zinc (II)-complex exhibited higher toxicity towards HepG-2 and MCF-7 cells and all the mixed ligand zinc (II) complexes showed antimicrobial activities against *E.coli*, *S.aureus*, *P.italicum* and *C.steinii*.

Yu et al. (2022) reported an unusual coordination of mixed imdazole-1,2,4–triazole ligands in a protonated cationic form in Zinc (II) and Cu(II) complexes. The flexible mixed donor ligand 1-(4-(1H-imidazol-1-yl)benzyl)-1H-1,2,4-triazole) was added to zinc nitrates in ethanol in the presence of HCl led to the formation of Zn (II) complex.

15.6.3 ZINC DINUCLEAR COMPLEXES

Two zinc dinuclear complexes with (i) a phenoxido and hydroxido bridges and (ii) an amine group attached to an alkyl chain were synthesized and found that they were able to capture the microvescicles isolated from human plasma (Figure 15.3).

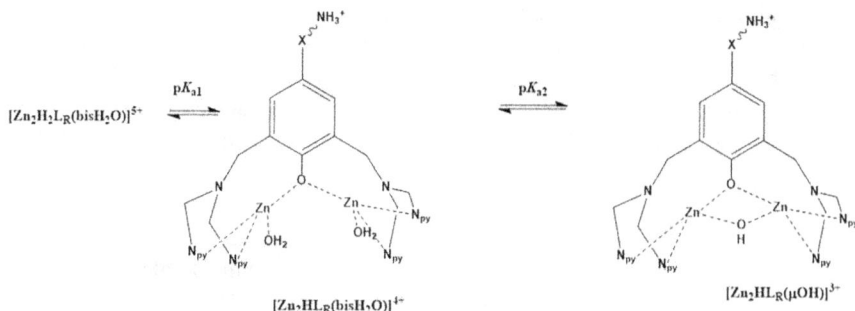

FIGURE 15.3 Proposed structure of zinc dinuclear complexes in H$_2$O:DMSO (70:30) (Van Der Heyden et al., 2023). Copyright permission obtained from Elsevier.

The latter are important markers in body fluids that reveal cellular activation in disease conditions such as thrombosis, inflammation and cancer (Van Der Heyden et al., 2023).

15.7 ZINC-BASED METAL-ORGANIC FRAMEWORKS (Zn-MOFs)

The term 'Metal-Organic Framework' refers to a family of new molecules generated using coordination constructions to produce porous three-dimensional structures that are utilized in the production of high-selectivity catalysts through the combination of suitable metals and organic ligands. Complexing transition metals with amine, hydroxyl and carboxylic acid-functionalized organic molecules via the host-guest interaction could lead to highly porous materials which have the potential sensor applications. Visible light-absorbing photocatalysts could be used in artificial photosynthesis for the purpose of CO_2 reduction, H_2 production and organic pollutant degradation. Due to the availability of organic ligands with high molar absorptivity, they are decent candidates for such applications.

While designing materials for specific applications such as sensing, catalysis, separation and storage, it is beneficial to consider self-assembling properties of difunctionalized organic molecules with transition metals to form 3D nanostructures. The flexible hierarchical porous structures have potential for use in the production of high-selectivity catalysts through the combination of suitable metals and organic ligands. Complexing transition metals with amine, hydroxyl and carboxylic acid-functionalized organic molecules via the host-guest interaction could result in the creation of uniformly porous materials with a large surface area that mimics the porosity of zeolites (He et al., 2015).

15.7.1 ZINC METAL-AZOLATE FRAMEWORK (MAF) AND APPLICATIONS

The Zn-based metal-azolate framework (MAF-6) may be synthesized by the reaction of $[Zn(NH_3)_4](OH)_2$ with 2-ethyl imidazole resulting in the formation of $[Zn(eim)_2]$ (He et al., 2015). MAF-6 had a larger surface area of 1695 $m^2.g-1$ and a wider pore size of 18.4 Å respectively (Figures 15.4a & 15.4b).

In a similar manner, adjusting the pore size might be accomplished by choosing suitable ligands. He and his coworkers (2015) were able to synthesize MAF-4 by combining Zn with 2-methyl imidazole. In comparison to MAF-6, MAF-4 exhibited a lower surface area and a more manageable pore size. The Zeolitic Imidazolate Frameworks (ZIFs) used in the construction of the MOFs have surface areas and porosity levels that are comparable to those of zeolites (Biswal et al., 2012).

MOF-74 structures have two-dimensional honeycomb structures that are prepared by combining 2,5-dihydroxy terephthalic acid with divalent metal ions such as zinc (II). For example, the visible absorption optoelectronic MOF-74 was obtained by adding a functionalized perylene diimide (PDI) derivative with the metal ions Zn (II), Mg(II) and Ni(II) (Figure 15.4b).

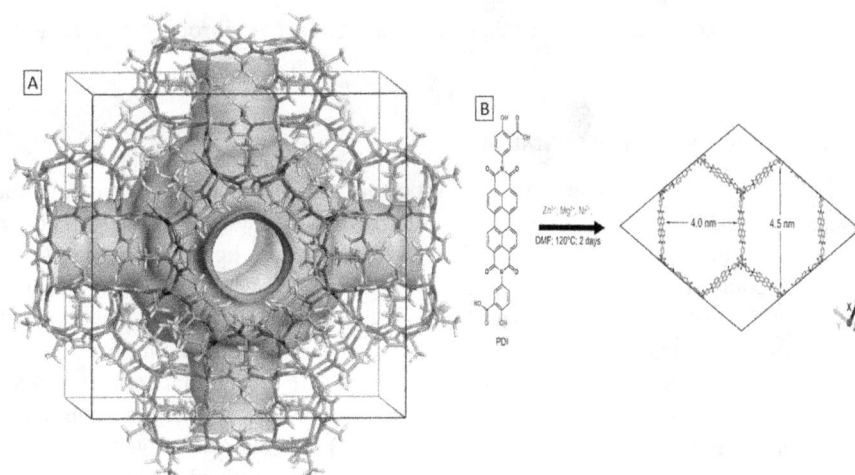

FIGURE 15.4 (a) MAF-6 structure (Zn – green, N-blue; C-grey) (He et al., 2015) Copy right obtained from ACS (b) PDI-MOF-74 framework (Scheurle et al., 2022). Copyright granted under the terms of the Creative Commons CC BY license.

The PDI-MOF-74 exhibited a highly crystalline character, with a significant optical bandgap of 2.11 eV and a considerable absorption in the visible region. Because of its high luminescence and excellent electrical conductivity, this product could be used in sensing or detector applications (Scheurle et al., 2022).

15.7.2 MOFs in Pollution Control

MOFs could be designed to have a wide range of applications including selective absorption of hazardous/greenhouse gases like carbon dioxide (CO_2) and volatile organic chemicals. In addition, a metal centre has the potential to perform catalytic organic conversion, which includes reduction of CO_2 and NO_2, generation of H_2 and other organic transformations. In addition, a MOF could be modified by selecting visible light-absorbing organic compounds such as naphthalene and perylene diimide-based conjugated organic molecules (Scheurle et al., 2022) and nanocomposites with metal oxide semiconductors such as ZnO, TiO_2, Fe_2O_3 and g-CN (Garcia-Salcido et al., 2022). Because of the availability of organic ligands with high molar absorptivity, the application of visible light-absorbing photocatalysts in artificial photosynthesis for the purpose of CO_2 reduction, H_2 production and organic pollutant degradation.

The functionalized organic ligands and metals included inside the MOFs make it feasible for the selective absorption of harmful contaminants such as heavy metals, organic pollutants and pharmaceutical effluents. For instance, a Zn-based MOF constructed from 1,3-benzene dicarboxylic acid demonstrated over cent percent cadmium recovery from the aqueous solution at pH 4. The buildup of heavy metals is an environmental concern (Nazari et al., 2017). Another Zn-MOF synthesised by depolymerizing bis(2-hydroxyethyl) terephthalate (BHET) exhibited significant cytotoxicity. The Zn-BHET-MOF demonstrated superior thermal stability, and its biocompatibility that led to an increase in fibroblast proliferation (Cabrera-Munguia et al., 2021).

15.7.3 MOFs in Photocatalysis

The MOFs play a significant part as photocatalysts in environmental remediation (Wang et al., 2020). Many newly discovered organic pollutants were destroyed by the Advanced Oxidation Process (AOP) which is driven by light. In recent years, it has been found that visible light-responsive photocatalysis is more eco-friendly. The role of MOF-based photocatalysts in the destruction of antibiotics and pesticides in aquatic environments was explored by some scientists. FMOFs constructed from the family of MIL-n, UiO, and ZIF with semiconductor metal oxides such as TiO_2, ZnO, Fe_2O_3, Fe_3O_4 and WO_3 as well as non-metal semiconducting materials such as $g-C_3N_4$ and RGO utilised visible light responsive photocatalysts for environmental applications. They were utilized in the degradation of organic pollutants such as textile dyes and pharmaceuticals besides oxidizing NO_x, reducing CO_2 and H_2 production (Garcia-Salcido et al., 2022).

The uses of binary and ternary MOF composites and the potential role of MOF-nanocomposites in the photocatalytic processes were reviewed. Nanocomposites are primarily being developed for the purpose of capturing visible light for photo-catalysis because of their high porosity and stability in water. The degradation of colours in textile wastewater was carried out by treating it with binary MOFs. Metal oxides such as TiO_2, ZnO, $BiVO_4$ and Ag_3PO_4 formed heterojunction nanocomposite and were used as visible light photocatalysts to treat the wastewater. Additionally, the coupling of MOF with a non-metal semiconductor such as $g-C_3N_4$ demonstrated higher visible light absorption and charge separation that led to an increase in the photocatalytic efficiency of the system (Mukherjee et al., 2022).

The UV and visible light-mediated AOPs are mainly catalysed by semiconductor metal oxides of transition metals such as Ti, Fe, Cr, Cu and Zn group elements. However, poor absorption of visible light and electron-hole recombination limit the use of metal oxides in visible light. Wang et al. (2018) have explored Zn-based MOFs comprising of 5-nitroisophthalic acid (H_2L) and 2,4,5-tri (4-pyridyl)-imidazole (Htpim) as organic ligands in the detection of acetone and UV-promoted degradation of Rhodamine B (RhB). In the UV-Vis studies, the Zn-MOF showed an optical bandgap in the range of 2.80 to 3.10 eV which is similar to that of ZnO and non-metal doped ZnO, which are in the 2.80–3.30 eV range (Vinitha et al., 2021). Furthermore, the degradation of RhB was increased to 92.3% when the Zn-MOF was used as a photocatalyst.

15.7.4 MOFs as Sensors

The detection of Al (III) up to 300 nM and methanol up to 0.7% (v/v) have been achieved by Zn-MOF containing 2-amino terephthalic acid (BDC) and 4,4'-bipyridine (BPy) ligands (Wiwasuku et al., 2020). The role of NH_2-functionalized Zn-BDC-MOF in sensing H_2 gas (BDC-1,4-dicarboxylic acid) which could be used in industries to avoid possible accidents due to hydrogen leakage was reported. The

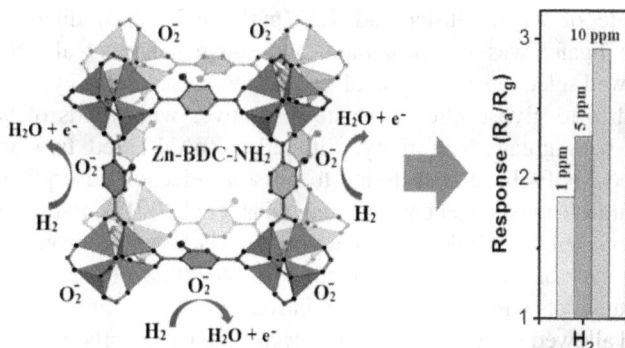

FIGURE 15.5 Zn-BDC-NH$_2$ framework and H$_2$ gas sensing (Yang et al., 2022). Copyright permission obtained from Elsevier.

sensor showed a lower detection limit, up to 1–10 ppm (Yang et al., 2022). The Zn-BDC-NH$_2$ was obtained by solvothermal method with 2-amino-benzene-1,4-dicarboxylic acid as ligand (Figure 15.5).

15.7.5 MOFs in Energy Production and Storage

MOF based on core shell nanosheets of ZnCo$_2$O$_4$-NiCo$_2$S$_4$-PPy synthesised on Ni foam is used in supercapacitor applications. The electrode showed maximum areal capacitance up to 3.75 F cm^{-2} with a specific capacitance of about 2507.0 F g^{-1}. In addition, the retention rate of capacitance is maintained to be 83.2% after 5000 cycles (Zhu et al., 2021). Thi et al. (2022) reported the use of nanocomposite Zn-MOF with r-GO in supercapacitor application. The MOF nanocomposite obtained by hydrothermal method revealed porous nanosheet-like structures with 82.5 F g^{-1} of specific capacitance and at 0.4 kW kg^{-1} showed an energy density of 7.1 Wh kg^{-1} (Thi et al., 2022). It has been reported that the semi-conductor hetero-junction formed by nanocomposites of MIL-n, UiO and ZIF with semiconductor metal oxides TiO$_2$, ZnO, Fe$_2$O$_3$, Fe$_3$O$_4$ and WO$_3$ could be used for H$_2$ production (García-Salcido et al., 2022).

15.7.6 MOFs in Selective Adsorption

MOFs possess tuneable porous structures by altering the type and functionalisation of organic ligands. There are examples which use amino group functionalised Zn-MOF for the selective adsorption of CO$_2$, NO$_x$, H$_2$, organic pollutants, sulphur compounds and phenols. The selective adsorption of CO$_2$ and NO$_x$ is the best instance of environmental remediation using Zn-MOF.

Kim et al. (2018) reported that nitrogen-rich three-dimensional Zn-MOF had enhanced selective adsorption of CO$_2$ (Kim et al., 2018). This was achieved at a relative pressure of 0.1–0.2 atm and 196° K which showed a maximum CO$_2$ adsorption up to 70–80 cm^3 g^{-1} when compared to N$_2$, H$_2$ and CH$_4$.

MOF made of Zn_4O cluster and 2,2'-(pyridine-2,5-diyl) diterephthalic acid (H4PDBDC) ligand was self-penetrable and stable (Zhou et al., 2018). Zn_4O-PDBDC showed selective adsorption of CO_2 and was used in the cycloaddition with epibromohydrin to give cyclic carbonate derivatives with yields of 90 and 99% respectively with greater selectivity. Similarly, Zn/Cd-based two amino group functionalized MOFs have been found to possess selective adsorption of CO_2 as well as useful in Knoevenagel condensation (Zhai et al., 2019). Pyrazole carboxylic acid functionalised Zn-MOF was used for the selective adsorption of cationic organic dyes such as rhodamine-6G, methylene blue, thiazole orange and Victoria blue. The study also revealed that the introduction of a linker increased pore diameter and allowed more cationic dye molecules to get adsorbed on the MOF (Liu et al., 2020).

15.7.7 MOFs in Organic Reactions

Zn-MOFs based on amide-functionalized pyridine dicarboxylic acid with 4,4'-[(pyridine-2,6-dicarbonyl)-bis(azanediyl)] dibenzoic acid (H2L) as organic ligand have been synthesized for one-pot tandem deacetylation and Knoevenagel condensation (Karmakar et al., 2020). By using Zn-MOF-1-NH2 as a heterogeneous catalyst for the selective absorption of CO_2 and cycloaddition with various substituted epoxides afforded cyclic carbonates in 90% yield with a conversion rate greater than 85%. Madasamy et al. (2019) reported bipyridyl (Bp) and 1,3,5-benzene tricarboxylic acid (BTC) functionalised Zn-Bp-BTC MOF for the Knoevenagel condensation reaction.

Synthesis of hierarchical mesoporous MOF, N-ZIF-8 was constructed from 2-methyl imidazole and Zn(II) by precipitation method. HP-N-ZIF-8 was obtained by spray drying technique using a polystyrene (PS) template-assisted method (Wang et al., 2022) and used for C-C bond formation in Knoevenagel condensation between benzaldehyde and malononitrile at room temperature. Metal azolate framework (Zn-MAF-6) was obtained from 2-ethyl imidazole and ZnO in the presence of ammonia. These catalysts were used in the depolymerisation of PET wastes with an ethylene-glycol-mediated glycolysis process. The PET conversion was achieved up to 92.4% with BHET yield of 81.7%.

15.7.8 MOFs in Drug Delivery Systems

In the drug delivery system, the role of the drug carrier is an essential component. According to Cai et al. (2020), amino-functionalized Zn-MOFs could be good drug carriers for curcumin. Zn-MOF-NH2, constructed from Zn(II) and 2-amino terephthalic acid (IRMOF-1), demonstrated 55.63% curcumin loading capacity but the identical analogue that lacked an amino group (IRMOF-3) showed just 49.30%.

15.7.9 Biological Applications of MOFs

The organic ligand Zn-MOF was synthesized by reacting Zn(II) with 2,6-pyridine dicarboxylic acid. The antimicrobial activity of this MOF was investigated and found to possess significant anti-fungal action making it an ideal candidate for fighting fungal diseases (Zeraati et al., 2022).

15.8 ZINC COMPLEXES WITH DRUG COMPOUNDS

15.8.1 ZINC COMPLEXES WITH CHLOROGENIC ACID

Metal complexes of plant phenolic compounds are being studied for non-toxic antioxidant materials from natural sources for plant protection, food industry and pharmaceutical applications. When plants are exposed to stress conditions, reactive oxygen species (ROS) are generated which damage the cellular components (Asada, 2006). Zinc chelation by chlorogenic acid, a plant-based compound, protected the plants and increased tolerance to stress. Zn(II)-chlorogenate's antioxidant and microbiological activities have been investigated and compared to ligand properties. The findings pointed its potential use in oxidative stress control and increasing plant resistance.

15.8.2 ZINC COMPLEXES WITH INDOMETHACIN

A study from the United Kingdom described the synthesis of a few zinc (II)-phenanthroline-indomethacin complexes and their anti-breast Cancer Stem Cell (anti-CSC) activity in a sub-micro and micro-molar range. Three zinc(II) complexes were found to possess significant inhibitory effects on mammosphere development and their reduced viability to the antibiotic salinomycin. Future studies on Metal-Phenanthroline-Non Steroidal Anti Inflammatory Drug complexes explored the cellular mechanism of zinc (II)-phenanthroline-indomethacin complexes in cancer stem cells and assessed their potential for translation into practical applications (Rundstadler et al., 2018).

15.8.3 ZINC COMPLEXES WITH NON-STEROIDAL ANTI-INFLAMMATORY DRUGS

NSAIDs are widely used as analgesics, antipyretics and anti-inflammatory agents but may cause gastrointestinal damage on long-term/improper use. Based on their chemical nature, they are characterized as carboxylic acids, furanones, oxicams and sulfonamides. Enhancing the therapeutic index of these drugs has become a priority in recent research. Metal complexes of NSAIDs have garnered significant attention in this regard (Psomas, 2020). In a study aimed to assess the impact of zinc complexation on the anti-inflammatory and ulcerogenic effects of ibuprofen and naproxen, rats were administered single and triple doses of the investigated complexes (zinc–ibuprofen and zinc–naproxen) together with their parent drugs and physical mixtures with zinc hydroaspartate (ZHA). The results showed a significant reduction in edema after carrageenan injection in the groups treated with the complexes, compared to control groups.

The combination of ZHA with ibuprofen and naproxen, administered three times into the stomach, exhibited a significant improvement in anti-inflammatory activity compared to both the control groups and the parent NSAIDs. In these animals, the growth of edema was reduced by an impressive 80.9%, surpassing the reductions achieved with zinc-naproxen (50.2%) or naproxen alone (47.9%). Furthermore, both complexes of NSAIDs with zinc and mixtures with ZHA effectively alleviated ulcerations caused by the parent NSAIDs. Notably, the mixtures of ibuprofen and naproxen with ZHA, administered three times, were found to cause the least damage. These findings provided valuable insights into the potential benefits of zinc supplementation during NSAID therapy which might help in preventing ulcers by reducing the required dosage of the parent drug while increasing its effectiveness (Jarosz et al., 2017).

The antioxidant properties of zinc (II)-NSAID complexes containing O-donor, N-donor and N, N'-donor co-ligands were studied and found to possess radical scavenging, inhibition of soybean lipoxygenase (LOX) and superoxide dismutase (SOD) activities (Psomas, 2020). The DPPH radical scavenging activities of these complexes were compared with the standard compounds such as nor-digydroguaric acid (NDGA) and butylated hydroxytoluene (BHT). Zn(II)-oxaprozin exhibited the highest DPPH scavenging activity among the metal-NSAID complexes.

15.8.4 Zinc Complex with Naproxen

A complex formed between zinc and naproxen demonstrated potent antibacterial properties against the Gram-positive *Staphylococcus aureus* and Gram-negative *Escherichia coli* (Sharma, 2003). Another modified zinc compound exhibited an enhanced effect specifically against *S. aureus*, surpassing its activity against *E. coli* (Abu Ali et al., 2015).

15.8.5 Zinc Complexes with Valproic Acid

Valproic acid is an anti-convulsant agent largely used to treat epilepsy and bipolar disorder. The structure and biological activity of mixed ligand zinc complexes based on sodium valproate and bioactive nitrogen-donor ligands were reported. 2,9-dimethyl-1,10-phenanthroline (dmphen) and valproate mixed ligand Zinc (II) complex exhibited enhanced antibacterial activity against gram-negative bacteria which arises due to the higher lipophilicity of the complex (Darawsheh et al., 2014).

15.8.6 Zinc Complexes with Curcumin and Cryptolepine

A series of zinc (II)-complexes with curcumin and cryptolepine were synthesized and investigated for photodynamic activity and cell imaging capabilities in T24 cancer cells. The addition of a glycosylated moiety to these compounds and their metal complexes led to improved bioavailability, water solubility and anticancer activity (Qin et al., 2021). Zhou et al. (2022) synthesized glycosylated zinc (II)-cryptolepine complexes [Zn(QA1)-Zn(QA3)], which were found to induce apoptosis and autophagy in SK-OV-3/DPP cancer cells. The study also revealed glycosylation enhanced antiproliferative activity of these zinc (II)-complexes. Further studies conducted in an *in vivo* model revealed tumor inhibition and confirmed anticancer effectiveness of the glycosylated zinc (II)-complexes.

15.9 CONCLUSION

This chapter highlights the significance of unique zinc complexes and their diverse applications in various fields. Zinc compounds particularly have promising prospects in drug research and therapeutic applications. The synergistic effects obtained by mixing zinc complexes with other metals or ligands open up interesting possibilities for the development of improved functional materials. In the rational design and optimization of zinc complexes, computational approaches and modelling techniques are important. This will facilitate the discovery of new complex structures with improved properties, enabling targeted and efficient applications. Finally, the study of unique zinc complexes and their applications is a rapidly evolving field with significant potential for scientific advancements and practical applications.

REFERENCES

Ali HA, Fares H, Darawsheh M, Rappocciolo E, Akkawi M, Jaber S. 2015. Synthesis, characterization and biological activity of new mixed ligand complexes of Zn(II) naproxen with nitrogen based ligands. Eur. J. Med. Chem. 89, 67–76. 10.1016/j.ejmech.2014.10.032.

Asada K. 2006. Production and scavenging of reactive oxygen species in chloroplasts and their functions. Plant Physiol. 141, 391–396. 10.1104/pp.106.082040.

Biswal BP, Panda T, Banerjee R. 2012. Solution mediated phase transformation (RHO to SOD) in porous Co-imidazolate based zeolitic frameworks with high water stability. Chem. Commun. 48, 11868. 10.1039/c2cc36651g.

Cabrera-Munguia DA, León-Campos MI, Claudio-Rizo JA, Solís-Casados DA, Flores-Guia TE, Cano Salazar LF. 2021. Potential biomedical application of a new MOF based on a derived PET: Synthesis and characterization. Bull. Mater. Sci. 44(4), 245. 10.1007/s12034-021-02537-9.

Cai M, Qin L, Pang L, Ma B, Bai J, Liu J, Dong X, Yin X, Ni J. 2020. Amino-functionalized Zn metal organic frameworks as antitumor drug curcumin carriers. New J. Chem. 44, 17693–17704. 10.1039/D0NJ03680C.

Darawsheh M, Ali HA, Abuhijleh AL, Rappocciolo E, Akkawi M, Jaber S, Maloul S, Hussein Y. 2014. New mixed ligand zinc(II) complexes based on the antiepileptic drug sodium valproate and bioactive nitrogen-donor ligands. Synthesis, structure and biological properties. Eur. J. Med. Chem. 82, 152–163. 10.1016/j.ejmech.2014.01.067.

Ermakova E A, Golubeva JA, Smirnova KS, Klyushova LS, Eltsov, IV, Zubenko AA, Fetisov LN, Svyatogorova AE, Lider EV. 2023. Bioactive mixed-ligand zinc(II) complexes with 1H-tetrazole-5-acetic acid and oligopyridine derivatives. Polyhedron 230, 116213. 10.1016/j.poly.2022.116213.

García-Salcido V, Mercado-Oliva P, Guzmán-Mar JL, Kharisov BI, Hinojosa-Reyes L. 2022. MOF-based composites for visible-light-driven heterogeneous photocatalysis: Synthesis, characterization and environmental application studies. J. Solid State Chem. 307, 122801. 10.1016/j.jssc.2021.122801.

He CT, Jiang, L, Ye ZM, Krishna R, Zhong ZS, Liao PQ, Xu J, Ouyang G, Zhang JP, Chen XM. 2015. Exceptional hydrophobicity of a large-pore metal–organic zeolite. J. Am. Chem. Soc. 137, 7217–7223. 10.1021/jacs.5b03727.

Jarosz M, Olbert M, Wyszogrodzka G, Młyniec K, Librowski T. 2017. Antioxidant and anti-inflammatory effects of zinc. Zinc-dependent NF-κB signaling. Inflammopharmacology 25, 11–24. 10.1007/s10787-017-0309-4.

Jiang Q, Zhu J, Zhang Y, Xiao N, Guo Z. 2009. DNA binding property, nuclease activity and cytotoxicity of Zn(II) complexes of terpyridine derivatives. BioMetals 22, 297–305. 10.1007/s10534-008-9166-3.

Karmakar A, Soliman MMA, Rúbio GM DM, Guedes Da Silva MFC, Pombeiro AJL. 2020. Synthesis and catalytic activities of a Zn(ii) based metallomacrocycle and a metal–organic framework towards one-pot deacetalization-Knoevenagel tandem reactions under different strategies: A comparative study. Dalton Trans. 49, 8075–8085. 10.1039/D0DT01312A.

Kim HC, Huh S, Lee DN, Kim Y. 2018. Selective carbon dioxide sorption by a new breathing three-dimensional Zn-MOF with Lewis basic nitrogen-rich channels. Dalton Trans. 47, 4820–4826. 10.1039/C7DT04134A.

Krężel A, Maret W. 2016. The biological inorganic chemistry of zinc ions. Arch. Biochem. Biophys. 611, 3–19. 10.1016/j.abb.2016.04.010.

Kumar N, Roopa Bhalla V, Kumar M. 2021. Beyond zinc coordination: Bioimaging applications of Zn(II)-complexes. Coord. Chem. Rev. 427, 213550. 10.1016/j.ccr.2020.213550.

Kusamoto T, Nishihara H. 2019. Zero-, one- and two-dimensional bis(dithiolato)metal complexes with unique physical and chemical properties. Coord. Chem. Rev. 380, 419–439. 10.1016/j.ccr.2018.09.012.

Liu XT, Chen SS, Li SM, Nie HX, Feng YQ, Fan YN, Yu MH, Chang Z, Bu XH. 2020. Structural tuning of Zn(II)-MOFs based on pyrazole functionalized carboxylic acid ligands for organic dye adsorption. CrystEngComm 22, 5941–5945. 10.1039/D0CE00798F.

Madasamy K, Kumaraguru S, Sankar V, Mannathan S, Kathiresan M. 2019. A Zn based metal organic framework as a heterogeneous catalyst for C–C bond formation reactions. New J. Chem. 43, 3793–3800. 10.1039/C8NJ05953E.

Marchetti F, Pettinari R, Verdicchio F, Tombesi A, Scuri S, Xhafa S, Olivieri L, Pettinari, C, Choquesillo-Lazarte D, García-García A, Rodríguez-Diéguez A, Galindo A. 2022. Role of hydrazone substituents in determining the nuclearity and antibacterial activity of Zn (II) complexes with pyrazolone-based hydrazones. Dalton Trans. 51, 14165–14181. 10.1039/D2DT02430F.

Marchetti F, Tombesi A, Di Nicola C, Pettinari R, Verdicchio F, Crispini A, Scarpelli F, Baldassarri, C, Marangoni E, Hofer A, Galindo A, Petrelli R. 2022. Zinc(II) Complex with Pyrazolone-Based Hydrazones is Strongly Effective against Trypanosoma brucei Which Causes African Sleeping Sickness. Inorg. Chem. 61, 13561–13575. 10.1021/acs.inorgchem.2c02201.

Mukherjee D, Van Der Bruggen B, Mandal B. 2022. Advancements in visible light responsive MOF composites for photocatalytic decontamination of textile wastewater: A review. Chemosphere 295, 133835. 10.1016/j.chemosphere.2022.133835.

Nazari Z, Taher MA, Fazelirad H. 2017. A Zn based metal organic framework nanocomposite: Synthesis, characterization and application for preconcentration of cadmium prior to its determination by FAAS. RSC Adv. 7, 44890–44895. 10.1039/C7RA08354H.

Öhrström L, Noa FMA 2021. Metal-Organic Frameworks. American Chemical Society. 10.1021/acs.infocus.7e4004.

Pellei M, Del Bello F, Porchia M, Santini C. 2021. Zinc coordination complexes as anticancer agents. Coord. Chem. Rev. 445, 214088. 10.1016/j.ccr.2021.214088.

Pola A, Tocci M., Goodwin FE. 2020. Review of microstructures and properties of zinc alloys. Metals 10, 253. 10.3390/met10020253.

Porchia M, Pellei M, Del Bello F, Santini C. 2020. Zinc complexes with nitrogen donor ligands as anticancer agents. Molecules 25, 5814. 10.3390/molecules25245814.

Psomas G. 2020. Copper (II) and zinc (II) coordination compounds of non-steroidal anti-inflammatory drugs: Structural features and antioxidant activity. Coord. Chem. Rev. 412, 213259. 10.1016/j.ccr.2020.213259.

Purcell KF, Kotz JC. 2010. Inorganic Chemistry. Cengage Learning India Private Limited.

Qin LQ, Liang CJ, Zhou Z, Qin QP, Wei ZZ, Tan MX, Liang H. 2021. Mitochondria-localizing curcumin-cryptolepine Zn(II) complexes and their antitumor activity. Bioorg. Med. Chem. 30, 115948. 10.1016/j.bmc.2020.115948.

Rundstadler T, Eskandari A, Norman, S, Suntharalingam K. 2018. Polypyridyl Zinc(II)-Indomethacin complexes with potent anti-breast cancer stem cell activity. Molecules 23, 2253. 10.3390/molecules23092253.

Scheurle PI, Biewald A, Mähringer A, Hartschuh A, Medina DD, Bein T. 2022. A novel electrically conductive perylene diimide-based MOF-74 series featuring luminescence and redox activity. Small Structures 3, 2100195. 10.1002/sstr.202100195.

Seyferth D. 2001. Zinc alkyls, Edward Frankland, and the beginnings of main-group organometallic chemistry. Organometallics 20, 2940–2955. 10.1021/om010439f.

Sharma J. 2003. Zinc–naproxen complex: Synthesis, physicochemical and biological evaluation. Int. J. Pharm. 260, 217–227. 10.1016/S0378-5173(03)00251-5.

Thi QV, Patil SA, Katkar PK, Rabani I, Patil AS, Ryu J, Kolekar G, Tung NT, Sohn D. 2022. Electrochemical performance of zinc-based metal-organic framework with reduced graphene oxide nanocomposite electrodes for supercapacitors. Synth. Met. 290, 117155. 10.1016/j.synthmet.2022.117155.

Van Der Heyden A, Chanthavong P, Angles-Cano E, Bonnet H, Dejeu J, Cras A, Philouze C, Serratrice G, El-Ghazouani FZ, Toti F, Thibon-Pourret A, Belle C. 2023. Grafted dinuclear zinc complexes for selective recognition of phosphatidylserine: Application to the capture of extracellular membrane microvesicles. J. Inorg. Biochem. 239, 112065. 10.1016/j.jinorgbio.2022.112065.

Vinitha V, Preeyanghaa M, Vinesh V, Dhanalakshmi R, Neppolian B, Sivamurugan V. 2021. Two is better than one: Catalytic, sensing and optical applications of doped zinc oxide nanostructures. Emergent Mater. 4, 1093–1124. 10.1007/s42247-021-00262-x.

Wang JJ, Zhang YJ, Chen Y, Si PP, Pan YY, Yang J, Fan RY, Li ZY. 2018. Three new Zn-based metal–organic frameworks exhibiting selective fluorescence sensing and photo-catalytic activity. CrystEngComm 20, 3877–3890. 10.1039/C8CE00598B.

Wang N, Wei Y, Chang M, Liu J, Wang JX. 2022. Macro-meso-microporous metal–organic frameworks: Template-assisted spray drying synthesis and enhanced catalysis. ACS Appl. Mater. Interfaces 14, 10712–10720. 10.1021/acsami.1c23297.

Wang Q, Gao Q, Al-Enizi AM, Nafady A, Ma S. 2020. Recent advances in MOF-based photocatalysis: Environmental remediation under visible light. Inorg. Chem. Front. 7, 300–339. 10.1039/C9QI01120J.

Wiwasuku T, Othong J, Boonmak J, Ervithayasuporn V, Youngme S. 2020. Sonochemical synthesis of microscale Zn (ii)-MOF with dual Lewis basic sites for fluorescent turn-on detection of Al 3+ and methanol with low detection limits. Dalton Trans. 49, 10240–10249. 10.1039/D0DT01175D.

Yang DH, Nguyen TTT, Navale ST, Nguyen LHT, Dang YT, Mai NXD, Phan TY, Doan, TLH, Kim SS, Kim HW. 2022. Novel amine-functionalized zinc-based metal-organic framework for low-temperature chemiresistive hydrogen sensing. Sens. Actuators B: Chem. 368, 132120. 10.1016/j.snb.2022.132120.

Yoshikawa Y, Yasui H. 2012. Zinc complexes developed as metallopharmaceutics for treating diabetes mellitus based on the bio-medicinal inorganic chemistry. Curr. Top. Med. Chem. 12, 210–218. 10.2174/156802612799078874.

Yu X, Gao E, Yao W, Fedin VP, Potapov, AS. 2022. Zinc (II) and cobalt (II) complexes with unusual coordination of mixed imidazole-1,2,4-triazole ligand in a protonated cationic form. Polyhedron 217, 115741. 10.1016/j.poly.2022.115741.

Zeraati M, Moghaddam-Manesh M, Khodamoradi S, Hosseinzadegan S, Golpayegani A, Chauhan NPS, Sargazi G. 2022. Ultrasonic assisted reverse micelle synthesis of a novel Zn-metal organic framework as an efficient candidate for antimicrobial activities. J. Mol. Struct. 1247, 131315. 10.1016/j.molstruc.2021.131315.

Zha, ZW, Yang SH, Lv YR, Du CX, Li LK, Zang SQ. 2019. Amino functionalized Zn/Cd-metal–organic frameworks for selective CO_2 adsorption and Knoevenagel condensation reactions. Dalton Trans. 48, 4007–4014. 10.1039/C9DT00391F.

Zhou HF, Liu B, Hou L, Zhang WY, Wang YY. 2018. Rational construction of a stable Zn 4 O-based MOF for highly efficient CO_2 capture and conversion. Chem. Commun. 54, 456–459. 10.1039/C7CC08473K.

Zhou Z, Du LQ, Huang XM, Zhu LG, Wei QC, Qin QP, Bian H. 2022. Novel glycosylation zinc (II) – cryptolepine complexes perturb mitophagy pathways and trigger cancer cell apoptosis and autophagy in SK-OV-3/DDP cells. Eur. J. Med. Chem. 243, 114743. 10.1016/j.ejmech.2022.114743.

Zhu J, Wang Y, Zhang X, Cai W. 2021. MOF-derived $ZnCo_2O_4$-$NiCo_2S_4$-PPy core–shell nanosheets on Ni foam for high-performance supercapacitors. Nanotechnology 32, 145404. 10.1088/1361-6528/abd20b

16 Role of Zinc Nutrition in Sustainable Vegetable Production

V. Kasthuri Thilagam, S. Manivannan, and R. Srinivasan

16.1 INTRODUCTION

Living organisms derive energy from carbohydrates, proteins, vitamins and minerals. Among these sources, vitamins and minerals are vital to human health which are available in fruits and vegetables. The nutritional quality of vegetables is directly influenced by the nutrient status of the soil. Micronutrients though required in small quantities, their role in crop production is well recognized. The deficiency of micronutrients will cause a reduction in the yield and quality of the crop. The green revolution has intensified agriculture, increased crop productivity, and depleted soil micronutrients. Micronutrients are also lost from the soil through erosion, runoff and leaching processes. Decreased organic inputs like farmyard manure, compost, green manures, and green leaf manures have made our soils deficient in micronutrients.

Zn is the third-most-limiting nutrient after N and P. Zinc deficiency is most prevalent in arid and semi-arid regions due to low solubility and high fixation (Donner et al., 2010). Soil Zn deficiency is an important factor for abiotic stress spread over 49% of cultivated lands globally, affecting sustainable agriculture (Hacisalihogluand Kochian, 2004). India, China, Pakistan, Iran and Turkey have widespread Zn deficiency in 50% and 70% of the cultivated land (Alloway, 2008).

Zinc deficiency is common in light-textured calcareous, peat, and high phosphorus soils (Rudani et al., 2018). Acidic soils are less deficient in Zn than soils with high pH (Shukla et al., 2019). Crops grown under these soils suffer from Zn deficiency disorders which drastically reduce the yield and quality of crops (Mousavi et al., 2007). Soils with marginal Zn deficiency can reduce crop yields and quality without expressing any Zn deficiency symptoms in plants called hidden deficiency (Alloway, 2009).

The introduction of high-yielding crops during the green revolution in India increased the demand for micronutrients. However, the indiscriminate application of primary nutrients like Nitrogen, phosphorus, and potassium through straight fertilizers and neglecting the micronutrients has led to micronutrient deficiency in most soils (Srivastava et al., 2017). Among the micronutrients, Zn is deficient in 43% of Indian soils, and the same may increase to 63% by 2025 (Sunithakumari, 2016).

DOI: 10.1201/9781003412472-16

Among the micronutrient fertilizers, $ZnSO_4$ consumption was highest, and 70% was applied to field crops and 30% to vegetable and fruit crops. Zn is involved in many activities of plants like nitrogen metabolism, energy transfer and protein synthesis. According to Havlin et al. (2014) the plant is Zn deficient when Zn concentration is <20 ppm in the leaf and becomes toxic when Zn levels are more than 400 ppm. The critical concentration level of Zn in cereals is 15 ppm, whereas in millets it varies from 15 to 20 ppm. In legume crops, the critical limit is 7–20 ppm. The highest critical Zn concentration is highest for vegetable crops (36 ppm) followed by oil seeds (12–25 ppm).

Zn is absorbed as Zn^{2+} form by most plants, but some plant roots absorb Zn in organic ligand–Zn complexes. Plants use two different strategies to uptake Zn in divalent form depending on the ligands secreted by the roots. The first approach involves the effluxes of reductants as organic acids (OA) and H+ ions that increase the solubility of Zn complexes (Zn phosphates and hydroxides), resulting in the release of Zn^{2+} ions for the absorption of the root epidermal cells. The strategy involves the efflux of phyto-siderophores (lower molecular weight compounds) that make stable complexes with Zn fostering Zn influx into the root epidermal cells in the cereal crops.

An important challenge for agriculture scientists is to feed the increasing population with nutrient-rich foods and reduce malnutrition. The wise option for managing the Zn deficiency is to apply micronutrients through integrated nutrient management with micronutrient fertilizers and organic manures.

16.2 ROLE OF Zn NUTRITION ON PLANTS

Zinc plays many roles in higher plants. Sommer and Lipman (1926) demonstrated the essentiality of Zn nutrition for the life and growth of higher green plants.

- Zn is a key structural constituent or regulatory co-factor of plant enzymes like auxins and proteins.
- It is an integral component of plant biomolecules like lipids and proteins and is essential for plant nucleic acid metabolism.
- It involves photosynthesis and the conversion of sugars to starch.
- Helps in pollen formation, maintaining the integrity of biological membranes, pest and disease resistance.
- It increases seed viability, and seedling vigor and gives protection to abiotic and biotic stresses.

Among the vegetable crops, beans and onion are highly sensitive, potato, tomato and sugar beet are mildly sensitive to zinc deficiency.

16.3 ZINC DEFICIENCY IN CROPS

Zn deficiencies are widespread worldwide and can be identified by distinctive visual symptoms in leaves, fruits and overall plant growth. Common symptoms include.

- Light green, yellow, or white areas between leaf veins, particularly in younger leaves
- Eventual tissue necrosis in chlorotic leaf areas
- Shortened stem or stalk internodes, resulting in stunted plants or bushy, rosette leaves
- Small, narrow, thickened leaves, often malformed by the growth of only part of leaf tissue
- Premature foliage loss
- Reduced root development
- Malformation of fruit, often with reduced or no yield.
- Zn deficiency leads to plants' iron (Fe) deficiency by preventing the transfer of Fe from root to shoot in zinc deficiency conditions
- Cause boron toxicity

16.4 ZINC IN SOILS

Zinc naturally occurs in the Earth's crust as rocks and minerals and the average content in soil depends on the nature of the parent material on which the soil is formed and soil mineralogy. The average Zn content of rocks in the Earth's crust is 80 mg kg^{-1} and a meager quantity of the total Zn is an exchangeable or soluble form (Lindsay, 1979). Zn exists in soil as seven discrete pools such as water-soluble Zn, exchangeable Zn, carbonate bound Zn, amorphous oxide bound Zn, organically bound Zn, complexed Zn and crystalline oxide bound Zn (Raja and Iyengar, 1986). The behavior of Zn in soils and Zn availability depends on the concentration of these Zn fractions to total Zn in soil (Liu et al., 2020).

Most plants cannot utilize the Zn in the soils as it is rapidly adsorbed into the soil colloids, forming chelates, and precipitated this reduces the bioavailability of Zn to plants. Zn concentrations in world soils range from 10 to 100 mg kg^{-1} (Mertens and Smolders, 2013). Zn content in Indian soils varied with soil type and climatic conditions. Arid and semi-arid soils have more Zn concentration compared to Humid and sub-humid tropics (Katyal and Vlek, 1985).

16.5 SOURCES OF ZINC NUTRITION TO PLANTS

Zn is supplied through various sources to correct Zn deficiency in crops that vary in Zn content, composition, price, and effectiveness.

16.5.1 ORGANIC SOURCES

Farmyard manure, compost and vermicompost can supply Zn nutrients to the plants throughout the crop growth period and also have a residual effect on the next crop. Plant available Zn in most animal wastes ranges from 0.01% to 0.05%; hence, a large quantity of manure must be applied to fulfill the plant Zn requirement. The primary benefit of organic manure is increased organic matter and associated natural chelation that increases Zn availability.

16.5.2 Inorganic Sources

Sulfates, oxides, carbonates, phosphates, chlorides and nitrates of Zn are the inorganic Zn fertilizers commonly available in the market. Crop management factors like type and application method can affect the efficiency of Zn fertilizers.

16.6 FACTORS AFFECTING Zn AVAILABILITY TO PLANTS

Zinc deficiency extends to every part of the world, and most crops respond positively to Zn application.

- Soil pH is crucial for the Zn availability in soil, high pH increases the adsorptive capacity of soil and decreases the Zn availability.
- Sandy, calcareous and alkaline soils are more prone to zinc deficiency than neutral and slightly acidic soils.
- The soils formed from gneisses, granites, sandstone and limestone are low in Zn.
- Quartz in the soils dilutes soil Zn as concentrations of Zn in quartz are very low.
- Soils with high phosphorus content or application of indiscriminate P fertilizers reduce the Zn uptake by the plants. This phenomenon is known as "P-induced Zn deficiency."
- Applying organic manure to soils facilitates the formation of soluble organic zinc complexes and makes the Zn available to plants.
- Peat and muck soils with high organic matter contents have low available Zn due to their inherent low Zn content or stable organic complex formation.
- Waterlogged soils are more deficient in Zn than dry soils due to reactions with free sulfide and precipitation.

16.7 Zn DEFICIENCY IN VEGETABLE CROPS

Zinc is an immobile element in the plant system; hence, it can not take Zn from older leaves for new leaves. Normally Zn deficiency is found in new leaves. The typical symptom of Zn deficiency is leaf color change from green to pale green, and yellow due to the less chlorophyll content. Necrotic leaf spots occur due to the leaf tissue's death in small concentrated areas. Bronzing of leaves also occurs due to Zn deficiency when the yellow areas turn bronze. Shortening of internodes and rosetting of leaves are also common symptoms of Zn deficiency in vegetable crops.

16.7.1 Beans (*Phaseolus vulgaris L.*)

Beans are susceptible to zinc deficiency. Regular application of soil test-based Zn is vital to maintain the beans' yield in the Zn deficient soil. Zn deficiency usually appears on the second set of trifoliate leaves, which become light green and mottled. Over time, the area between the leaf veins becomes pale green and turns yellow near the tips and outer edges. A severe deficiency causes graying or browning, shredding of leaves, and aborted blossoms.

16.7.2 POTATO (*SOLANUM TUBEROSUM L.*)

Deficiency of Zn in potatoes results in stunted plant growth, leaf rolling, rigid plants, and shorter internodes, also known as "little leaf" or "fern leaf". Deficiencies will occur when Zn content in the leaf is less than 15–20 mg kg^{-1}. Intense interveinal necrosis will be caused as the deficiency progresses and the main veins remain green.

16.7.3 TOMATO (*LYCOPSERICUM ESCULENTUM MILL*)

Zn deficiency in tomatoes will affect the vegetative and reproductive growth of plants. It also causes interveinal chlorosis in leaves and curls downwards. In severe conditions, the petioles curl like a corkscrew. It is followed by the brownish-orange colouration of the older leaves which often show necrotic spots. Under severe deficiency conditions, brown-colored fluid will be oozed out from the leaves, affecting the fruit set and yield.

16.7.4 SWEET POTATO (*IPOMOEA BATATAS*)

Zn deficiency in sweet potatoes causes the thickening of leaves with limited plant growth. Chlorosis in young leaves leads to complete bleaching, purple pigmentation, and narrowing of the leaf blade, also caused due to Zn deficiency. In severe deficiency, brown discolouration will occur in sweet potato flesh.

16.7.5 CAULIFLOWER (*BRASSICA OLERACEA VAR BOTRYTIS L*)

Young leaves of cauliflower plants become small and curled. Curd will turn to a brown color and sparsed, which reduces the economic value of the produce. In severe Zn deficiency conditions, leaves turned light-purple, and the older leaves became chlorotic, tough and leathery.

16.7.6 ONION (*ALLIUM CEPA L.*)

Zinc deficiency in onions shows up as irregular leaves will be formed with narrow tips. Stunting, twisting and bending tips are also symptoms (Figure 16.1).

16.7.7 CARAT (*DAUCUS CAROTA L.*)

Carrots grown on sandy calcareous soils and soil with high organic matter are prone to zinc deficiency. Carrots grown in Zn-deficient soils will be small in size with stunted, chlorotic leaves.

FIGURE 16.1   Zn deficiency symptoms in different vegetable crops.

16.7.8   Chili (*Capsicum sp.*)

Chilies require a large amount of zinc compared with other vegetable crops, Zn deficiency will affect the plant growth severely. Reduced crop foliage with interveinal chlorosis and stunted growth is observed in severe deficiency.

16.7.9   Peas (*Pisum sativum*)

Leaves will be narrow, pointed and curled inwards. Yellow mottling on the lamina started from the midrib's base and moved upwards. Poor pod set and seed development.

16.7.10   Lettuce (*Lactuca Sativa L.*)

Lettuces grown under sandy and clay loam soils will show symptoms like stunted growth, slow maturity and reduced weight of heads.

16.8   RESPONSE OF VEGETABLE CROPS TO Zn NUTRIENT

Crops vary widely in their tolerance Zn deficiency concerning uptake and utilization due to their physiological and molecular mechanisms (Zeng et al., 2021). Understanding these mechanisms will improve the Zn use efficiency, Zn deficiency resistance, and Zn accumulation in plants (Table 16.1).

TABLE 16.1

Percent Yield Increase in Vegetable Crops with Zn Application

S.No	Crop	Response to Zn in terms of Yield increase (%)	Reference
1.	Potato	22	Ahmad et al., 2012
		37	Dwivedi, 1991
		36.7 (Zn +B)	Sarker et al., 2018
2.	Cabbage	7.3	Shukla and Tiwari, 2016
		38.4	Singh and Singh, 2017
3.	Carat	85.0 (Zn Nano fertilizer)	Awad et al., 2021

16.9   STRATEGIES FOR Zn NUTRIENT MANAGEMENT IN VEGETABLE CROPS

Regular soil or plant testing is required to determine the Zn status and ensure the optimal application of Zn fertilizers. Zinc deficiency in soils can be identified by soil testing or analyzing the plant leaves. The deficiency can be corrected by applying soil test-based applications and followed by maintenance doses through the foliar application (Zia et al., 2006). The critical concentrations of Zn are site-specific, and one critical limit may not represent every soil type or crop. Most crops respond to the Zn application when soil Zn content is 0.5 ppm or less.

16.9.1   Zn Application Method

Soil test-based Zn application is the most economical approach for zinc fertilization in vegetable crops. In this approach, deficient soils can be supplemented with the required quantity of Zn specific to the vegetable crop. Zinc fertilizers usually applied to the soil and incorporated as mere broadcasting will not be helpful for plant uptake.

Foliar applications of zinc have not been consistently effective in correcting Zn deficiencies. Zn chelates will help to increase the Zn use efficiency.

Integrated nutrient management with organic and inorganic sources significantly enhances soil Zn availability. INM can improve soil's physical and chemical properties, increasing fertilizer use efficiency.

- Soil organic matter acts as a factor in the distribution of micronutrients between soil colloids and solution by controlling the sorption of micronutrients through the active functional groups, high specific surface area, cation exchange capacity and capability to form soluble complexes.
- Arbuscular mycorrhiza undoubtedly can increase the accumulation of many nutrients, including Zn (Srinivasagam et al., 2013).
- 2.0% Zn-enriched urea (ZEU) as hydrated Zn sulfate will increase the Zn uptake in crops (Singhal et al., 2012).

- Identification and selection of Zn-efficient crops and varieties that can grow and yield better even under Zn-deficient soils
- Application of superphosphate, a source of P that contains varying amounts of Zn will also add Zn to the soil simultaneously.

16.10 CHALLENGES

Zn use efficiency is low and hardly exceeds 5% in most crops. Farmers do not realize the role of micronutrients in crop production, and balanced fertilization is lacking. The growing population in all countries poses many challenges to farmers, scientists and policymakers in feeding healthy and nutritious food. The present scenario of intensive agriculture reduced the usage of organic manures also a major concern in Zn nutrition.

- Creating awareness among the farmers on Zn fertilization through meetings, training, Kisan melas and celebrating Zinc Days
- GST on bulk fertilizers is now 5%, but for micronutrients, it remains 12%, whereas for micronutrient mixture is 18%. Hence the GST has to be reduced for micronutrients to ensure balanced fertilization by the farmers.
- Coating Zn with other major nutrients is an effective way to increase micronutrient use efficiency.
- Many research studies demonstrated the role of nano zinc fertilizers in enhancing nutrient use efficiency. Therefore, scientists should develop cost-effective, customized nano fertilizers without compromising crop requirements.
- Breeding vegetable genotypes that accumulate high levels of Zn in edible portions will increase the nutritional value.
- An interdisciplinary approach integrating soil scientists, breeders, agronomists, and nutritionists promises to improve plant and human nutrition.

16.11 CONCLUSION

In the scenario of a burgeoning population and emerging pandemic diseases on Earth, zinc deficiency in soil and crops is a growing concern for plant and human health. Zinc deficiency causes significant loss in vegetable quality and yield. As most soils are deficient in Zn nutrients, farmers should give due importance to Zn management for vegetable production. Efficient Zn management should consider soil availability, cropping systems, residual availability, sources of nutrients, and Zn use efficiency. Soil application of zinc fertilizers is the best method, whereas foliar application is a complementary approach for increasing the Zn availability to vegetable crops. Zn chelates also can be used to improve the use efficiency in vegetable crops. Zn-sensitive crops must be given due attention for vegetable production. As Zn plays a major role in human health, current research programs seek to improve crop yields and Zn concentrations in vegetable crops' edible portions to address food security and health.

REFERENCES

Ahmad W, Watts MJ, Imtiaz M, Ahmed I, Zia MH. 2012. Zinc deficiency in soils, crops and humans. Agrochimica 56, 65–97.

Alloway BJ. 2008. Zinc in Soils and CropNutrition, Second ed. IZA and IFA, Paris, France.

Alloway BJ. 2009. Soil factors associated with zinc deficiency in crops and humans. Environ. Geochem. Health 31, 537–548.

Awad AAM, Rady MM, Semida WM, Belal EE, Omran WM, Al-Yasi HM, Ali EF. 2021. Foliar nourishment with different zinc-containing forms effectively sustains carrot performance in zinc-deficient soil. Agronomy 11, 1853. 10.3390/agronomy11091853.

Donner E, Broos K, Heemsbergen D, Warne MSTJ, Mclaughlin MJ, Hodson ME, Nortcliff S. 2010. Biological and chemical assessments of zinc ageing in field soils. Environ. Poll. 158, 339–345.

Dwivedi GK. 1991. Mode of application of micronutrients to potato in acid soil for Garhwal Himalaya. Indian J. Hortic. 45, 258–263.

Hacisalihoglu G, Kochian LV. 2004. How do some plants tolerate low levels of soil zinc? Mechanisms of zinc efficiency in crop plants. New Phytol. 159, 341–350.

Havlin JL, Tisdale SL, Nelson WL, Beaton JD. 2014. Soil Fertility and Fertilizers - An Introduction to Nutrient Management. Eight ed. Indian Edition, Prentice Hall of India Ltd.

Katyal JC, Vlek PLG. 1985. Micronutrient problems in tropical asia. Fertil. Res. 7, 69–94.

Lindsay WL. 1979. Chemical Equilibria in Soils. John Wiley & Sons, Inc. New York.

Liu YM, Liu DY, Zhao QY, Zhang W, Chen XX, Xu SJ, Zou CQ. 2020. Zinc fractions in soils and uptake in winter wheat as affected by repeated applications of zinc fertilizer. Soil Tillage Res. 200, 104612.

Mertens J, Smolders E. 2013. Zinc, in: Alloway, BJ (Ed.), Heavy Metals in Soils 22. Springer, Netherlands.

Mousavi SR, Galavi M, Ahmadvand G. 2007. Effect of zinc and manganese foliar application on yield, quality and enrichment on potato (*Solanum tuberosum* L.). Asian J. Plant Sci. 6, 1256–1260.

Raja EM, Iyengar BRV. 1986. Chemical pools of Zinc in some soils as influenced by sources of applied Zinc. J. Ind. Soc. Soil Sci. 34, 98–105.

Rudani K, Patel V, Prajapati K. 2018. The importance of Zinc in plant growth – A review. Int. Res. J. Nat. Appl. Sci. 5, 38–48.

Sarker MMH, Moslehuddin AZM, Jahiruddin M, Islam MR. 2018. Effects of micronutrient application on different attributes of potato in floodplain soils of Bangladesh. SAARC J. Agric. 16, 97–108.

Shukla AK, Sanjib KB, Satyanarayana T, Majumdar K. 2019. Importance of micronutrients in Indian agriculture. Better crops- South Asia 11, 6–10.

Shukla AK, Tiwari PK. 2016. Micro and secondary nutrients and pollutant elements research in India. Progress Report 2014-16. AICRP-MSPE, ICAR-IISS, Bhopal. pp. 1–196.

Singh N, Singh R. 2017. Effect of zinc nutrition on yield, quality and uptake of nutrients in cabbage (*Brassica oleracea*). Ann. Plant Soil Res. 19, 299–302.

Singhal SK, Sharma VK, Singh T. 2012. Integrated effect of enriched compost and fertilizer on yield and uptake of micronutrient by maize (*Zea mays* L.). Agric. Sci. Digest 32, 43–47.

Sommer AL, Lipman CB. 1926. Evidence on the indispensable nature of Zinc and boron for higher green plants. Plant Physiol. 1, 231–249.

Srinivasagam K, Natarajan B, Raju M, Kumar SR. 2013. Myth and mystery of soil mycorrhiza: a review. Afr. J. Agric. Res. 8, 4706–4717.

Srivastava PP, Pandiaraj T, Das S, Sinha AK. 2017. Assessment of micronutrient status of soil under Tasar host plant growing regions in Jashpur district, Chhattisgarh State. Imp. J. Interdiscip. Res. 3, 1080–1083.

Sunithakumari K, Padmadevi SN, Vasandha S. 2016. Zinc solubilizing bacterial isolates, from the agricultural field of Coimbatore, Tamil Nadu, India. Current sci. 110, 196–205.

Zeng H, Wu H, Yan F, Yi K, Zhu Y. 2021. Molecular regulation of zinc deficiency responses in plants. J. Plant Physiol. 261, 153419.

Zia MH, Ahmad R, Khaliq I, Ahmad A, Irshad M. 2006. Micronutrients status and management in orchards soils: Applied aspects. Soil Environ. 25, 6–16.

17 Soil-Plant Nutrition with Zinc in Tropical Tuber Crops

K. Susan John

17.1 INTRODUCTION

In light of the current state of the environment, it is important to focus on soil productivity, food security, high-quality tuber starch production and climate resistance. Millennium crops include tropical tubers such as cassava (*Manihot esculenta* Crantz), sweet potato (*Ipomoea batatas* L. Lam), yams (*Dioscorea sp.*) and aroids such as elephant foot yam (*Amorphophallus peaoniifolius*), taro (*Colocasia* sp.) and tannia (*Xanthosoma* sp.). Due to their role as a staple food for many nations and as a primary or supplementary food for one-fifth of the world's population, they are recognised as the third most important food crop after cereals and grain legumes. These crops are powerful due to their increased biological efficiency, availability of a low-cost energy source, particularly for the weaker segments of the population, and capacity to grow in challenging soil and climate conditions. These qualities make these crops perfect for growing in the less developed nations of Asia, Africa and Latin America, where they are now regarded more and more as sources of employment and income in addition to being crops for food security.

Cassava, popularly known as tapioca, is a tropical root and tuber crop native to South America that ranks first in terms of area and production worldwide. Manure and fertilizer applications are effective in tropical tuber crops, such as cassava (John et al., 2005). These crops have exceptional productivity compared to field crops and result in very high nutrient uptake, implying significant nutrient loss from the soil. This leads to nutrient replenishment from organic or inorganic sources. Zinc (Zn) is one of the essential micronutrients for these crops, particularly in lowering the cyanogenic glucoside level that gives cassava tubers their bitter taste besides tuber productivity and quality. These unique properties of Zn are related to its function in the enzymes necessary for photosynthesis, root respiration, oxidation, translocation and transport in plants. According to John et al. (2016), removal of Zn from the soil by cassava, sweet potato, yams, elephant foot yam, taro and Chinese potato yielding 30,18,18–30,33,17 and 26 t ha^{-1} is equivalent to 0.81, 0.46, 0.33–0.53, 0.49,0.65 and 0.40 kg ha^{-1} respectively.

DOI: 10.1201/9781003412472-17

17.2 CASSAVA

The most significant tropical tuber crop is cassava (*Manihot esculenta* Crantz; Family: Euphorbiaceae) which has a large area under cultivation, high productivity, tolerance to abiotic and biotic stresses, good quality storage roots (tuber) and starch. This can be used to make many high-value industrial products, such as food, feed, adhesives, pharmaceuticals, ethanol, super absorbent polymers and biodegradable plastics. On a global scale, 25.59 million ha area of cassava are found, yielding 277.81 million tonnes at a productivity of 11.30 t ha^{-1} (FAO, 2019). The tuber yield can be raised by 32% by managing low soil fertility, which is thought to be one of the restrictions in cassava production (Henry and Gottret, 1996).

17.3 Zn APPLICATION ON TUBER YIELD AND TUBER QUALITY

In acid laterite soils, NPK @ 100 kg ha^{-1} and ZnSO$_4$ @ 12.5 kg ha^{-1} increased tuber yield, tuber starch (fresh weight basis) and tuber cyanogens as 29.4 t ha^{-1}, 29.6% and 90.3 μg g^{-1} respectively. This indicates a 16% and 7% increase in tuber yield and starch respectively and a 24.5% decrease in HCN content as a result of Zn application.

17.4 EFFECT OF SOIL Zn UNDER DIFFERENT NUTRIENT MANAGEMENT PRACTICES

The Permanent Manurial Trial (PMT) at ICAR-CTCRI, where a cassava crop was fertilized with various combinations of N, P and K both alone and in conjunction with farm yard manure (FYM) continuously for 14 years starting from 1977, revealed the need for research to maintain the status of soil Zn (Kabeerathumma et al., 1993; John et al., 1998). Treatment with Phosphorus alone (0.14 μg g^{-1}) and combination treatments without FYM and K had the lowest level of available Zn. When cassava is continuously grown in the same field with major nutrients alone, the soil Zn status decreases from an initial level of 0.82 μg g^{-1} to 0.47, 0.49, 0.55 μg g^{-1} in the case of N alone, P alone and K alone treatments, with absolute control recording a Zn content of 0.22 μg g^{-1} (John et al., 2005). This indicates the need to replenish Zn. Based on this result, other crops also underwent Zn nutrition studies.

17.4.1 Continuous Application of Blanket Doses of Zn

After observing depletion of Zn due to continuous application of major nutrients, ZnSO$_4$ @12.5 kg ha^{-1} alone and together with MgSO$_4$ @20 kg ha^{-1} along with Package of Practices (PoP) was included as treatments in the second phase of the Long Term Fertilizer Experiment (LTFE) (1990–2005) in cassava. It was found that, independent application of Zn was better in tuber yield, quality and the soil's Zn status. As a result of the continuous ZnSO$_4$ alone treatment, the soil's Zn status was raised over the critical level of 0.6 μg g^{-1}, but no discernible improvement in tuber production was noticed (Figure 17.1).

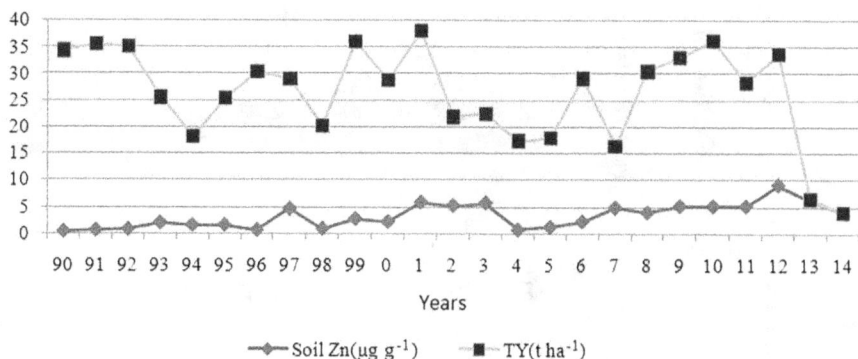

FIGURE 17.1 Relationship between tuber yield and soil Zn status over a period of 25 years in an Ultisol of Kerala, India.

Even though no discernible impact of Zn application with blanket recommendation (BR) on the characteristics of the quality of the tubers, soil Zn, leaf Zn, stem Zn and total plant Zn uptake were all significantly higher (Table 17.1) (John et al., 2015).

17.4.2 SOIL TEST-BASED APPLICATION OF Zn

The Zn dose to be applied was determined based on its soil status, soil critical level and plant requirement and included in the PoP recommendations of Kerala Agricultural University, Kerala, India (KAU, 2011, 2016). This was done using ten years of data under LTFE on Zn nutrition of cassava.

According to John et al. (2010), the application rates of the same were $ZnSO_4$ @ 12.5,10,7.5, 5.0 and 2.5 kg ha^{-1} when the soil Zn status was 0.2, 0.2–0.3, 0.3–0.4, 0.4–0.6 and >0.6 µg g^{-1}. In the third phase, a maintenance dose of $ZnSO_4$ @ 2.5 kg ha^{-1} was applied, when the Zn level was above 0.6 µg g^{-1}, based on the results of the second phase (non-response to BR of Zn). While there was no significant impact on tuber yield under STB recommendation (T3), the overall effect during these years clearly demonstrated that, Zn application could result in a significant reduction in HCN and an increase in the starch content of cassava tubers (T1 and T2 are the same as in the previous second-phase treatments). In the pilot study conducted at Ultisol in Kerala, India, it was shown that, the overall Zn uptake and the same in the leaves, stems and tubers had greatly increased (Table 17.1).

Based on the above study, $ZnSO_4$ (7.5–10 kg ha^{-1}) was recommended for the low and uplands of Kerala's nine main cassava-growing soils (John et al., 2014a) to improve the status of nutrient to increase the production and quality of the tuber.

Field validation trials on soil test-based Zn application in the districts of Pathanamthitta and Kollam of Kerala, five hectares of fields were used to show the effect of Zn. The diverse cropping pattern, variation in the kind and quantity of manures and fertilizers applied, and difference in the type of crops produced by farmers prior to soil sampling were all related to the selected areas' significant variation in Zn status. Its status ranged from very low to low with values between 0.235 and 0.851 µg g^{-1}. According to the technique proposed by Howeler (1996), it was necessary to apply Zn

TABLE 17.1

Response of Cassava to Zn Nutrition Over Years under Blanket and Soil Test-based Application (Pooled mean) in an Ultisol of Kerala, India

Duration	Treatments	TY (t ha^{-1})	HCN (µg g^{-1})	Starch (%) (Fresh weight basis)	Soil Zn (µg g^{-1})	Leaf Zn (µg g^{-1})	Stem Zn (µg g^{-1})	Tuber Zn (µg g^{-1})	Plant Zn uptake (g ha^{-1})
1990–2004 (BR)	T1	28.56	54.0	23.51	1.04	62.0	27.2	23.7	520
	T2	24.82	62.5	22.28	0.98	58.9	26.8	19.9	463
	T3	26.16	50.9	23.96	2.57	76.9	36.6	24.4	686
	CD	2.443	NS	NS	0.883	13.47	5.13	NS	105.3
2005–2014 (STB)	T1	27.87	61.1	20.93	3.67	76.0	41.5	25.2	823
	T2	26.68	81.2	21.42	2.69	66.8	35.2	21.8	732
	T3	24.89	46.9	23.08	4.47	78.6	45.0	29.6	754
	CD	NS	NS	1.473	0.959	NS	6.40	NS	NS
1990–2014 (Overall)	T1	28.27	56.93	21.96	2.14	67.8	33.69	24.35	646
	T2	25.59	70.30	21.76	1.69	62.2	30.59	20.71	575
	T3	25.63	49.25	23.43	3.36	77.6	40.41	26.62	714
	CD	1.740	15.032	1.110	0.636	9.45	3.921	4.632	88.4

(Susan John et al., 2019)

BR: Blanket Recommendation; STB: Soil Test Based; TY: Tuber Yield; HCN: Cyanogenic Glucosides

CD: Critical Difference.

based on the Zn requirements of cassava as well as its condition in these soils. Table 17.2 outlines the tuber yield, cyanogens and starch content that were found under trial settings based on a soil test. It was evident that STB applications of Mg (T3) (NPK @ 82:6.3:68 kg ha^{-1}+FYM @6.8 t ha^{-1}+MgSO$_4$ @14.50 kg ha^{-1}) and Zn (T4) (NPK @82:6.3:68 kg ha^{-1}+FYM @ 6.8 t ha^{-1}+ZnSO$_4$@ 10.50 kg ha^{-1}) were better than other treatments (Table 17.2).

A low-input management strategy involving soil test-based application of Zn, application of N, P, K, MgSO$_4$ and ZnSO$_4$ depending on soil test results, was one of the items in the low input management (LIM) method that was established. In the first year, they were applied at 106: 0: 83: 20: 2.5 kg ha^{-1}, and in the second year, at 106: 0: 94: 10: 2.5 kg ha^{-1}. The tuber yield under the management practises was 28.34, 27.53, 29.61 and 28.73 t ha^{-1} for PoP, soil test-based NPK +FYM, PoP+ biofertilisers and low input management strategy respectively. The LIM strategy with soil test-based administration of Zn was comparable to other management practises. The percentage increase in input costs under the above practises over the low input practise varied significantly up to 55% with a BC ratio of 4.41. The latter practise could save P, K, Mg and Zn to the tune of 100%, 11.5%, 62.5% and 80%, respectively (Shanida, 2017; Shanida et al., 2015; John et al., 2016, 2018 and 2019).

TABLE 17.2

Tuber Yield, Quality and Economic Parameters as Influenced by the Treatments (Mean of 13 Locations)

Treatments	Treatment description	Tuber yield (t ha^{-1})	Tuber quality parameters			Economics	
			Dry matter (%)	Starch (%) (Fresh weight basis)	Cyanogenic glucosides (ppm)	Net income (Rs/ha)	BC Ratio
T1	Farmers practice	28.946	35.532	21.477	56.115	5914	1.09
T2	Package of Practices	33.184	35.412	22.957	53.577	17448	1.27
T3	Soil test based NPK + FYM + Mg	38.844	37.88	24.195	37.885	37099	1.62
T4	Soil test based NPK + FYM + Zn	42.187	38.031	23.081	34.077	45125	1.75
T5	Soil test based NPK+FYM	34.625	35.276	21.156	43.615	26871	1.45
CD (0.05)			3.348	0.601	1.108	3.751	–	–

(Susan John et al., 2010, 2018, 2019).

Zn-enriched cassava starch factory solid waste (*thippi*) compost was made in nutrient recycling by combining FYM, *Glyricidia*/cassava leaves, Mussoriphos, and rock powder with vermincomposting. Chithra et al. (2016) and John et al. (2018) made *thippi* (cassava starch factory solid residue) compost with a very high C:N ratio of 8:1 and Zn content 89.93 $\mu g\ g^{-1}$ higher than *thippi* having a C: N ratio of 82:1 and Zn content of 7.8 $\mu g\ g^{-1}$. In cassava, the effect was that, the same @ 3.87 t ha^{-1} along with soil test-based application of NPK @ 78:0:48 kg ha^{-1} and FYM @ 5 t ha^{-1} (24.436 t ha^{-1}) can replace external application of $ZnSO_4$ @2.5 kg ha^{-1} along with these applications.

17.4.3 Diagnosis and Correction of Zn Deficiency

Although Zn deficiency rarely affects the cassava crop, it is undoubtedly caused by its ongoing monoculture (John et al., 2005). In the permanent manurial experiment, young leaves under N and NP treatments at about 7–8 months stages of the crop showed well-developed Zn deficiency signs as indicated by the reduction of leaf lobes and a widening of the angle of lobes. As a result, the stem eventually appeared very thin with only a few rudimentary leaves at the apex (Kabeerathumma et al., 1993; John et al., 1998). The senescence of the damaged leaves occurred more significantly and quickly.

The younger leaves developed a distinctive interveinal chlorosis due to Zn shortage. Between the veins, little white or light yellow chlorotic patches were formed and additional leaves with smaller size and more chlorotic areas were produced. The borders of the leaf lobes cupped upward and grew narrow, with a light green to white colour. Under nutrient-stress conditions, leaf tips turned necrotic. Frequently, the basal lobes turned away from the stem. Zn deficiency significantly decreased plant growth and yield because it most significantly influenced the growing point. According to Nair et al. (1988), it may be caused by the decrease of Zn availability in soil and Zn-P antagonism. The concentration of Zn falls below the usual limit of 35 $\mu g\ g^{-1}$ in the index leaf, which was the youngest fully expanded leaf (YFEL) at higher P levels. $ZnSO_4$. $7H_2O$ @12.5 kg ha^{-1} could be used to treat the Zn deficiency under these circumstances. In order to effectively control Zn shortage, John et al. (2006) suggested dipping stakes in a 2–4% solution of $ZnSO_4$ for 15 minutes before planting or applying 1–2% solution of $ZnSO_4$ topically to the leaves.

17.4.4 Critical Level of Zn in Soil and Plants

According to Howeler (1983), the critical Zn level in soil for cassava production was set as 0.5, 0.5–1.0, 1–5, 5–50 and >50 $\mu g\ g^{-1}$, which correspond to very low, low, medium, high and very high respectively. The critical Zn content was fixed as 25, 30–60 and >120 $\mu g\ g^{-1}$ as the deficiency and sufficiency ranges as well as toxicity levels for Zn in cassava plants, according to Asher et al. (1980). The YFEL petiole or leaf blade at 3–4 month stage is the best indicator of the nutritional status of the crop.

17.5   Zn STATUS OF CASSAVA-GROWING SOILS OF KERALA

An assessment of the nutrient status of Kerala's cassava-growing soils was conducted between 2003 and 2006. The Zn status and other soil chemical parameters of the State's major cassava growing districts viz. Thiruvananthapuram, Kollam, Pathanamthitta, Alappuzha, Kottayam, Idukki, Ernakulam and Palakkad (both lowland and uplands) were also determined.

Based on the figures from the Farm Guide (2004), the Zn status of the soils used to grow cassava in the nine districts with an area of over 5000 hectares spread across 76 panchayats of the 45 blocks was examined as a preliminary step prior to soil test based micronutrient recommendations. Based on the findings, the soils were categorized into sufficient, very low, low, medium, high and very high soils, keeping the general soil critical level of Zn as 0.6 μg g^{-1} (Dev, 1997).

The Zn status of soils used to grow cassava in the aforementioned sites ranged from 0.619 to 2.836 μg g^{-1}, and there were few differences across the districts or between the upland and lowland soils within the same district. The lowest soil available Zn status (0.619 μg g^{-1}) and highest (2.836 μg g^{-1}) were found in Ernakulam and Idukki districts respectively. Based on the examination of 21, 953 soil samples from the districts of Pathanamthitta and Kottayam. John et al. (2012 and 2013) found that, 56 panchayats in Pathanamthitta had high levels of accessible Zn, with 90% of the samples having enough Zn levels ranging from 1.4 to 31.6 μg g^{-1} (with a mean value of 5.6 μg g^{-1}). By examining the Zn status of the 7 panchayats, a nutrient management plan was created for one block (Elanthur) in the Pathanamthitta district and found that, the Zn status ranged from 2.90 to 4.30 μg g^{-1} which was sufficient (John et al., 2012; 2014b). Out of the 11,605 samples evaluated in the Kottayam district, 95% had sufficient available Zn levels (Geetha et al., 2013). The Kerala State Planning Board (KSPB, 2013) and Rajasekharan et al. (2014) revealed that, the Zn deficiency was minor in Kerala as 87% of the samples indicated appropriate soil Zn levels based on the field data of roughly 1,54, 531 soil samples collected from 14 districts. Soil lacking in Zn was only found on the eastern Wayanad plateau (Figure 17.2).

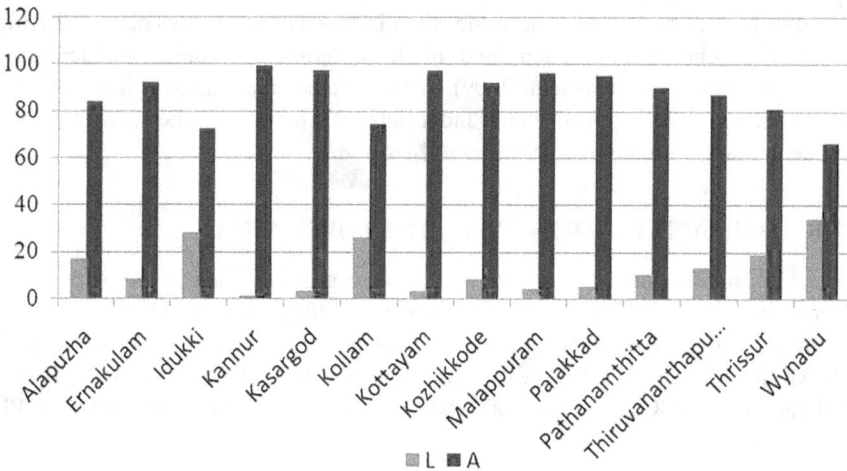

FIGURE 17.2   Percentage of the tested samples with low and adequate Zn in the different districts of Kerala, India.

17.6   EFFECT OF Zn IN THE MANAGEMENT OF CASSAVA MOSAIC DISEASE (CMD)

Both under controlled conditions and in the field, the impact Zn in managing CMD was evaluated. In the most vulnerable variety, Sree Visakham, soil application of $ZnSO_4$ @1 g per plant within a month of top dressing and foliar application (0.1%) every other week from 3MAP to 6MAP under controlled conditions showed that Zn had an impact on tolerance to CMD. The PCR examination of the viral load in the treated plants at various times showed a similar pattern, with Zn producing the greatest outcomes. As regards to other nutrients, Ca, Mg, K, Cu, Zn, B, Mn and Si, Zn application had the best response and the lowest incidence of CMD. Among the several nutrients tested, Zn treatment produced the maximum tuber production. Field tests with the most susceptible cassava variety H226 in 2016–17 showed better results in lowering the infection to 68% from 86% under absolute control by applying Zn (CTCRI, 2017).

17.7   SWEET POTATO

Sweet Potato (*Ipomoea batatas* L. Lam; Fam: Convolvulaceae), is another major crop mostly grown for its carbohydrates. Protein, minerals, vitamins and antioxidants are abundant in the crop. One week of sprouting until 6MAP on alternate weeks, Zn application, increased tuber output by 16%, reaching 24.93 t ha^{-1}, placing Zn next to boron (tuber yield; 26.68 t ha^{-1}). Zn controlled the amount of antioxidants in the orange and purple-fleshed sweet potatoes. The tubers are safe for diabetics due to the starch's low glycaemic index.

In the humid tropical and subtropical regions of the world, sweet potatoes are a widely grown resilient and nutrient-rich staple crop. Subsequent thickening of the upper portion of some feeder roots produces edible roots or tubers that have enormous potential for use in the food industry. With a 77% area share and an 82% production share, it is mostly grown as a rainfed crop in Eastern India particularly in Odisha, West Bengal, Uttar Pradesh, Bihar and Jharkhand (Edison et al., 2009). It is placed fifth in terms of economic value, sixth in terms of dry matter output, seventh in terms of energy production, and ninth in terms of protein production in developing nations (Lobenstein, 2009). After potatoes and cassava, it is the third most significant tuber crop in India. India ranks 12, 8 and 5 in the world for area, production and productivity, respectively.

17.8   SOIL APPLICATION OF Zn ON TUBER YIELD

In general, soils sufficient in Zn are sweet potato-growing laterites. An experiment with various levels of Zn was done to standardise the best application rate. Based on tuber yield statistics, the optimum was determined to be Zn @ 6.3 kg ha^{-1} (30 kg ha^{-1} $ZnSO_4$). The best results were obtained with an independent application of $ZnSO_4$ @ 20 kg ha^{-1}, which could begin after one week of NPK top dressing (after 45 DAP).

17.9   FOLIAR APPLICATION OF Zn ON TUBER YIELD

Foliar nutrition is typically applied to crops either to correct nutritional abnormalities or enhance crop growth and output. At the ICAR-CTCRI, foliar application of Zn in various combinations was undertaken regulating in greater tuberization with 0.1% Zn EDTA @ 500–750 g ha^{-1}, depending on the crop's needs as determined by a soil test or symptom manifestation. As a standard guideline, two sprays of 0.1% Zn EDTA spaced at 20 days apart was adopted during the crop's peak vegetative development stage (starting at 1MAP) followed by two sprays spaced 20 days apart during the tuber bulking stage (starting at 65 DAP). These may also be used in conjunction with other major nutrients N, P and K as a foliar application of 19:19:19 containing 19% each as well as ZnEDTA as 1% @ 625 litres ha^{-1} during the peak vegetative growth stage twice and at the tuber bulking stage twice at an interval of 15–20 days (CTCRI 2018; CTCRI, 2019). The soil and foliar treatments were successful in raising tuber production by up to 15–25% over (PoP) (Aparna, 2017; CTCRI, 2017).

17.10   Zn DEFICIENCY AND MANAGEMENT

The index leaf is the youngest fully expanded leaf (YFEL) at one month after planting (MAP) and 11 µg g^{-1} of Zn is stated to be the essential concentration (Sullivan et al., 1997). ICAR-CTCRI carried out a sand culture experiment under controlled conditions to simulate Zn deficiency symptoms resulting in thickened but not distorted leaves with serious plant growth limitation, including shortening of internodes, chlorosis of young leaves, etc., that were nearly the same as what was noticed under field conditions (Pillai et al., 1986). The Zn concentration in the plant tissues was 33.6 µg g^{-1} compared to control Zn deficiency under field settings. Treatment with foliar spraying of 1–2% $ZnSO_4.7 H_2O$ solution and soaking cuttings in 2–4% of the same for 15 minutes prior to planting offered good results (John et al., 2006; 2018).

17.11   YAMS

Yams are regarded as the most nutritious of the tropical root crops (Wanasundera and Ravindran, 1994) containing four times as much protein as cassava, as well as higher levels of essential amino acids, fibre, minerals and vitamins A and C. They also have lower levels of anti-nutritional factors like phytate and trypsin inhibitors. White yam (*Dioscorea rotundata* Poir), greater yam (*Dioscorea alata* L.) and lesser yam [(*Dioscorea esculents* (Lour.) Burk.)] are the three most popularly grown edible yams species. Due to their superior drought tolerance and high biological efficiency in arid areas, they are the main food crop in African nations. The mucilage found in yams is hypoglycaemic and has the power to lower blood sugar levels. Yams with purple flesh are high in anthocyanins.

17.12   Zn CONTENT IN DIFFERENT PARTS OF YAMS

Kabeerathumma et al. (1987) determined the critical nutrient content of various yam species in an Ultisol. The study showed that, the first fully mature leaf could be used as an index leaf, Zn concentrations in *D. alata*, *D. rotundata*, and *D. esculenta* were 24, 22 and 21 μg g^{-1}, respectively. The study also revealed that, the Zn contents in the leaves, vines, roots and tubers of the above species as values at 3,5,7 and 9 MAP with a significant decline with crop maturity. According to Kabeerathumma et al. (1987), *D. esculenta* and *D. rotundata* showed a minor drop in leaf and vine Zn at 7 MAP and an increasing trend in all of the plant parts at the harvest stage. The build up may be caused by senescence and the subsequent transfer of Zn from aging leaves to growing portions, whereas the low concentration of Zn during vigorous vegetative development may be due to dilution impact. Zn uptake was highest in *D. esculenta* (0.531 kg ha^{-1}), followed by *D. rotundata* (0.456 kg ha^{-1}) and *D. alata* (0.327 kg ha^{-1}) with tuber uptake of Zn as low as 0.286 kg ha^{-1} in *D. rotundata* 0.278 kg ha^{-1} in *D. esculenta* and 0.134 kg ha^{-1} in *D. alata* (John et al., 2018).

17.13   ELEPHANT FOOT YAM (EFY)

Due to therapeutic qualities and the presence of phenols and alkaloids, elephant foot yam (*Amorphophallus peaoniifolius;* Family: Araceae) is important among the aroids. It is a crucial component in numerous Ayurvedic medicines. The crop's high yield (30–70 t ha^{-1}) and tolerance of shade made it an intercrop in plantations. It is known as the 'King of Tuber Crops' due to the large size of its corms, which can weigh up to 20–30 kg. It is perennial and herbaceous originating from South East Asia. Many nations of South East Asia including the Philippines, Java, Indonesia, Sumatra, Malaysia, Bangladesh, India, China, it is a native staple cuisine. In India, is grown in northern and eastern States including Andhra Pradesh, West Bengal, Gujarat, Kerala, Tamil Nadu, Maharashtra, Uttar Pradesh and Jharkhand.

The primary edible 'corm' is a stem alteration that produces little cormels, the quantity of which is a varietal characteristic. Corm, cormels, immature petioles and unopened inflorescence of cultivated species can be used as vegetables. In addition to flavonoids and fibre, they are high in protein, carbohydrates and minerals such as Ca, Fe and P besides vitamins A, B and C.

17.14   EFFECT OF SOIL AND FOLIAR APPLICATION OF Zn ON TUBER YIELD

The nutrient level experiment was carried out in EFY under open conditions as sole crop, revealed that, the ideal rate of Zn administration was 4.73 kg ha^{-1} (ZnSO$_4$ @ 22.5 kg ha^{-1}) that might go up to 20–30 kg ha^{-1} as soil application based on its state and critical Zn level. The studies carried out at ICAR-CTCRI suggested that, foliar application of Zn EDTA (0.5%), twice at an interval of one month during peak vegetative growth and tuber bulking stage be used to correct the disorder or improve crop growth and yield (CTCRI, 2017; CTCRI, 2018). Based on this advice, soil, soil + foliar and foliar applications with PoP (NPK @ 100:50:150 kg ha^{-1}) reported yields of 3.998, 3.135 and 3.765 kg plant^{-1} respectively which were 10–12% higher than PoP alone treatment (CTCRI, 2017).

17.15   Zn CONTENT IN DIFFERENT PLANT PARTS OF EFY

The threshold tissue Zn content of the index leaf was found to be 121 $\mu g\ g^{-1}$ and the same in leaf, pseudo-stem, tuber and root of the plant at growth stages of 3, 5, 7 and 9 MAP showed a decline with maturity due to the active uptake in the early stages. The content of the tuber was 60 $\mu g\ g^{-1}$ and its removal via the entire crop was 0.49 kg ha^{-1} with 0.085 kg ha^{-1} through the crop residue (Kabeerathumma et al., 1987).

17.16   TARO

This is a herbaceous plant, also known as 'eddo' or 'dasheen' (*Colocasia esculenta*; Family: Araceae) and cultivated for its edible leaves, corms and cormels. It originated from South Eastern Asia and moved to the Pacific islands where it established itself as a major crop. It is a well-liked tuberous vegetable produced as an insurance crop in plantations to earn decent revenue and is a great source of many minerals, fibre and carbohydrates.

17.17   Zn CONTENT IN DIFFERENT PLANT PARTS AND Zn UPTAKE

The index leaf has a threshold Zn content of 22 $\mu g\ g^{-1}$ (acceptable range of 22–50 μg g^{-1}) (Sullivan et al., 1996). The percentage removal of Zn by its various parts (leaf, petiole, root and tuber) at different growth stages (30, 60, 90, 120 and 150 days) after planting showed a tendency for Zn content to increase with the increase in crop age, with the peak concentration being observed at harvest. The root has the highest Zn concentration (107 $\mu g\ g^{-1}$) and the leaves the lowest. All the other plant parts showed decreased Zn removal as the plants grew. With maturation, the percentage elimination of Zn also grew, reaching a peak between 120 and 150 DAP. Zn absorption by plants as a whole was estimated to be 647 g ha^{-1} (Kabeerathumma et al., 1987).

17.18   CHINESE POTATO

Chinese potato or Coleus (*Solenostemon rotundifolius*; Family: Lamiaceae) is a herb with aromatic leaves and succulent stalks that clusters tubers at the base of the stems. The tubers are aromatic, alkaloid and polyphenol-rich as well as high in minerals including Ca and Fe. It is said to be an Indian native and produces 10–20 t ha^{-1} of tubers in 4–6 months. As per the experiments on the Zn concentration of the crops' various plant components (leaf, vine, root and tuber), the Zn content of leaves dropped with age but was higher than that of vine and root throughout the first two MAP. The Zn concentration of the stem and root did not alter significantly as the plant grew. Tuber Zn was somewhat lower at the beginning phase and rose until harvest from 3 MAP. The total Zn intake was 397 g ha^{-1} and its content in the tubers was 31 $\mu g\ g^{-1}$ (Kabeerathumma et al., 1987).

17.19   CONCLUSION

Zinc is one of the most important of the 17 essential plant nutrients for crop growth, development and yield as it is a component of many enzymes and co-enzymes of both plant and animal systems. It is found in the body cells and essential to the immune system besides cell proliferation, cell division, wound healing and carbohydrate breakdown. It is needed for taste and smell as well as to power numerous metabolic processes in all crops. The absence of Zn in enzymes of plant tissues affects crop growth, development and yield. In plants lacking Zn, production of carbohydrates, proteins and chlorophyll is dramatically reduced.

Plants with low Zn levels have stunted development, fewer tillers, chlorosis, smaller leaves, longer crop maturation times, spikelet sterility and lower quality of harvested produce. This chapter discussed the role of Zn in tropical tuber crops' growth and productivity, quality of the tubers and soil productivity with regard to its enrichment. The application of Zn, particularly in bitter cassava cultivars having high cyanogenic glucosides decreased the latter and boosted starch content. The technique that not only increased in the tuber yield but also decreased symptom manifestation of cassava mosaic disease in leaves may be considered as socially acceptable practice. An academic approach for future students, planners and policymakers in the implementation of greater Zn-centric research and extension activities connected to crop and soil productivity is the need of the hour in cassava growing lowland and uplands of Kerala.

However, further studies are required to examine the effects of Zn delivery through various modalities and fortification. With more practical programmes, the objectives can be implemented quickly as this is an urgent necessity since the tuber crops serve as a mainstay for the nation's nutritionally challenged citizens.

REFERENCES

Aparna S. 2017. Secondary and micronutrients on soil health and plant growth under sweet potato. MSc. Thesis, Mahatma Gandhi University, Kerala, India, p. 89.

Asher CJ, Edwards DG, Howeler RH. 1980. Nutritional Disorders of Cassava (*Manihot esculenta* Crantz). University of Queens land, St. Lucia, Australia, p. 48.

Chithra S, Susan John K, Sreekumar J. 2016. Low cost traditional cassava starch factory solid waste (thippi) composting: A possible strategy for organic nutrient management and economic security for tribal farmers. J. Root Crops. 42, 52–58.

CTCRI. 2017. ICAR-Central Tuber Crops Research Institute, Thiruvananthapuram, Kerala, Annual Report, pp. 46, 47, 62.

CTCRI. 2018. ICAR-Central Tuber Crops Research Institute, Thiruvananthapuram, Kerala, Annual Report, pp. 40–41.

CTCRI. 2019. ICAR- Central Tuber Crops Research Institute, Thiruvananthapuram, Kerala, Annual Report, pp. 40.

Dev G. 1997. Soil fertility evaluation for balanced fertilization. Fert. News 42, 3–34.

Edison S, Hegde V, Makesh Kumar T, Srinivas T, Suja G, Padmaja G. 2009. Sweet potato in the Indian sub continent, in: Lobenstein G, Thottappill G (Eds.), The Sweet Potato. Springer Science Publishers, pp. 391–414.

FAO. 2019. Agricultural Outlook 2019-2028. FAO Statistics Division, Food and Agriculture Organization, Rome, p. 12.

Farm guide. 2004. Area under crops, 2001-02. Farm Information Bureau, Government of Kerala.

Geetha K, Susan John K, Dhanyamol K, Praveena N, Radhika TA, Manikantan Nair M, Joseph J, Paul S, Anuroopa S. 2013. Soil fertility: Kottayam district, in: Rajasekharan P,

Nair KM, Rajasree G, Suresh Kumar P, Narayanan Kutty MC (Eds.), Soil Fertility Assessment and Information Management for Enhancing Crop Productivity in Kerala. Kerala State Planning Board, Thiruvananthapuram, Kerala, pp. 284–304.

Henry G, Gottret V. 1996. Global Cassava Trends: Remaining the Crops Future. CIAT working document No.157. Centro Internacional de Agricultura Tropicale, Cali, Columbia.

Howeler RH. 1983. Analisis del tejidovegetalen el diagnostico de problems nutritionales de algunas cultivos tropicals. Centro International de Agricultura Tropicale (CIAT), Cali, Colombia, ISBN 84-89206-35-X.

Howeler RH. 1996. Diagnosis of nutritional disorders and soil fertility management of cassava, in: Kurup GT, Palaniswami MS, Potty VP, Padmaja G, Kabeerathumma S, Santha VP (Eds.), Tropical Tuber Crops: Problems, Prospects and Future Strategies, Oxford and IBH Publishing Co Pvt., Ltd., New Delhi, pp. 181–193.

John KS, Mohan Kumar CR, Ravindran CS, Prabhakar M. 1998. Long term effect of manures and fertilizers on cassava production and soil productivity in an acid Ultisol, in: Swarup A, Reddy DD, Prasad RN (Eds.), Proceedings of the National Workshop on Long term Soil Fertility Management through Integrated Plant Nutrient Supply. Indian Institute of Soil Science, Bhopal, India, pp. 318–325.

John KS, Ravindran CS, George J. 2005. Long Term Fertilizer Experiments- Three Decades Experience in Cassava. Technical Bulletin Series No. 45, Central Tuber Crops Research Institute, Thiruvananthapuram, Kerala, India.

John KS, Suja G, Edison S, Ravindran CS. 2006. Nutritional Disorders in Tropical Tuber Crops. Technical Bulletin Series No. 48, Central Tuber Crops Research Institute, Thiruvananthapuram, Kerala, India.

John KS, Nair MM, Paul S, Anuroopa S, Geetha K Dhanyamol K, Praveena N, Radhika TA. 2013. Soil fertility: Pathanamthitta district, in: Rajasekharan P, Nair KM, Rajasree G, Kumar PS and Kutty MCN (Eds.), Soil Fertility Assessment and Information Management for Enhancing Crop Productivity in Kerala. Kerala State Planning Board, Thiruvananthapuram, Kerala, pp. 377–399.

John KS, Ravindran, CS, Suja G, Prathapan K. 2010. Soil test based fertilizer cum manurial recommendation for cassava growing soils of Kerala. J. Root Crops 36, 44–52.

John KS, Bharathan R, Manikantan Nair M, Suja G. 2012. Soil based nutrient management plan for Pathanamthitta district of Kerala state. J. Root Crops 38, 51–59.

John KS, Ravindran CS, Manikantan Nair M, George J. 2014a. Nutrient Status of Cassava Growing Soils of Kerala. Technical Bulletin Series No.56, Central Tuber Crops Research Institute, Thiruvananthapuram, Kerala, India.

John KS, Ravindran CS, Manikantan Nair M, George J. 2014b. Soil Test Based Fertilizer cum Manurial Recommendation for Cassava Growing Soils of Kerala. Technical Bulletin Series No. 58, Central Tuber Crops Research Institute, Thiruvananthapuram, Kerala, India.

John KS, Ravindran CS, George J. 2015. Soil test and plant analysis as diagnostic tools for fertilizer recommendation for cassava in an Ultisol of Kerala, India. Commun. Soil Sci. Plant Anal. 46, 1607–1627.

John KS, George J, Shanida Beegum SU, Shivay YS. 2016. Soil fertility and nutrient management of tropical tuber crops-An overview. Ind. J. Agron. 61, 263–273.

John KS, Anju PS, Suja G, Shanida Beegum SU, Chithra S, George J, Ravindran CS, Manikantan Nair M, Mathew J. 2018. Zinc nutrition of tuber crops in ultisols of Kerala. Ind. J. Fert. 14, 50–57.

John KS, Anju PS, Suja G, Jeena Mathew, Shivay YS. 2019. Zinc nutrition in tropical tuber crops: A review. Ind. J. Agron. 64, 100–109.

Kabeerathumma S, Mohan Kumar B, Nair PG. 1987. Nutrient Uptake and their Utilization by Yams, Aroids and Coleus. Technical Bulletin Series No.10, Central Tuber Crops Research Institute, Thiruvananthapuram, Kerala.

Kabeerathumma S, Mohan Kumar CR, Nair GM, Nair G. 1993. Effect of continuous cropping of cassava with organics and inorganics on the secondary and micronutrient elements status of an acid Ultisol. J. Ind. Soc. Soil Sci. 41, 710–713.

KAU. 2011. Kerala Agricultural University. Package of Practices (POP) Recommendations: Crops, Fourteenth ed. Directorate of Extension, Kerala Agricultural University, Thrissur.

KAU. 2016. Kerala Agricultural University. Package of Practices (POP) Recommendations: Crops, Fifteenth ed. Directorate of Extension, Kerala Agricultural University, Thrissur.

KSPB, 2013. Kerala State Planning Board. Soil Fertility Assessment and Information Management for Enhancing Crop Productivity in Kerala, Rajasekharan P, Nair KM, Rajasree G, Suresh Kumar P, Narayanan Kutty MC (Eds.), Kerala State Planning Board, Thiruvananthapuram, Kerala.

Lobenstien G. 2009. Origin, distribution and economic importance, in: Lobenstein G, Thottappilly G (Eds.), The Sweet Potato. Springer Science Publishers, pp. 9–12.

Nair PG, Mohan Kumar BM, Prabhakar M, Kabeerathumma S. 1988. Response of cassava to graded doses of phosphorus in acid laterite soils of high and low P status. J. Root Crops 14, 1–9.

Pillai NG, Mohan Kumar B, Kabeerathumma S, Nair PG. 1986. Deficiency symptoms of micronutrients in sweet potato (*Ipomoea batatas* L). J. Root Crops 12, 91–95.

Rajasekharan P, Nair KM, Susan John K, Suresh Kumar P, Narayanan Kutty MC, Ajith RN. 2014. Soil fertility related constraints to crop production in Kerala. Ind. J. Fert. 10, 56–62.

Shanida BSU, Susan John K, Sheela MN, Sreekumar J. 2015. Low cost cassava production strategy through nutrient use efficient genotypes integrated with low input management. J. Root Crops 41, 42–48.

Shanida BSU. 2017. Low input management strategy in cassava: Implications of rhizosphere dynamics and carbon sequestration under changing global environment. Ph.D Thesis, Kerala University.

Sullivan JN, Asher CJ, Blamey FPC. 1996. Diagnostic criteria for nutrition disorders of root crops in the South Pacific, in: Proceedings of Australian Centre for International Agricultural Research Workshop, Nuku'alofa, Kingdom of Tonga, 17-20 April 1995, Canberra, Australia.

Sullivan JN, Asher CJ, Blamey FPC. 1997. Nutrient Disorders of Sweet Potato, in: Australian Centre for International Agricultural Research Monograph No. 48, Department of Agriculture, The University of Queensland, ISBN I 863202102.

Wanasundera JPD, Ravindran G. 1994. Nutritional assessment of yam (*Dioscorea alata*) tubers. Plant Foods Human Nutr. 46, 33–39.

18 Enrichment of Zinc by Encapsulation for Food Applications

F. Magdaline Eljeeva Emerald, B. G. Seethu,
P. Aditya Sukumar, Heartwin A. Pushpadass,
P. Abhinash, and Annamalai Manickavasagan

18.1 INTRODUCTION

Minerals that are essential for normal functions of the human body are categorized as micro- and macro-minerals depending upon the quantities required. Zinc is the second-most important microelement next to iron, and its role is considered as vital as it is needed by the body to perform primary metabolic activities and cellular metabolism, including catalytic, structural and regulatory roles. More specifically, it plays a vital role in cell growth, healing wounds, immune system function, blood clotting, mineralization of bones, brain development and cognitive functions (Mayo-Wilson et al., 2014). It is the only element and micronutrient that is a co-factor to more than 300 enzymes in the body (Ali et al., 2018) besides being a powerful therapeutic tool to manage many illnesses (Bhowmik et al., 2010). The role of zinc starts from white matter development in neonates and goes on to the entire life span of a human being. Hence, people deficient in zinc might experience increased susceptibility to diseases (Prasad, 1998). Its Recommended Dietary Allowance (RDA) is 11 and 8 mg/day for men and women respectively (D'Imperio et al., 2022).

As there is no systemic storage of zinc in the body, it needs to be restocked daily through dietary intake. The daily diet should supply an adequate amount of zinc to the body for better health, but its absorption rate depends on the origin of the food (Gibson, 2012). Though it is found in an expansive range of foods, it is present in higher concentrations in meat, seafood, poultry, eggs, nuts, seeds, legumes and whole-grain cereals (Sangeetha et al., 2022). Some vegetarian sources are also high in zinc content, but its absorption is a worriment due to native phytates in such foods. As zinc is carried through the body by serum albumin, its deficiency also leads to malabsorption. Furthermore, as most soils are becoming increasingly deficient in zinc, the grains harvested from such soils often lack it. Hence, nearly 31% of the global population is at the peril of deficiency in zinc (Das et al., 2019). People in South Asian countries are facing severe zinc deficiency in their population to the tune of 19 to 45% (Akhtar, 2013). Deficiency of zinc is thus a major

DOI: 10.1201/9781003412472-18

malnutrition problem now, particularly among young children. Supplementation of zinc is a key approach to address this malnutrition issue, predominant in most developing countries. Baldelli et al. (2021) demonstrated that microcapsules of zinc produced by spray-drying could be used for fortification of foods with the aim of alleviating its deficiency.

18.2 ADVANTAGES OF ENCAPSULATION

Direct addition of zinc to milk and other foods as a means of supplementation imparts bitter, metallic and astringent after-taste, and hence, a suitable vehicle is needed for its delivery. Micro- and nanoencapsulation could be used to wrap zinc inside wall materials, thereby masking its metallic taste besides enhancing its stability and shelf-life. In addition, the method also could help in the controlled release of zinc (Singh et al., 2010), which enables its better absorption in the gastrointestinal tract and bioavailability. Madene et al. (2006) also opined that encapsulation by physical methods is mainly preferred for the protection and efficient delivery of bioactives and minerals in food applications. A recent study concluded that microencapsulation was an attractive method to deliver zinc by fortification with better stability, without affecting the organoleptic qualities of fortified milk (Polekkad et al., 2021). Baldelli et al. (2023) reported that the absorption of iron in the body increased by 38% after microencapsulation.

The major methods of microencapsulation are spray and freeze drying, *in-situ* polymerization, extrusion, coacervation, liposome entrapment and electrospinning. The choice of a particular method is influenced by the properties of the active ingredient to be encapsulated (core) and coating material (encapsulant). It is also based on the release mechanism desired, prevailing processing conditions and the particle size of microcapsules (Choudhury et al., 2021).

Spray drying transforms the atomized liquid droplets of feed into a dry powder. It is effective in micro-encapsulation of the core ingredient and reduction of its biological and chemical degradation. The lower water activity of a dried product ensures good microbiological stability and a lower rate of chemical and/or biological degradation (Gharsallaoui et al., 2007). Also, the spray-dried microcapsules possess superior functional and reconstitution properties, which could be engineered by controlling the process parameters such as inlet temperature of hot drying air, feed rate liquid product and drying air, solid content in the atomized feed, ratio of carrier material to core ingredient, etc. Therefore, microcapsules with a mean diameter ranging from 2–15 µm are ideal for fortification of foods because this size allows uniform blending of the microcapsules in those foods (Gonçalves et al., 2016). In addition, the microscale dimensions of microcapsules produced by spray drying could facilitate in enhanced bioavailability of the encapsulated minerals or bioactive compounds (Arredondo et al., 2006). However, for successful microencapsulation of any bioactive such as zinc, an appropriate wall material or a combination of them needs to be selected and the spray drying conditions need to be optimized (Homayoonfal et al., 2022).

18.3   WALL MATERIALS FOR ENCAPSULATION OF ZINC

The encapsulant used as coating material should have adequate stability, good emulsifying ability and proclivity to form a stable and firm network so that separation of core material during spray-drying does not occur. The stability of microcapsules and encapsulation efficiency is thus decided by the protection offered by wall material. The wall material should be able to release the core material actively at the targeted site, and possess good rheological properties at higher concentrations so as to form a stable emulsion for drying in the spray dryer and should also be cheap, edible and possess "Generally Recognized AsSafe" (GRAS) status. Some common wall materials for encapsulation of minerals in a spray dryer are carbohydrates like starches, maltodextrin, chitosan, cellulose derivatives and proteins like whey protein isolate (WPI), gelatin, etc. (Diosady et al., 2002; Cho et al., 2015; Bai et al., 2017; Pratap-Singh and Leiva, 2021; Kaul et al., 2022).

18.4   MICROENCAPSULATION OF ZINC

For microencapsulation, the dispersion of zinc sulfate heptahydrate ($ZnSO_4.7H_2O$) was made using different wall materials namely, maltodextrin, HiCap-100 and WPI with biopolymer to zinc ratios of 10:1 and 20:1 and spray-dried. Gum Arabic was added to all three wall materials (encapsulants) at 0.5% (w/w) to enhance the binding between wall materials and zinc. Each wall material and zinc sulfate was dispersed individually in double-distilled water so that the total solid content in the feed was maintained at 30% (w/w). Prior to spray drying, the dispersion was homogenized using a high-shear homogenize rate of 15,000 rpm for 15 min.

The microcapsules of zinc sulfate were produced in a lab-scale co-current spray dryer (LU-222 Adv., LabUltima Process Technologies Pvt. Ltd., Mumbai) fitted with a twin-fluid atomizer, drying chamber, cyclone separators, hot air system, pre-filters and a bag filter. The experimental design consisted of spray drying 18 samples and the morphology and microstructural properties of zinc microcapsules were analyzed together with the physicochemical properties and encapsulation efficiency of zinc. The effects of three independent factors are, namely, drying-air temperature, nature of coating material (wall material) and ratio of wall material to zinc sulfate on the bulk density of microcapsules and encapsulation efficiency of zinc also were determined.

18.5   CHARACTERIZATION OF SPRAY-DRIED ZINC
MICROCAPSULES

The content of zinc in the microcapsules obtained by spray drying was determined usingInductively-Coupled Plasma Optical Spectroscopy (ICP-OES) (Optima 8000, PerkinElmer, Inc., Waltham, USA), while encapsulation efficiency was estimated by the method of Abbasi and Azari (2011) with slight modifications. The whiteness index of microencapsulated zinc was calculated after acquiring the color parameters of the microcapsules in a Konica Minolta CM-5 color spectrophotometer (Konica Minolta, Tokyo, Japan) at ambient conditions.

The morphological characteristics of the microcapsules of zinc wereexaminedin Field-Emission Gun Scanning Electron Microscope (FEGSEM) (Ultra 55, Zeiss Gemini, Jena, Germany). The elements in the spray-dried zinc microcapsules were identified and quantified using X-Max Energy-Dispersive X-ray Spectroscopic (EDX) detector operating at 5 kV (Polekkad et al., 2021). The average particle diameter of microcapsules was measured from the SEM images using Fiji software. The innate characteristics of zinc-loaded microcapsules were examined using Transmission Electron Microscopy (TEM) (Tecnai Spirit G^2, FEI, Netherlands). The 2-D and 3-D topographical features of zinc microcapsules were studied using Atomic Force Microscopy (AFM) (ScanAsyst, Bruker, Santa Barbara, USA).

The crystalline nature and physical state of zinc and microcapsules were examined using X-ray diffractometry(SmartLab, Rigaku Corporation, Tokyo, Japan). The spectra of pure zinc sulfate and zinc microcapsules were obtained by a Fourier transform infrared (FTIR) spectrometer (Frontier, Perkin Elmer, Singapore). The moisture content of spray-dried microcapsules was estimated by the method 927.05 of AOAC (AOAC, 2016), while dispersibility, wettability and insolubility index were determined as per A/S Niro Atomizer (1978). The bulk and tapped densities were measured by ASTM method D7481 (ASTM, 2009), while particle density was measured as proposed by Pushpadass et al. (2014). The true and tapped densities were used to calculate the porosity of the microcapsules. The influence of each process factor on the encapsulation efficiency of zinc and bulk density of microcapsules was evaluated using Analysis of Variance (ANOVA) at a significance of 0.05.

18.6   FORTIFICATION OF MILK WITH ZINC MICROCAPSULES

Exactly 5 g of zinc microcapsules based on the three wall materials was added to 500 mL milk. The fortified milk was pasteurized and evaluated for its physicochemical and sensory properties. The pH was determined using pH meter, while for estimation of acidity fortified milk was titrated against 0.1 N NaOH with phenolphthalein as an indicator. Ostwald viscometer was used to measure the viscosity of fortified milk at 25°C, and the heat stability was evaluated by "clot on boiling" test. The fuzzy-logic approach was employed to assess the organoleptic quality of milk fortified with zinc microcapsules (Franklin et al., 2019). Twenty experienced and trained judges were involved. The set of perceptions from the sensory panel was subjected to fuzzy modeling using MATLAB software and the ranking of samples and quality attributes was performed based on their similarity values.

18.7   CHARACTERISTICS OF SPRAY-DRIED ZINC MICROCAPSULES

All three wall materials were found to be suitable for the production of encapsulated zinc microcapsules by spray drying. The spray-dried powder was granular in all cases, with moisture content in the range of 2.64 to 3.53%. The microcapsules of HiCap-100-zinc had the highest whiteness index ranging from 75.26 to 81.26, followed by those of maltodextrin and WPI ranging from 75.01 to 80.0 and 69.91 to

78.21, respectively. The whiteness index of microcapsules showed an upward trend with increasing wall material ratio and decreasing drying air temperature (Table 18.1).

The morphological features of zinc microcapsules were examined by SEM and the microcapsules based on all three wall materials were more or less spherical in shape with minor dents on the surface (Figure 18.1a). The absence of major cracks on the surface of spherical microcapsules in the study confirmed the surface integrity of spray-dried zinc suggesting that they possessed excellent characteristics for food applications, wherein they would be intact and less likely to experience leaching during storage. The surface of microcapsules was smooth, with some wall material residues attached to it and resembling brittle flakes. Hence it was concluded that the spray-chilling technique successfully produced 2-acetyl-1-pyrroline zinc chloride microcapsules with desirable morphological characteristics.

The sphericity of WPI, maltodextrin and HiCap-100-based microcapsules, as calculated from the SEM images using Fiji software, was 0.88, 0.84 and 0.81, respectively. It was observed that the concentration of wall material had a

TABLE 18.1

Whiteness Index and Encapsulation Efficiency of Zinc Microcapsules (Polekkad et al., 2021)

S. No.	Temperature (°C)	Wall material	Wall material to $ZnSO_4$ ratio	Whiteness index	Encapsulation efficiency (%)
1	170	Maltodextrin	10:1	74.63	76.86
2	170	Maltodextrin	20:1	78.21	84.37
3	185	Maltodextrin	10:1	72.79	82.80
4	185	Maltodextrin	20:1	74.63	88.00
5	200	Maltodextrin	10:1	69.91	79.42
6	200	Maltodextrin	20:1	75.40	86.54
7	170	HiCap-100	10:1	77.96	85.00
8	170	HiCap-100	20:1	80.00	87.73
9	185	HiCap-100	10:1	76.88	87.60
10	185	HiCap-100	20:1	78.42	92.65
11	200	HiCap-100	10:1	75.26	86.83
12	200	HiCap-100	20:1	78.31	88.90
13	170	Whey protein isolate	10:1	78.59	79.83
14	170	Whey protein isolate	20:1	81.26	83.62
15	185	Whey protein isolate	10:1	78.28	82.60
16	185	Whey protein isolate	20:1	79.27	87.90
17	200	Whey protein isolate	10:1	75.01	80.20
18	200	Whey protein isolate	20:1	77.92	84.07

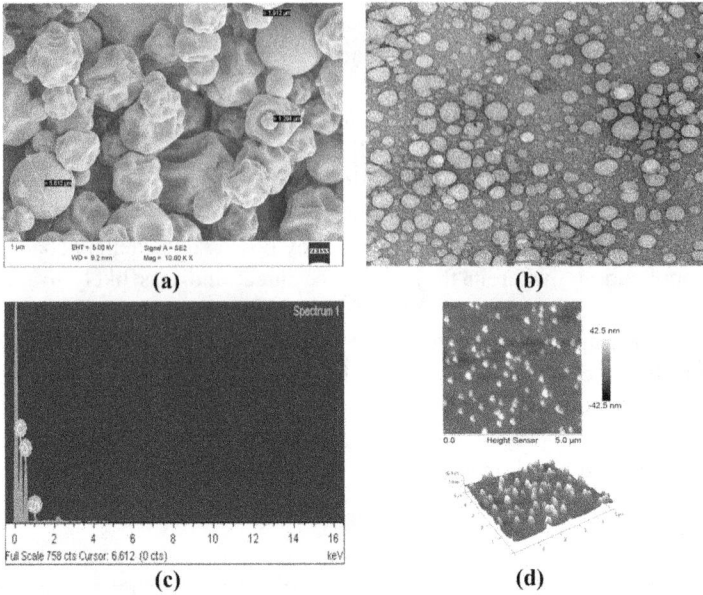

FIGURE 18.1 (a) SEM (b) SEM-EDX (c) TEM and (d) AFM images of spray-dried zinc microcapsules prepared using HiCap-100 as wall materials.

significant positive influence in producing spherical microcapsules with fewer surface indentations and cracks. The spherical shape of microcapsules without flaky and sharp edges in the SEM images signified the successful encapsulation of zinc regardless of the wall material used. Baldelli et al. (2021) evaluated several compounds of zinc (such as zinc oxide, zinc gluconate, zinc citrate, zinc chloride and zinc sulfate) as core materials to produce microcapsules by spray drying for fortification of foods. The maximum solid weight percentage was limited to 16 (w/w) due to the viscosity limitations of the atomizer used. In addition to solid weight percentage, the effects of various ratios of ZnO and maltodextrin (0.3 to 1.8), ZnO and pea protein (0.1 to 2.3), maltodextrin and pea protein (0.3 to 12) and zinc oxide and titanium dioxide (0.3 to 2.5) on the morphological properties of spray-dried particles were evaluated. It was concluded that these ratios had a significant impact on the formation and characteristics of dried particulates, consequently affecting the morphology and chemical properties of encapsulated zinc.

The diameter of microcapsules was obtained from the SEM images and was found to be in the range of 3.13–4.5, 2.77–5.15 and 4.9–9.39 μm for maltodextrin, HiCap-100 and WPI-based microcapsules, respectively (Figure 18.1a). It should be noted that apart from wall material, particle size is influenced by other factors such as inlet temperature of hot air, concentration of encapsulant, feed flow rate and atomization. Microcapsules produced from feed solution containing 1 mM of zinc had a particle size of 100 nm, whereas those containing 5 and 10 mM of zinc had a particle size as high as 1 μm.

The skin-forming property of WPI presumably provided a ballooning effect by entrapping more air as compared to other wall materials, which increased the occluded air content within the microcapsules (Kwapińska and Zbiciński, 2005), thereby increasing the diameter. The TEM images also manifested the spherical shape of WPI and maltodextrin-based microcapsules were also evident in the TEM images, corroborating the results of SEM (Figure 18.1b). The spherical shape of micro-capsules increases their stability due to a reduction in surface area to volume ratio.

The size of microcapsules produced at the inlet air temperature of 170°C was relatively small, but as temperature increased, the former also increased. As the residence time in the lab-scale spray dryer was shorter, the microcapsules experienced inadequate drying at 170°C, which led to the formation of wrinkled surfaces and concavities upon cooling (Nijdam and Langrish, 2005). As inlet air temperature increased, the microcapsules became increasingly spherical-shaped.

The EDX spectra of microcapsules revealed that they primarily had elements such as carbon, oxygen and zinc (Figure 18.1c). The weight percentage of zinc in the microcapsules was in the range of 1.28 to 1.83%. As 10% zinc was spray-dried along with the wall materials, the much lower values obtained in the study suggested that the material was encapsulated inside the sub-layer of wall materials. The 2-D and 3-D AFM images also showed the surface topography of zinc microcapsules (Figure 18.1d). The maltodextrin, HiCap-100 and WPI-based microcapsules had surface roughness of 237, 243.7 and 238 nm, respectively. The highest surface roughness was thus observed in HiCap-100-based microcap-sules. These findings mirror those of SEM images. The heterogeneous topography of zinc microcapsules was attributed to the variations in the properties of wall materials (Figure 18.1d). However, the absence of sharp pyramid-like peaks in the AFM images suggested that the outer crust of zinc microcapsules was mostly amorphous-like, which was typical of the wall materials us.

Among the wall materials, HiCap-100 yielded the maximum mean encapsulation efficiency of 88.12%, while WPI provided the least value of 81.24% (Table 18.1). Encapsulation efficiency was significantly influenced ($p<0.001$) by the type of encapsulant, wall-to-core material ratio and inlet temperature of drying air. The encapsulation efficiency was highly influenced by the type of encapsulant, followed by the ratio of wall material to zinc and inlet air temperature(Figure 18.2a and 18.2b).

The rate and thickness of the semi-permeable membrane that encircled the zinc sulfate core increased with the quantity of wall material, which ensured adequate encapsulation of zinc inside the microcapsules. The encapsulation efficiency of zinc was found to be increased with drying air temperature owing to expedited drying of droplets resulting in a rapid rate of crust formation in the microcapsules and locking zinc inside the wall materials. The optimal conditions of spray drying to achieve the highest encapsulation efficiency of zinc were HiCap-100 as coating material, coating material to zinc ratio of 20:1 and inlet air temperature of 185°C. Thus, the desired zinc microcapsules were produced by selecting suitable wall materials and inlet air temperature.

In the present investigation, the bulk densities of maltodextrin, HiCap-100 and WPI-based zinc microcapsules were observed to be 536.4, 541.2 and 437.4 kg/m^3 respectively (Figure 18.2c). The type of wall material used for encapsulating zinc

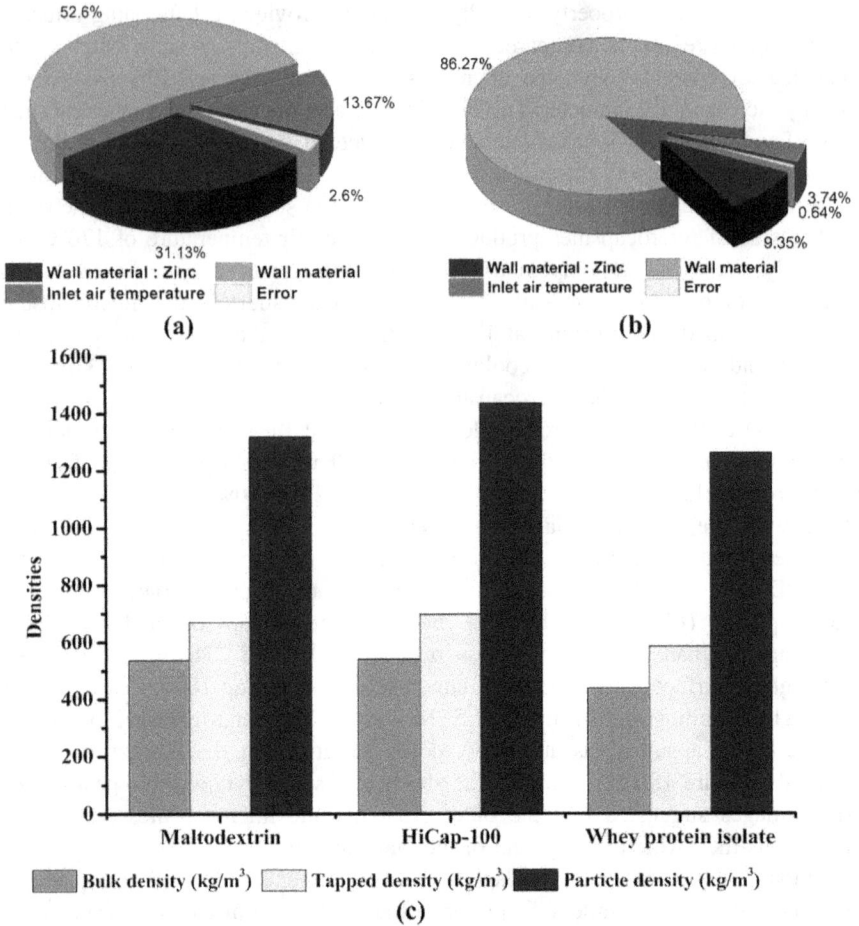

FIGURE 18.2 Contribution percentage of independent factors on (a) Encapsulation efficiency of zinc (b) Bulk density of microcapsules and (c) Densities of zinc microcapsules prepared using different wall materials.

sulfate had the greatest influence ($p<0.001$) on the above property, followed by the ratio of wall material ($p<0.001$) to air temperature ($p<0.001$). The highest bulk density of HiCap-100-based microcapsules was primarily due to their smaller size caused by shrinkage during drying (Sanchez-Reinoso et al., 2017) and the relatively higher molecular weight of HiCap-100. In comparison, the WPI-based zinc microcapsules had a much higher diameter, and hence, had the lowest density. Zinc microcapsules based on maltodextrin, HiCap-100 and WPI had particle densities of 1319.88, 1436.50 and 1263.25 kg/m³ respectively, while their corresponding porosities were 49.20, 51.40 and 53.60%. The characteristics of WPI and the reasons discussed earlier could be assigned to the higher porosity of WPI-based microcapsules. Because the particle size was small, zinc microcapsules produced with HiCap-100 as wall material exhibited superior properties in terms of flowability and bulk density.

The reconstitution properties of microcapsules are important for consumers' acceptance and fortification in food systems. In general, when spray-dried powders come into contact with water, they dissolve quickly and do not segregate into individual particles. In this study, the mean wettability time of maltodextrin, HiCap-100 and WPI-based zinc microcapsules was 120, 102 and 91 s respectively. The WPI-based microcapsules had the least wetting time owing to their relatively much bigger particle size, higher porosity and spherical shape, which influenced the contact angle with water. Microcapsules prepared using HiCap-100 had the highest dispersibility as compared to the other two. It is defined as the ease with which agglomerates dissociate into individual particles when dissolved in water. In comparison, the microcapsules prepared using maltodextrin, HiCap-100 and WPI had insolubility index of 0.8, 1.1 and 1.3 mL respectively. The higher solubility of maltodextrin-based microcapsules was due to the inverse relationship between insolubility index and solubility.

The XRD peaks ofpure zinc sulfate at 16.67°, 20.13°, 21.50°, 33.67°, 36.2°, 41.37°, 47.5° and 53.61°which are typical of its crystalline structure (Figure 18.3a). In comparison, the zinc-loaded microcapsules from all three wall materials were amorphous, manifesting only a broader peak at 22° and burying the typical highly intense crystalline peaks of zinc sulfate (Figure 18.3a). The disappearance of typical zinc-related peaks in the XRD implied that it was successfully residing inside the wall materials. The amorphous property possessed by zinc microcapsules may be advantageous in food systems because crystalline products are difficult to handle and they release the core materials easily (Drusch, 2007). The zinc microcapsules are easy to fortify in food systems due to the highly water-soluble nature of amorphous ingredients.

The FTIR spectrum of pure zinc sulfate is illustrated in Figure 18.3b. The characteristic peaks at 3082 cm^{-1}, 1614 cm^{-1} and 757.81 cm^{-1} were representative of the symmetric OH stretching, bending vibrations and libration mode of innate water molecules respectively. These are water molecules crystallized as the heptahydrate. From the peaks observed at 1092 and 627 cm^{-1}, the triply degenerate symmetric stretching modes of sulfate groups were identified, whereas the peaks at 978 and 464 cm^{-1} were indicative of the non-degenerate mode and doubly degenerate mode vibrations respectively. In comparison, the FTIR spectra of zinc microcapsules resembled quite similar to that of wall materials (not shown in Figure) rather than that of zinc sulfate. The peaks of HiCap-100-based zinc microcapsules were observed at 3394, 2918, 1626, 1421, 1355, 1150, 1083, 1010, 831 and 772 cm^{-1} (Figure 18.3b). Thus, it is obvious that the typical peaks of zinc sulfate were absent in the spectra of microcapsules. The absence of major spectral changes such as new bond formation or shifting of peaks indicated the absence of strong structural interactions between zinc sulfate and the functional groups of wall materials after micro-encapsulation. It could be stated that spray drying is a suitable method for microencapsulating zinc without altering its inherent properties.

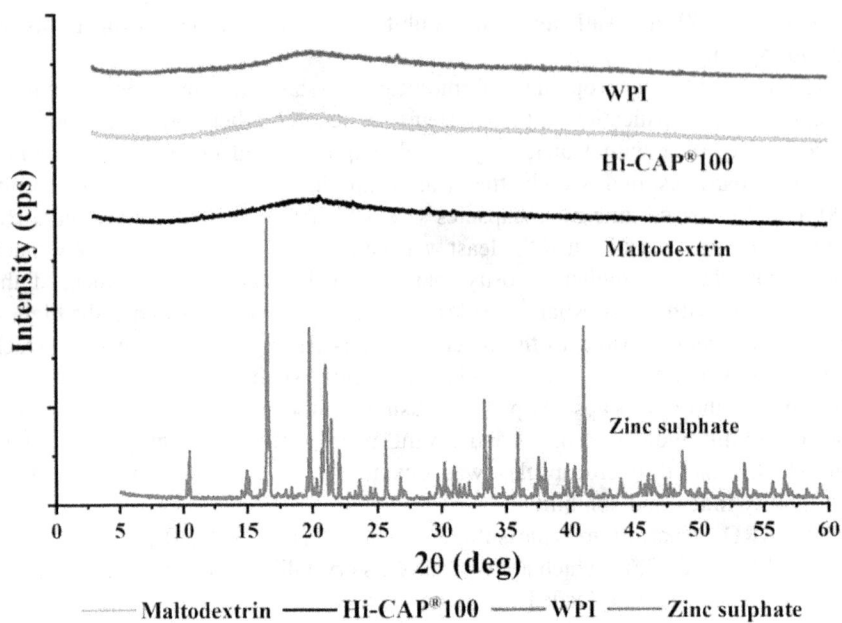

(a)

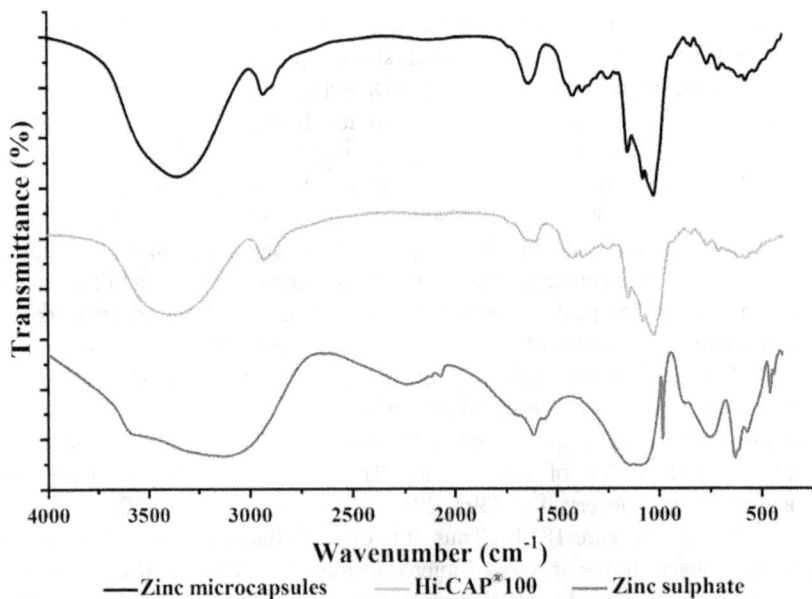

(b)

FIGURE 18.3 (a) X-ray diffractograms of zinc sulphate and microcapsules prepared using different wall materials (b) FTIR spectra of zinc sulphate, HiCap-100 and microcapsules prepared using HiCap-100 as wall material.

18.8 PROPERTIES OF ZINC-FORTIFIED MILK

One of the vital aspects of food fortification is the selection of a suitable vehicle for the delivery of zinc. Some minerals and vitamins can be fortified simply by blending or mixing. This method is easier and cheaper but may fail to protect and deliver the micronutrient effectively. It also ignores the negative impact on physicochemical and organoleptic changes in foods, which may be a major concern for consumer acceptance. Öner et al. (1988) reported that zinc cannot be added directly to milk as it has an unpleasant and bitter metallic after-taste that is not preferred by consumers.

In this study, microencapsulated zinc that was prepared using three wall materials was added to milk at 1% (w/w), and the fortified milk was pasteurized and analyzed. The physicochemical qualities of milk did not change significantly after fortification (Table 18.2). The titratable acidity and pH of fortified milk were similar to those of normal milk (control), while the viscosity was higher than that of the latter due to higher solid contents and protein-polysaccharide interactions of the wall material with innate milk proteins. Similarly, there were no adverse changes in the organoleptic quality. The *in-vitro* release pattern was the same as normal but as the wall material ratio increased, the time scale for release of zinc from the microcapsules got extended.

For the evaluation of organoleptic quality, the membership function of the standard fuzzy scale and overall membership function of organoleptic scores were used. The overall ranking of all samples was such that normal milk (Very Good) > milk fortified with HiCap-100 based microcapsules (Very Good) > milk fortified with maltodextrin-based microcapsules (Good) > milk fortified with WPI-based microcapsules (Satisfactory). The panel of judges perceived similarity in the acceptability of both milk fortified with HiCap-100-based microcapsules and normal pasteurized milk. The organoleptic acceptance of both normal milk and milk fortified with HiCap-100-based zinc microcapsules by them was similar. The similarity values were: Aroma (Highly Important) > Taste (Highly Important) > Color and appearance (Highly Important) > Overall acceptability

TABLE 18.2

Physicochemical Properties of Milk Fortified with Zinc Microcapsules (Polekkad et al., 2021)

Characteristic	Control	Maltodextrin-based	HiCap-100-based	Whey protein isolate-based
pH	6.70	6.56	6.58	6.54
Titratable acidity (%LA)	0.19	0.21	0.20	0.22
Viscosity (cP)	1.60	1.84	1.86	1.72
Clot on boiling test	Negative	Negative	Negative	Negative
Zinc content (ppm)	5.57	22.04	23.02	21.19

(Highly Important). All quality attributes were found to have significant roles in the consumer acceptance of milk fortified with microencapsulated zinc. Earlier studies confirmed that microencapsulation facilitated the sustained release of zinc and made it bioavailable in the small intestine rather than in the gastric conditions of the stomach.

18.9 CONCLUSIONS

Microencapsulation of zinc by spray drying with HiCap-100 as wall material had entrapment efficiency as high as 92.7%. It was observed that the drying conditions and nature of wall material had a significant influence on the physicochemical and microstructural properties of dried microcapsules. After encapsulation, no major structural changes were observed in the FTIR and XRD spectra of zinc microcapsules implying that there were no strong interactions between zinc and wall materials during spray drying and zinc sulfate was encapsulated and protected as such in microcapsules. Milk was found to be a suitable vehicle for fortification of microencapsulated zinc for its wider consumption by the majority population and had the least adverse changes in physicochemical and organoleptic qualities.

REFERENCES

Abbasi S, Azari S 2011. Efficiency of novel iron microencapsulation techniques: fortification of milk. Int. J. Food Sci. Technol. 46(9), 1927–1933.

Akhtar S 2013. Zinc status in South Asian populations-an update. J. Health Popul. Nutr. 31(2), 139.

Ali A, Phull AR, Zia M 2018. Elemental zinc to zinc nanoparticles: Is ZnO NPs crucial for life? synthesis, toxicological, and environmental concerns. Nanotechnol. Rev. 7(5), 413–441.

AOAC. 2016. Official Methods of Analysis International. twentieth ed., Association of Official Analytical Chemists, Rockville, Maryland.

Arredondo M, Martinez R, Nune MT, Ruz M, Olivares M 2006. Inhibition of iron and copper uptake by iron, copper and zinc. Biol. Res. 39(1), 95–102.

ASTM D7481-09. 2009. Standard Test Methods for Determining Loose and Tapped Bulk Densities of Powders using a Graduated Cylinder. ASTM International, West Conshohocken, PA.

A/S Niro Atomizer. 1978. Analytical Methods for Milk Products, third ed., Copenhagen, Denmark.

Bai K, Hong B, He J, Hong Z, Tan R 2017. Preparation and antioxidant properties of selenium nanoparticles-loaded chitosan microspheres. Int. J. Nanomed. 12, 4527.

Baldelli A, Liang DY, Guo Y, Pratap-Singh A 2023. Effect of the formulation on mucoadhesive spray-dried microparticles containing iron for food fortification. Food Hydrocoll. 134, 107906.

Baldelli A, Wells S, Pratap-Singh A 2021. Impact of product formulation on spray-dried microencapsulated zinc for food fortification. Food Bioproc. Technol. 14(12), 2286–2301.

Bhowmik D, Chiranjib KP, Kumar S 2010. A potential medicinal importance of zinc in human health and chronic. Int. J. Pharma Bio Sci. 1(1), 05–11.

Cho HY, Kim B, Chun JY, Choi MJ 2015. Effect of spray-drying process on physical properties of sodium chloride/maltodextrin complexes. Powder Technol. 277, 141–146.

Choudhury N, Meghwal M, Das K 2021. Microencapsulation: An overview on concepts, methods, properties and applications in foods. Food Front. 2(4), 426–442.

Das S, Chaki AK, Hossain A 2019. Breeding and agronomic approaches for the biofortification of zinc in wheat (*Triticum aestivum* L.) to combat zinc deficiency in millions of a population: A Bangladesh perspective. Acta Agrobot. 72(2), 1–13.

D'Imperio M, Durante M, Gonnella M, Renna M, Montesano FF, Parente A, Mita G, Serio F 2022. Enhancing the nutritional value of *Portulaca oleracea* L. by using soilless agronomic biofortification with zinc. Food Res. Int. 155, 111057.

Diosady LL, Alberti JO, Mannar MV 2002. Microencapsulation for iodine stability in salt fortified with ferrous fumarate and potassium iodide. Food Res. Int. 35(7), 635–642.

Drusch S 2007. Sugar beet pectin: A novel emulsifying wall component for micro-encapsulation of lipophilic food ingredients by spray-drying. Food Hydrocoll. 21(7), 1223–1228.

Franklin ME, Pushpadass HA, Kamaraj M, Muthurayappa M, Battula SN 2019. Application of D-optimal mixture design and fuzzy logic approach in the preparation of chhana podo (baked milk cake). J. Food Proc. Eng. 42(5), e13121.

Gharsallaoui A, Roudaut G, Chambin O, Voilley A, Saurel R 2007. Applicationsof spray-drying in microencapsulation of food ingredients: An overview. Food Res. Int. 40(9), 1107–1121.

Gibson RS 2012. Zinc deficiency and human health: Etiology, health consequences, andfuture solutions. Plant Soil 361(1-2), 291–299.

Gonçalves A, Estevinho BN, Rocha F 2016. Microencapsulation of vitamin A: A review. Trends Food Sci. Technol. 51, 76–87.

Homayoonfal M, Malekjani N, Baeghbali V, Ansarifar E, Hedayati S, Jafari SM 2022. Optimization of spray drying process parameters for the food bioactive ingredients. Crit. Rev. Food Sci. Nutr. 1–41.

Kaul S, Kaur K, Mehta, N, Dhaliwal SS, Kennedy JF 2022. Characterization and optimization of spray dried iron and zinc nanoencapsules based on potato starch and maltodextrin. Carbohydr. Polym. 282, 119107.

Kwapińska M, Zbiciński I 2005. Prediction of final product properties after cocurrent spray drying. Dry. Technol. 23(8), 1653–1665.

Madene A, Jacquot M, Scher J, Desobry S 2006. Flavour encapsulation and controlled release-A review. Int. J. Food Sci. Technol. 41(1), 1–21.

Mayo-Wilson E, Junior JA, Imdad A, Dean S, Chan XHS, Chan ES, Jaswal A, Bhutta ZA 2014. Zinc supplementation for preventing mortality, morbidity, and growth failure in children aged 6 months to 12 years of age. Cochrane Database Syst. Rev. 5, CD009384.

Nijdam JJ, Langrish TAG 2005. An investigation of milk powders produced by a laboratory-scale spray dryer. Dry. Technol. 23(5), 1043–1056.

Öner L, Kaş HS, Hincal AA 1988. Studies on zinc sulphate microcapsules (1): Microencapsulation and *in vitro* dissolution kinetics. J. Microencapsul. 5(3), 219–223.

Polekkad A, Franklin MEE, Pushpadass HA, Battula SN, Rao SN, Pal DT 2021. Microencapsulation of zinc by spray-drying: Characterisation and fortification. Powder Technol. 381, 1–16.

Prasad AS 1998. Zinc and immunity, in: Pierce GN, Izumi T, Rupp H, Grynberg A (Eds.), Molecular and Cellular Effects of Nutrition on Disease Processes. Springer, Boston, MA, pp. 63–69.

Pratap-Singh A, Leiva A 2021. Double fortified (iron and zinc) spray-dried microencapsulated premix for food fortification. LWT-Food Sci. Technol. 151, 112189.

Pushpadass HA, Emerald FME, Chaturvedi B, Rao KJ 2014. Moisture sorption behavior and thermodynamic properties of gulab jamun mix. J. Food Proc. Preserv. 38(6), 2192–2200.

Sanchez-Reinoso Z, Osorio C, Herrera A 2017. Effect of microencapsulation by spray drying on cocoa aroma compounds and physicochemical characterisation of microencapsulates. Powder Technol. 318,110–119.

Sangeetha VJ, Dutta S, Moses JA, Anandharamakrishnan C 2022. Zinc nutrition and human health: Overview and implications. eFood, 3(5), e17.

Singh MN, Hemant KSY, Ram M, Shivakumar HG 2010. Micro encapsulation: A promising technique for controlled drug delivery. Res. Pharm. Sci. 5(2), 65.

19 Impact of Different Processing Techniques on the Retention of Zinc in Foods

Nagamaniammai Govindarajan, Safreena Kabeer, and Navin Venketeish

19.1 INTRODUCTION

Zinc is a vital micronutrient that is mandatory for the body to stay healthy. It is the mineral part of some foods and acts as a dietary supplement. But modern processing techniques lead to the depletion of micronutrients such as zinc and increase the anti-nutritional factors that in turn, decrease their bioabsorption and bioavailability while the conventional processing methods do not have these adverse outcomes. This chapter discusses the impact of different processing methods and ways to increase the bioavailability of zinc with its high retention in processed foods.

19.2 ZINC DEPLETION AND BIOAVAILABILITY IN FOODS

Malnutrition of zinc can cause adverse health issues and increase the morbidity rate. Zinc depletion affects every part of the human body resulting in metabolic dysfunctions. Zinc deficit is a principal issue related to poor growth, decreased immunity, severe infections and several neural abnormalities. In 1961, zinc deficiency in humans was recognized as a malnutrition issue worldwide. It affected those who consumed low animal foods and high amounts of cereals (Roohani et al., 2013). The ensuing malnourishment in developing countries is due to an imbalanced supply of nutrients through diet. Among the nutrients, the role of micronutrients is significant as micronutrient malnutrition is a general problem widely prevalent in developing countries. It can disturb all age groups, but infants and pregnant women are always at high risk (Black, 1998). Different approaches to regulate zinc deficiency include supplementation, fortification and dietary modification (Carmona et al., 2020).

DOI: 10.1201/9781003412472-19

Bioavailability is the total intake of nutrients absorbed and utilized by an individual and is an essential aspect of the deficiency of minerals. The mineral zinc is bound with several inhibitory factors (Bailey et al., 2015). Infants had an intake of unrefined cereals, which might have affected the absorption of zinc as anti-nutritional factors present in them reduced zinc's absorption. Phytate is a major inhibiting factor that reduces zinc's bioavailability (Sandberg, 1991). In poor regions, the diet primarily comprises of cereals and legumes containing phytate (Calhoun et al., 1974). Polyphenols and phytate are dominant factors in crops such as legumes and cereals. Strong chelates are formed by the inhibitory compounds and get complexes with macro and microelements leading to lowering their absorption and causing malnutrition among people (Gharibzahedi and Jafari, 2017).

In a study of pregnant women, it was found that 55% had lower zinc status than average serum zinc concentration (Hotz and Brown, 2004). Another study revealed that 35% of infants had low serum zinc concentration. Several traditional processing methods are used to boost the bioavailability of zinc, including thermal processing, mechanical processing, milling, cooking, fermentation, soaking, germination and other non-thermal processing techniques (Hotz and Gibson, 2007).

The bioavailability technique should be considered to better understand the interaction between food components and minerals in gastrointestinal tracts.

For zinc homeostasis, loss and adsorption of zinc in the small intestine are contemplated in a common mechanism. Plant-based foods consist of more phytic acid compared to animal-sourced. Protein content in the small intestine and stomach degrades the enzymatic activity due to phytic acid that is present in large amounts in plant foods than animal products (Kies et al., 2006).

19.3   ANTINUTRITIONAL FACTORS AND PROCESSING

Fibers, phenolic compounds and phytates are recognized as Anti-Nutritional Factors (ANF) that cause less bioavailability and bioaccessibility of zinc and iron in plant foods. Processing technologies such as heating, germination, mechanical treatment and soaking help to improve the bioaccessibility of zinc and iron in plant food (Raes et al., 2014). Several processing methods are used to decrease the ANF, which supports enhancing the bioavailability of minerals.

Legumes and cereals possess rich mineral sources but the bioavailability of these minerals is less due to the presence of ANF such as polyphenols and phytate. Phytic acid is a critical anti-nutrient as it is present in cereals and has more capacity to complex with metal ions making them inaccessible for human consumption. Modest traditional methods such as cooking, soaking, germination and roasting are used in cereal processing to enrich the nutritional quality (Nadeem et al., 2010). Phytates are the chief loading formula of phosphorous and they are present in legumes and cereals (Reddy et al., 1989). Nutrients such as zinc, calcium, magnesium and iron are chelating agents which complex with phytate phosphorus and reduce their bioavailability (Sandberg, 2002). In order to remove ANF from cereals, germination is used as an important technique.

19.4   ANTINUTRITIONAL FACTORS IN BIOAVAILABILITY OF ZINC

Wild yam tubers contain phytate and depending on the location, irrigation methods and soil types, the content varies. High phytate content in yam reduces the availability of minerals such as calcium and zinc as phytic acid forms calcium phytate with calcium which in turn inhibits the absorption of zinc (Srikka, 1997).

Lentils are high in minerals but their bioavailability is low due to oxalic acid, phytate and phenolic compounds. Lectins found in legumes and trypsin inhibitors (which act as a protein source) can also be anti-nutrients and they obstruct digestion. Other important anti-nutrients are flavonoids and polyphenolic compounds. Metals such as iron and zinc get chelated with these compounds and reduce nutrient absorption. Almost all foods contain anti-nutrients in different amounts and processing techniques do not completely remove them (Aykroyd et al., 1982). A study report mentions that phytates too cannot be completely removed by cooking or allowing to stand in the boiling water (Ranhotra et al., 1974).

19.5   FACTORS AFFECTING THE BIOAVAILABILITY OF ZINC

19.5.1   INTAKE OF ZINC

A low zinc diet increases its absorption better while the excess is counter-productive. This is so in all age groups and is considered the universal behavior of zinc absorption. The increased consumption also diminishes the uptake of zinc's efficiency (Solomons, 2001).

19.5.2   PROTEIN QUANTITY AND SOURCE

The quantity of protein in the food is related to the absorption of zinc in the body. Several studies with different protein sources and varying amounts proved that there is a linear proportion relationship between protein and zinc. Thus, increased dietary protein is the reason for an increase in the bioavailability of zinc (Sandström et al., 1990). The protein source in the diet also affects zinc intake. The absorption is higher if the protein source is animal-based rather than of plant origin. Animal-based protein interacts with the inhibitory factors and enhances the intake of zinc (Sandström and Sandberg, 1992). When animal protein sources are added to vegetable-based foods, the bioavailability of zinc improves significantly. Protein breaks down to form peptides and amino acids after digestion to produce low molecular weight compounds which in turn combine with zinc and form complexes that enhance the solubility and bioavailability of zinc (Regvar et al., 2011).

19.5.3   PHYTATE AND DIETARY FIBER

As discussed earlier, phytate has an inhibitory action of zinc absorption. Foods such as corn, cereal, rice, legumes, pulses, etc., contain high amounts of phytate. A study conducted with soy-based infant formula and human milk proved that the bio-availability of zinc was very low in the former. When the phytate was removed from

TABLE 19.1

Zinc and Phytate Content in Selected Plant-based Food

Food	Zinc content (mg/100 g)	Phytate content (mg/100 g)
Lentils	3.03–4.02	749–961
Sesame seeds	2.48	1525
Durum wheat	2.4–4.8	460–952
Sweet potato	0.30	31–37
Passion fruit	0.41–0.48	77.2–86.8
Refined wheat flour	0.52	37

Source: Maares and Haase, 2020.

soy protein, a drastic difference in zinc intake was observed with an improvement in zinc absorption (Lönnerdal et al., 1988).

Fiber content also has an impact on zinc absorption. This again is due to the presence of phytate as most fiber-rich foods contain phytate. A study was conducted on phytate-free bread with white bread that contained low fiber. In both cases, there was a considerable increase in the intake of zinc, signifying that fiber alone has little or no effect on zinc absorption (Halsted et al., 1972). Table 19.1 shows the zinc and phytate contents in some selected plant-based foods.

19.5.4 MICRONUTRIENTS

The presence of micronutrients also affects the absorption of zinc. Minerals such as calcium, iron and copper compete for absorption since they have similar chemical characteristics. It was found that high doses of iron decrease the absorption of zinc. Undesirable effects of both inorganic and heme iron on zinc absorption were studied by *in vivo* methods and the effect was better when the iron was used as an aqueous solution than with a meal (Solomons and Jacob, 1981).

Calcium, too, has adverse effects on the absorption of zinc and disturbs the intake due to the presence of phytate-containing compounds. Calcium easily forms a complex with phytate affecting the absorption of zinc (Lonnerdal et al., 1984). Copper does not have any influence on zinc intake and absorption (August et al., 1989).

19.6 EFFECTS OF VARIOUS PROCESSING TECHNIQUES ON ZINC

19.6.1 CONVENTIONAL METHODS

19.6.1.1 Thermal and Mechanical Processing

There are various processing techniques that affect the bioavailability of zinc. Different heat processing methods such as pressure cooking, roasting, microwave cooking, spray drying and drum drying were studied for zinc's bioavailability.

a. **Pressure Cooking** of grains reduced zinc's bioavailability, especially in pulses. A study conducted by Gayathri et al. (2004) found that this method reduced the bioavailability of zinc by 57% in rice and 63% in finger millet. The decrease was due to the interaction of zinc with the proteins thereby hindering its absorption. In raw pulses, the presence of phytic acid decreases Zn bioavailability. The former is stable and mostly unaffected by the heat treatment resulting in low bioavailability (Hemalatha et al., 2007).

b. **Roasting** denatures protein and causes zinc to improve its bioavailability (Frontela et al., 2008). This is considered as a less expensive method to increase Zn bioavailability (Jaiswal and Lakshmi, 2022).

c. **Microwave cooking** reduces the bioavailability of zinc in rice (39%) and maize (19%). Soaking, de-husking and sprouting of rice followed by microwave and pressure cooking increase soluble zinc content and also decrease anti-nutritional factors (Kaur and Kawatra, 2002).

d. **Spray drying** is used to preserve the nutritional quality of dried products and causes the depletion of inhibitors such as phytic and oxalic acids thereby improving protein and carbohydrate availability. This improves bioavailability of zinc and releases the entrapped zinc also (Jaisal and Lakshmi, 2022).

e. **Drum drying** is used to obtain ready-to-use products on a large scale. The products with rich phytate content are dried and made into slurry prior to drying. This enhances the bioavailability of zinc as phytate is degraded and improves protein digestibility (Tonin et al., 2018).

19.6.1.2 Milling

Milling is a process that grinds the grains into flour which enhances the bioavailability of zinc. Milling of rice increases zinc content in the grains which depends on the degree of milling. As the latter increases, phytate content is significantly reduced (Dipti et al., 2017). The highest reduction in the phytate content occurred between 0% and 2% degrees polishing of milling. Normally, the outer part of cereals and pulses contains more anti-nutrients that could be compacted by milling (Oghbaei and Prakash, 2016). Harland and Morris (1995) stated that reducing phytate, tannin and phenolic acid through milling helps to recover the availability of minerals and digestibility of proteins and carbohydrates.

19.6.1.3 Cooking

Phytic acid content was found to decrease in legumes by soaking and cooking methods (Vadivel and Biesalksi, 2012). Reduced anti-nutrient content was observed during boiling of the food thereby enhancing the nutritional value (Rehman and Shah, 2005). Studies showed that boiling/cooking enhanced the nutritional composition and reduced anti-nutritional factors such as trypsin inhibitors and tannins (Patterson et al., 2017).

19.6.1.4 Fermentation

During conventional food processing techniques such as fermentation, phytic acid content decreases in vegetables and cereals with an increase in the phytase activity.

To process wheat germs, traditional methods such as fermentation and germination are done before consumption. Fermented food is a source of protein. The process involves chemical changes that increase the organoleptic properties and sugar content and mineral bioavailability (Samtiya et al., 2021) and help breakdown of some endogenous compounds. Fermentation lowers the ANFs' level in food grains and enhances the extractability of minerals. During fermentation, components such as fibers, lignin and cellulose are minimally lowered depending upon the involvement of microorganisms.

19.6.1.5   Soaking

Soaking is a process of immersion of grains in water or acidified solution at a certain temperature and time period to cook and enhance nutritional value (Hurrell and Egli, 2010). Soaking is mostly used for seeds, grains and beans. In the process, cells take up the water which helps in activating the endogenous enzymes. It also helps in softening the cell wall and improving the extraction of water-soluble components. Its effect depends upon the temperature, run time and pH. Higher bioavailability of zinc and iron has been found after the soaking process. The reduction of phytate also triggers phenolic content in food material. Soaking increases the bio-accessibility of minerals by degrading ANFs such as phenolic compounds, pectin and phytates. In cereals, naturally occurring phytases are activated by soaking. In wheat bran, within an hour of soaking, about 95% phytate hydrolysis was complete and took two hours to degrade completely. It was observed that phytate content was lowered below 0.5 mol/gram to enhance the solubility of iron if the promoting factors were absent (Sandberg et al., 1991).

19.6.1.6   Germination

Germinated grains are widely used in infant and geriatric foods as they increase the nutrient content and digestibility. Germination is used to decrease the viscosity by breaking down starch and reducing the bulkiness which decreases intestinal motility and digestion in aged people (Devadas, 1998).

The effect of germination on micronutrients showed that the accessibility of zinc in rice (87%) and wheat (80%) has increased due to germination. The process also decreased phytate content and enhanced the bioavailability of zinc (Lestienne et al., 2005). Ungerminated grains had high amounts of phytic acid that in turn reduced the zinc's bioavailability (Konietzny et al., 1995).

19.6.2   Modern Methods

Non-thermal processing is a modern method that includes pulsed electric field, cold plasma, microwave, supercritical technology, ultrasonication, etc. Food exposed to thermal processing improves shelf-life but causes deterioration of nutritional and sensory qualities. In non-thermal processing techniques, food is exposed to a minimum temperature for a short period that causes no damage to nutritional composition (Jadav et al., 2021). Some studies have proved that these methods can capture enzyme activity, reduce ANFs and increase organic acid contents. The latter binds to minerals and forms soluble ligands which stop the development of phytate (Cilla et al., 2019).

19.7 CONCLUSION

Zinc is an important micronutrient and its absence leads to several undesirable health issues. By supplying nutrients through natural food sources, the bioavailability and digestibility of zinc could be improved. Most of the existing processing methods increase the number of anti-nutritional factors and do not improve the digestibility of minerals. The conventional processing methods lower the anti-nutritional factors such as phenolic compounds and phytic acid and improve the bioavailability of zinc. Phytate content should be inactivated to enhance the bioaccessibility of zinc content in food.

REFERENCES

August D, Janghorbani M, Young V 1989. Determination of zinc and copper absorption at three dietary Zn-Cu ratios by using stable isotope methods in young adult and elderly subjects. Am. J. Clin. Nutr. 50, 1457–1463. 10.1093/ajcn/50.6.1457

Aykroyd WR, Doughty J, Walker AF 1982. Legumes in Human Nutrition. Food & Agriculture Org.

Bailey RL, West JrKP, Black RE 2015. The epidemiology of global micronutrient deficiencies. Ann. Nutr. Metab. 66, 22–33. 10.1159/000371618

Black MM 1998. Zinc deficiency and child development Am. J. Clin. Nutr., 68(2), S464–S469. 10.1093/ajcn/68.2.464S

Calhoun NR, Smith JC, KL Becker. (1974). The role of zinc in bone metabolism. Clin. Orthop. 103, 212–234.

Cilla A, Barberá R, López-García G, Blanco-Morales V, Alegría A, Garcia-Llatas G 2019. Impact of Processing on Mineral Bioaccessibility/Bioavailability. Elsevier Inc., Amsterdam, The Netherlands. 10.1016/B978-0-12-814174-8.00007-X.

Carmona YR, Gutiérrez ED, Uribe ES, Aguirre PM, Flores M, Salmerón J 2020. Zinc supplementation and fortification in Mexican children. Food Nutr. Bull. 41, 89 101. Doi: 10.1177/0379572119877757.

Devadas RP 1998. Local strategies to support child nutrition. Nutr. Res. 18, 233–239. 10.1016/S0271-5317(98)00015-3

Dipti SS, Hotz C, Kabir KA, Bipul M 2017. Changes in the zinc content of selected Bangladeshi rice varieties through modified parboiling and milling methods. SAARC J. Agric. 15, 31–43. 10.3329/sja.v15i2.35153

Frontela, C, García-Alonso, FJ, Ros, G, Martínez, C 2008. Phytic acid and inositol phosphates in raw flours and infant cereals: The effect of processing. Journal of Food Composition and Analysis, 21, 343–350. https://www.foodinfotech.com/effect-of-various-food-processing-techniques-on-zinc-bioavailability/

Gayathri, GN, Platel, K, Prakash, J, Srinivasan, K 2004. Influence of antioxidant spices on the retention of β-carotene in vegetables during domestic cooking processes. Food Chemistry, 84, 35–43. https://www.foodinfotech.com/effect-of-various-food-processing-techniques-on-zinc-bioavailability/

Gharibzahedi SMT, Jafari SM 2017. The importance of minerals in human nutrition: Bioavailability, food fortification, processing effects and nanoencapsulation. Trends Food Sci. Techno. 62, 119–132. 10.1016/j.tifs.2017.02.017

Halsted J, Ronaghy H, Abadi P, Copper RL, Johnston KE, DuBard MB, Hauth JC 1972. Zinc deficiency in man: The Shiraz experiment. Am. J. Med. 53, 277–284. 10.1016/0002-9343(72)90169-6

Harland BF, Morris ER 1995. Phytate: a good or a bad food component. Nutr. Res. 15, 733–754. 10.1016/0271-5317(95)00040-P

Hemalatha S, Platel K, Srinivasan K 2007. Influence of heat processing on the bioaccessibility of zinc and iron from cereals and pulses consumed in India. J. Trace Elem. Med. Biol. 21, 1–7. 10.1016/j.jtemb.2006.10.002.

Hotz, C, Brown, KH 2004. Assessment of the risk of zinc deficiency in populations and options for its control. APA format.

Hotz C, Gibson RS 2007. Traditional food-processing and preparation practices to enhance the bioavailability of micronutrients in plant-based diets. J. Nutr. 137, 1097–1100. 10.1093/jn/137.4.1097

Hurrell R, Egli I 2010. Iron bioavailability and dietary reference values. Am. J. Clin. Nutr. 91, 1461–1467. 10.3945/ajcn.2010.28674F

Jadhav, H. B., Annapure, U. S., Deshmukh, R. R. (2021). Non-thermal Technologies for Food Processing. Frontiers in Nutrition, 8, 657090. 10.3389/fnut.2021.657090.

Jaiswal, A., Jyothi Lakshmi, A. (2022). Maximising the bioaccessibility of iron and zinc of a complementary food mix through multiple strategies. Food Chemistry, 372, 131286. https://www.foodinfotech.com/effect-of-various-food-processing-techniques-on-zinc-bioavailability/

Kaur M, Kawatra BL 2002. Effect of domestic processing on zinc bioavailability from rice bean (Vigna autamte) diets. Plant Food Human Nutr. 57, 307–318. 10.1023/A:102184 8916175

Kies AK, De Jonge LH, Kemme PA, Jongbloed AW 2006. Interaction between protein, phytate, and microbial phytase. In vitro studies. J. Agric. Food Chem. 54, 1753–1758. 10.1021/jf0518554

Konietzny U, Greiner R, Jany KD 1995. Purification and characterisation of a phytase from spelt. J. Food Biochem. 18, 165–183. 10.1111/j.1745-4514.1994.tb00495.x

Lestienne ICM, Rivier CI, Verniere I, Rochette S, Treche. 2005. The effects of soaking of whole, dehulled and ground millet and soybean seeds on phytate degradation and Phy/Fe and Phy/Zn molar ratios. Int. J. Food Sci. Tech. 40, 391–399. 10.1111/j.1365-2621. 2004.00941.x

Lönnerdal B, Bell JG, Hendrickx AG, Burns RA, Keen CL. 1988. Effect of phytate removal on zinc absorption from soy formula. Am. J. Clin. Nutr. 48, 1301–1306. 10.1093/ajcn/48.5.1301

Lönnerdal B, Cederblad Å, Davidsson L, Sandström B 1984. The effect of individual components of soy formula and cow's milk formula on zinc bioavailability. Am. J. Clin. Nutr. 40, 1064–1107. 10.1093/ajcn/40.5.1064

Maares M, Haase H 2020. A guide to human zinc absorption: General overview and recent advances of in vitro intestinal models. Nutrients. 12. 10.3390/NU12030762

Nadeem M, Anwar AA, Hussain A, Khan S 2010. Performance of winter cereal-legumes fodder mixtures and their pure stand at different growth stages under rainfed conditions of Pothowar. J. Agric. Res. (03681157) 48(2).

Oghbaei M, Prakash J 2016. Effect of primary processing of cereals and legumes on its nutritional quality: A comprehensive review. Cogent Food & Agriculture 2(1), 1136015. 1080/23311932.2015.1136015

Patterson CA, Curran J, Der T 2017.Effect of processing on antinutrient compounds in pulses. Cereal Chem. 94, 210. 10.1094/CCHEM-05-16-0144-FI

Raes, K, Knockaert, D, Struijs, K, & Van Camp, J (2014). Role of processing on bioaccessibility of minerals: Influence of localization of minerals and anti-nutritional factors in the plant. Trends in Food Science & Technology, 37, 32–41. 10.1016/j.tifs. 2014.02.002. ISSN 0924-2244.

Ranhotra GS, Lowe RJ, Puyat LV 1974. Phytic acid in soyabeans and its hydrolysis during bread making. J. Food Sci. 39, 1023–1025.

Reddy NRMD, Pierson SK, Sathe DK, Salunkhe. 1989. Occurrence, Distribution, Content and Dietary Intake of Phytate. CRC Press, Boca Raton, FL, USA. pp. 39–56.

Regvar M, Eichert D, Kaulich B, Gianoncelli A, Pongrac P, Vogel-Mikuš K, Kreft I 2011. New insights into globoids of protein storage vacuoles in wheat aleurone using synchrotron soft X-ray microscopy. J. Exp. Bot. 62, 3929–3939. 10.1093/ jxb/err090.

Rehman ZU, Shah WH 2005. Thermal heat processing effects on antinutrients, protein and starch digestibility of food legumes. Food Chem. 91, 327–331. 10.1016/j.foodchem. 2004.06.019

Roohani N, Hurrell R, Kelishadi R, Schulin R 2013. Zinc and its importance for human health: An integrative review. J. Res. Med. Sci. 18, 144–157. PMID: 23914218.

Samtiya M, Aluko RE, Puniya AK, Dhewa T 2021. Enhancing micronutrients bioavailability through fermentation of plant based Foods: A concise review. MDPI. 7, 63. 10.3390/ fermentation7020063Fermentation

Sandberg AS 1991. The Effect of Food Processing on Phytate Hydrolysis and Availability of Iron and Zinc. Nutritional and Toxicological Consequences of Food Processing. Springer Science, Business Media, New York. 10.1007/978-1-4899-2626-5_33

Sandberg AS 2002. Bioaccessibility of minerals in legumes. Br. J. Nutr. 88 (Suppl), S281–S285.

Sandström B, Davidsson L, Eriksson R, Alpsten M 1990. Effect of long-term trace element supplementation on blood trace element levels and absorption of Se, Mn, and Zn. J. Trace Elem. Electrolytes Health Dis. 4, 65–72. 10.1007/978-1-4613-0723-5_253

Sandström B, Sandberg AS 1992, Inhibitory effects of isolated inositol phosphates on zinc absorption in humans. J. Trace Elem. Elec. Health Dis. 6, 99–103. PMID: 1422186.

Srikka P 1997. Myoinositol phosphates: Analysis, content in foods and effects in nutrition. Lebensm. Wiss. U. Technol. 30, 633–647. 10.1006/fstl.1997.0246

Solomons NW 2001. Dietary sources of zinc and factors affecting its bioavailability. Food Nutri. Bull. 22, 138–154. 10.1177/156482650102200204.

Solomons NW, Jacob RA 1981. Studies on the bioavailability of zinc in humans: effect of heme and nonheme iron on the absorption of zinc. Am. J. Clin. Nutr. 34, 475–482. 10. 1093/ajcn/34.4.475

Tonin IP, Ferrari CC, da Silva MG, de Oliveira KL, Berto MI, da Silva VM, Germer SPM 2018. Performance of different process additives on the properties of mango powder obtained by drum drying. Drying Technol. 36, 355–365. 10.1080/07373937.2017. 1334000

Vadivel V, Biesalksi HK 2012. Effect of certain indigenous processing methods on the bioactive compounds of ten different wild type legume grains. J. Food Sci. Technol. 49, 673e684. 10.1007/s13197-010-0223-x

20 Zinc Enhances Immunity in Marine Finfishes

B. Deivasigamani and A. Queen Elizabeth

20.1 INTRODUCTION

With the increased world population, there is a demand for animal proteins, and hence, their production through aquaculture has intensified in many countries (Ellis et al., 2017; Osmundsen et al., 2020). The unique functions of zinc on immunological response and antioxidant activities are effectively utilized in aquaculture projects for substantial output and health of aquatic organisms such as fish (Kruse-Jarres, 1989). Finfishes *Oreochromis niloticus* and *Mugil cephalus* are examples of an aquatic ecosystem with a high capacity for tolerance and growth in various environmental conditions yielding high profits commercially (EI-Sayed, 2006).

The first study conducted about 90 years ago reported the importance of zinc in the development and growth in rat models (Todd et al, 1933). Later studies revealed novel insights into the relationship between diet zinc and inflammation, oxidative stress and growth in insulin as well as lipid metabolic activities. Zinc is the main source of zinc-metalloenzyme action and the active central ion includes alcohol dehydrogenase, superoxide dismutase, carbonic anhydrase and alkaline phosphatase enzymes (Rink and Gabriel, 2000; Miranda and Dey, 2004; Hara et al., 2017). It inhibits the oxidative capacity physiologically (Salgueiro et al., 2000; Ogawa et al., 2011) and its cofactor with copper promotes the antioxidant action (Chalaux et al., 1999). Reports have stated that oxidation resistance and immunity activate ALP, SOD and phenoloxidase in white Pacific shrimp (Lin et al., 2013) and *Penaeus monodon* grass shrimp. Zinc dietary supplement offers immunity, bioavailability, oxidative resistance and growth of marine finfishes *O.niloticus* and *M.cephalus* as well as other species (Lin et al., 2013; Katya et al., 2016). However, the effectiveness of dietary zinc supplements in finfishes remains to be studied in detail.

In this chapter, the effect of a zinc complex (Glu-Zn) as a diet supplement in growth performance, antioxidant activity and immune response of two marine finfishes are discussed.

20.2 ZINC ENHANCEMENT IN MARINE ORGANISMS

Zinc is the major constituent of cell organelles and is found in a variety of aquatic organisms. The important role of zinc is based on the catalyst part of metallo-enzymes

DOI: 10.1201/9781003412472-20

activity (Ikem and Egiebor, 2005; Marín-Guirao et al., 2008), the elevation of which becomes toxic to the aquatic species. Many researches showed that the dietary supplement of zinc is the basic cause of its increase in marine fishes (Xu and Wang, 2002). Elevated levels of zinc can also affect the pancreatic tissues and protein metabolism, disrupting their function and leading to arteriosclerosis (Demirezen and Uruc, 2006; Ahmed et al., 2016).

20.3 EXPERIMENTAL STUDIES

20.3.1 MATERIALS

Formulated raw Gluconate Zinc nutrition (Glu-Zn) and lysozyme measuring kit were obtained from Hi-Media Chemicals, Mumbai. The formulation consists of 5.0 g of Glu-Zn dissolved in 10 ml distilled water. The diet was prepared from this stock solution on a daily basis. Five concentrations of experimental diet were formulated *viz.* 42.4 (diet base), 52.4, 62.4, 72.4 and 82.4 mg/L (Shi et al., 2021).

20.3.2 METHODOLOGY

The active live species of *O. niloticus* and *M. cephalus* were collected from Annan Kovil Landing Centre, Parangipettai, Cuddalore District of Tamil Nadu (India) and acclimatized in a fibre-glass tank (20 L) for two weeks fed with a commercial diet to adjust to the conditions of experimental diets. After this period, 50 each of *O. niloticus* (Wt. 79.56 ± 0.03 g) and *M. cephalus* (Wt. 76.18 ± 0.03 g) were separately divided into five groups (n = 10) and fed with an experimental diet. A separate set of ten fishes served as a control fed with a commercial diet (non-zinc). The tanks were maintained in 12 h day and 12 h night cycles for 60 days with feeding at regular intervals. The diet feed was prepared according to the 4% body weight of the fish and adjusted to a subsequent level every two weeks with the weight gain of the fish (Luo et al., 2021). All the fish were fed with fullness. The leftover and the faeces were removed from the tank periodically to avoid contamination. The tanks were supplied with constant aeration and regular flow-through water. All physio-chemical parameters were maintained throughout the experimental period. The water quality was maintained in the experimental tank uniformly (Temperature 30.82 ± 1.820°C; pH 8.1 ± 0.27; Dissolved Oxygen 5.92 ± 0.43 mg/l; Salinity 13.92 ± 5.78 ppt; Alkalinity 170.6 ± 4.26) and checked regularly.

After 60 days, all fishes were fasted for 24 h before sample collection and ten fish from each tank were taken and observed for weight gain (WG%), feeding efficiency (FE%), condition factor (CF g/cm^3) and survival rate (%). Five fish were selected randomly from each tank and blood was collected from the caudal vein without disturbing the fish (Dawood et al., 2016). A further five samples each were collected from the control and treated groups and anaesthetized using clove oil. 1 ml of blood samples were withdrawn from the caudal vein with disposable heparinized syringes and stored in Eppendorf vials coated with EDTA to analyze haematological and biochemical changes. The blood was left to at 40°C for 15 min and centrifuged at 3000 rpm for 20 min to get fresh serum. It was analyzed for Zn

TABLE 20.1
Effect of Dietary Glu-Znon Serum Immune Parameters of Experimental Finfishes

Parameters	Control Commercial Diet	*O.niloticus* Glu-Zn Fed	Control Commercial Diet	*M.cephalus* Glu-Zn Fed
lysozyme(U/ml)	8.23 ± 1.02^b	9.08 ± 1.02^c	7.22 ± 1.02^a	8.61 ± 1.16^c
IgM(g/L)	0.32 ± 0.09^a	0.39 ± 0.13^a	0.43 ± 0.18^b	0.51 ± 0.22^b
C3(g/L)	0.77 ± 0.12^c	0.84 ± 0.18^a	0.70 ± 0.20^c	0.89 ± 0.22^a
C4(g/L)	0.80 ± 0.09^a	0.83 ± 0.09^b	0.78 ± 0.17^b	0.81 ± 0.12^b

IgM- immunoglobulin M; C3, complement 3; C4, complement 4.

Values are mean±SEM of three replicates. $p<0.05$ among treatments and a,b,c represent the values in the same row.

source and immune assays (lysozyme, IgM, C3 and C4) from two species separately. Values of three replicates with mean±SEM were obtained. One-way ANOVA followed by Tukey's test was used to compare the mean values to find out the significance ($p<0.05$) between various groups.

20.3.3 RESULTS

The results of the study indicating growth performances and immunological parameters are presented in Tables 20.1 and 20.2. The immune levels of *O. niloticus* and *M. cephalus* fed with Glu-Zn diets were increased as compared to those fed with non-zinc commercial feed (control). The supplement (Glu-Zn) significantly increased lysozyme and immunological parameters also compared to control fishes.

20.3.3.1 Bioaccumulation of Zinc

Zinc bio-accumulates in the organs and body tissues such as the hepato-pancreas depending upon its concentration. The rate of its deposition is enormously elevated with dietary zinc concentration. Finfishes fed with 42.4 mg/L had a minimum deposit of Zn whereas 82.4 mg/L diet base had a maximum concentration of deposition. Bioaccumulation in the muscle, gill and liver was compared with the fishes not fed with zinc supplements. Glu-Zn accumulates copper, iron and manganese also in the liver, muscle and intestine of treated fishes and plays a positive role in increasing the accumulation of minerals in tissues.

20.3.3.2 Zinc Enhances Immunity

Zinc enhanced the immunity of the fishes. The intake of zinc feed resulted in a significant increase in immunological activity as compared to those in unfed groups. In both the species of finfishes, Glu-Zn increased lysozyme, IgM, C3 and C4 when compared to unfed fishes (Table 20.1). Zinc promotes the inflammasome through C3 and C4 action complement activation. Phagocytosis is one of the major activities in innate immunity. Antibodies such as IgM produce adaptive immunity. Our study

TABLE 20.2

Effect of Dietary Glu-Zn on Growth, Feeding and Morphologic Characteristics of Experimental Finfishes

Parameters	Oreochromisniloticus	Mugil Cephalus
IBWT(g/fish)	79.56±0.03	76.18±0.03
FBWT(g/fish)	256.45±31.34	264.86±34.45
WG(%)	263.67±53.95	245.91±51.94
FI(g/fish)	223.43±9.56	214.37±9.52
FE(%)	93.17±18.21	92.14±17.23
CF(g/cm^3)	2.02±0.11	2.03±0.13
SR (%)	100	100
VSI(%)	7.67±1.08	7.51±1.01
HSI(%) crude protein (%)	2.52±0.49	2.17±0.47
Ash(%)	16.34±1.13	16.06±1.33
Moisture(%)	70.13±0.52	71.43±0.56
Crude lipid(%)	18.93±1.37	19.93±1.32
Crude protein(%)	57.09±0.93	56.09±0.84

IBWT- Initial Body Weight, FBWT- Final Body Weight, WG- Weight Gain, FI- Feed Intake, FE-Feeding Efficiency, CF- Condition Factor, SR- Survival Rate, VSI-Vicero Somatic Index, HIS- Hepato Somatic Index. Values are three replicates of mean ± SEM variation (n =5) with $P < 0.05$.

demonstrated that the fishes fed with Glu-Zn could increase immunity as observed in the serum protein and influence the action of immunity in *O. niloticus* and *M. cephalus* (Figure 20.1).

20.3.3.3   Hematology and Hormonal Changes

In an earlier study, biochemical changes were reported with an increase in blood glucose, lipid, ALT and AST levels in the treated groups of Nile tilapia with BIO-ZnONPs and CH-ZnONPs (El-Saadony et al., 2021). The same observations were noted in the finfishes when treated with Glu-Zn. High modulations in the LYZ and NBT cell organelles as well as significant hormonal changes such as FSH, GH, testosterone and cortisol were also recorded. Hence, it may be inferred that zinc supplements change the blood biochemistry, hematology and hormone levels and also influence the development of fish species.

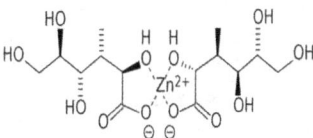

FIGURE 20.1   Structure of Gluconate-Zinc (Glu-Zn).

20.3.3.4   Growth Performance

The results on the effect of Glu-Zn supplement on the growth performance of two finfishes are presented in Table 20.2 which indicate that the diet improved zinc deposition and absorption. The differences between pre and post-treatments in the body weights, weight gain, feeding efficiency, condition factor, survival rate, viceroy somatic index, hepatic somatic index, ash, moisture, lipid crude and crude protein were recorded. A minimal WG ratio was observed from one of the tried feed regimens and the optimal ratio was found to be 68.6 mg/L in addition to increased development in the species and feed admission.

20.3.3.5   Antioxidant Ability

Dietary Zn supplement is a good source to increase the antioxidant activities in organisms. In the present study, the antioxidant levels in the finfishes were significantly raised with the Glu-Zn supplement and the level of MDA in hepato-pancreas was found to decrease thereby supporting the claim that a zinc-containing diet enhances antioxidant activity.

20.3.3.6   Zinc Signaling

Zinc transporters (ZNTs) mediate the signaling. Salute Carrier 39 level (SLC39) family carries Zn^{2+} and enters the biological membranes. Extracellular Zn^{2+} activates SLC 39 and intracellular Zn^{2+} binds to ALP and SOD to regulate the action of the antioxidant and immune response. Excess amount stored in mitochondria maintains homeostasis of the cell to enumerate intra-cellular Zn^{2+}. Zinc signaling helps to study the physiological process of the cell organelles and its role in the mechanism of zinc homeostasis in organisms. It also confirms its crucial role in them which are regulated by ZNTs specifically by zinc-binding proteins.

20.4   DISCUSSION

Earlier studies revealed the beneficial effects of organic minerals on immunity and oxidation resistance in *Ictalurus punctatus* fish (Li and Robinson), *Oncorhynchus mykiss* rainbow trout (Apines-Amar et al., 2004) and Pacific white shrimp *Litopenaeus vannamei* (Yuan et al., 2018 and 2020). Usually, zinc feeds contain its inorganic salts (carbonate and sulfate) and inorganic zinc compounds were found to pollute water resources and the environment in aquaculture. Further, they become toxic in water bodies due to concentration and poor absorption capacity at low levels (NRC, 2011; Kumar et al., 2017).

Administration of inorganic zinc diets has led to higher zinc concentrations in the intestines of organisms forming insoluble substances in the digestive canal (Predieri et al., 2005). In *O.niloticus* and some other species, zinc additives improved the bioavailability of zinc (Do Carmo e Sá et al., 2005). Previous investigations have also indicated the beneficial effects of Glu-Zn in the growth, antioxidation, bioavailability and immune responses of finfishes (Yuan et al., 2020; Luo et al., 2021). A few studies demonstrated zinc supplements' role in non-specific immunological responses in Nile tilapia (Huang et al., 2015; Awad et al., 2015).

This species showed an immense response and resisted bacterial infections on zinc supplementation. The present study demonstrated the beneficial effect of Glu-Zn in increasing immunity as observed in serum protein and its influence on immunity in the finfishes investigated.

A similar immune stimulation response was also reported by Elseweidy et al. (2017). Many researchers found that Glu-Zn is a biomarker for inflammatory action and oxidation (Ansari et al., 2019) and indicated that the diet stimulated the action of innate and adaptive responses in *O. niloticus* and *M. cephalus*. In the present investigation, it was observed that the immune response in finfishes due to Glu-Zn feeding offered a maximum role in the innate and adaptive immune responses. The immune response and modulating capacity were also found to be influenced by zinc signaling. The action of zinc was high in mRNA expression in hepato-pancreas (Moazenzadeh et al., 2018). This study also revealed the lysozyme functional status of the immune response of the studied fishes by the enhancement of zinc action in the pathways of innate and adaptive immunity which act together to face the challenges of immune response and constitute different defense mechanisms.

20.5 CONCLUSION

The effect of dietary zinc supplement Glu-Zn has been studied in two marine finfishes *O. niloticus* and *M. cephalus*. The growth performance, antioxidant activity and immune responses were found to be enhanced in them on supplementation. The dietary zinc could activate the minerals by mediating and signaling molecules. The relationship between the zinc source supplements and pathways gave novel insights into the role of zinc in fish health and diseases. The study also suggested that zinc enhancement is required to increase fish production besides achieving quality fish culture.

ACKNOWLEDGEMENT

I would like thanks to authorities of Annamalai University Annamalai Nagar Chidambaram Tamil Nadu India for the support and help.

REFERENCES

Ahmed M, Ahmad T, Liaquat M, Abbasi K, Abdel-Farid I, Jahangir M 2016. Tissue specific metal characterization of selected fish species in Pak. Environ. Monit. Assess. 188, 212. doi.org/10.1007/s10661016-5214-6.

Ansari MM, Ahmad A, Mishra RK, Raza SS, Khan R 2019. Zinc gluconate-loaded chitosan nanoparticles reduce the severity of collagen-induced arthritis in Wistar rats. ACS Biomater. Sci. Engg. 5, 3380–3397. doi.org/10.1021/acsbiomaterials. 9b00427.

Apines-Amar MJS, Satoh S, Caipang CMA, Kiron V, Watanabe T, Aoki T 2004. Amino acid-chelate: A better source of Zn, Mn, and Cu for rainbow trout, *Oncorhynchus mykiss*. Aquaculture. 240, 345–358. doi.org/10.1016/j.aquaculture.2004.01.032.

Awad A, Zaglool AW, Ahmed SAA, Khalil SR 2015. Transcriptomic profile change, immunological response, and disease resistance of *Oreochromis niloticus* fed with conventional and Nano-Zincoxide dietary supplements. Fish Shellfish Immunol. 93, 336–343. doi.org/10.1016/j.fsi.2019.07.067.

Chalaux E, López-Rovira T, Rosa JL, Pons G, Boxer LM, Batrons R, Ventura F 1999. A zinc-finger transcription factor induced by TGF-beta promotes apoptotic cell death in epithelial Mv1Lu cells. FEBS Lett. 457, 478–482. doi.org/10.1016/S0014-5793:01051-0.

Dawood, MAO, Koshio S, Ishikawa M, Yokoyama S, El Basuini MF, Hossain MS, Nhu TH, Dossou S, Moss AS 2016. Effects of dietary supplementation of Lactobacillus rhamnosus or/ and Lactococcus lactis on the growth, gut microbiota, and immune responses of red sea bream, Pagrus major. Fish Shellfish Immunol. 49, 275–285. doi.org/10.1016/j.fsi.2015.12.047.

Demirezen D, Uruç K 2006. Comparative study of trace elements in certain fish, meat and meat products. Meat Sci. 74, 255–260. doi.org/10.1016/j.meatsci.2006.03.012.

Do Carmo e Sá MV, Pezzato LE, Barros MM, De Magalháes Padilha P 2005. Relative bio availability of zinc in supplemental inorganic and organic sources for Nile tilapia Oreochromisniloticus fingerlings. Aquac. Nutr. 11, 273–281.

Ellis RP, Urbina MA, Wilson RW 2017. Lessons from two high CO2 worlds -future oceans and intensive aquaculture. Glob. Chang. Biol. 23, 2141–2148. doi.org/10.1111/gcb.13515.

El-Saadony MT, Alkhatib FM, Alzahrani SO, Manal E, Shafi, Shereen El, Abdel- Hamid, Taha F Taha, Salama M, Aboelenin, Mohamed. M, Soliman, Norhan H, Ahmed. 2021. Impact of my cogeniczincnano particle son performance, behavior, immune response, and microbialload in Oreochromisniloticus. Saudi J. Biol. Sci. doi.org/10.1016/j.sjbs. 2021.04.066.

EI-Sayed AFM 2006. Tilapia culture in saltwater: Environmental requirements, nutritional implications, and economic potentials. Avances in Nutricion Acuicola. Tilapia Culture, Page: 1–277. doi.org/10.1079/9780851990149.0000.

El - Seweidy MM, Ali AA, Elabidine NZ, Mursey NM 2017. Effect of zinc gluconate, and sage oil on inflammatory patterns and hyper glycemiain zinc-deficient diabetic rats. Biomed. Pharmacother. 95, 317–323. doi.org/10.1016/j.biopha.2017.08.081.

Hara T, Takeda T, Takagishi T, Fukue K, Kambe T, Fukada T 2017. Physiological roles of zinc transporters: Molecular and genetic importance in zinc homeostasis. J. Physiol. Sci. 67, 283–301. doi.org/10.1007/s12576-017-0521-4.

Huang F, Jiang M, Wen H, Wu F, Liu W, Tian J, Yang C 2015. Dietary zinc requirement of adult Nile tilapia (Oreochromis niloticus) fed semi-purified diets, and effects on tissue mineral compositionand antioxidant responses. Aquaculture. 439, 53–59. doi.org/10. 1016/j.aquaculture.2015.01.018.

Ikem A, Egiebor N 2005. Assessment of trace elements in canned fishes (mackerel, tuna, salmon, sardines, and herrings) marketed in Georgia and Alabama (United States of America). J.Food Compos. Anal. 18, 771 787. doi.org/10.1016/j.jfca.2004.11.002.

Katya K, Lee S, Yun H, Dagoberto S, Browdy CL, Vazquez-Anon M, Bai SC 2016. Efficacy of inorganic and chelated trace minerals (Cu, Zn, and Mn) premixsources in Pacific white shrimp, Litopenaeus vannamei (Boone) fed plant protein-baseddiets. Aquaculture 459, 17–123.

Kumar N, Krishnani KK, Kumar P, Jha AK, Gupta SK, Singh NP 2017. Dietary zinc promotes immuno-biochemical plasticity and protects fish against multiple stresses. Fish Shellfish Immunol. 62, 184–194. doi.org/10.1016/j.fsi.2017.01.017.

Kruse-Jarres JD 1989. Review the significance of zinc for humoral and cellular immunity. J. Trace Elem. Electrolytes Health Dis. 3, 1–8.

Lin SM, Lin X, Yang Y, Li, FJ, Luo L 2013. Comparison of chelated zinc and zinc sulfate as zinc sources for growth and immune response of shrimp (Litopenaeus vannamei). Aquaculture 406–407, 79–84. doi.org/10.1016/j.aquaculture.2013.04.026.

Luo F, Fu Z, Wang M, Ke Z, Wang M, Xiong W, Hasan WM, Shu X 2021. Growth performance, tissue mineralization, antioxidantactivity, and immune response of

Oreochromis niloticus fed withconventional and gluconic acid zinc dietary supplements. Aquac. Nutr. 27, 897–907. doi.org/10.1111/anu.13234.

Marín-Guirao L, Lloret J, Marin A 2008. Carbon and nitrogen stable isotopes and metal concentration in food webs from a mining-impacted coastal lagoon. Sci. Total Environ. 393, 118–130. doi.org/10.1016/j.scitotenv.2007.12.023.

Miranda ER, Dey CS 2004. Effect of chromium and zinc on insulin signaling in skeletal muscle cells. Biol. Trace Elem. Res. 101, 19–36. doi.org/10.1385/bter:101:1:19.

Moazenzadeh K, RajabiIslami H, Zamini A, Soltani M 2018. Effects of dietary zinc level on performance, zinc status, tissue composition, and enzyme activities of juvenile Siberian sturgeon, Acipenserbaerii (Brandt 1869). Aquacult. Nutr. 24, 1330–1339. doi.org/10.1111/anu.12670.

National Research Council (NRC) 2011. Committee on the Nutrient Requirements of Fish and Shrimp, Nutrient Requirements of Fish and Shrimp. National Academies Press, Washington, D.C., USA, pp. 176–179. dels.nas.edu/banr.

Ogawa D, Asanuma M, Miyazaki I, Tachibana H, Wada J, Sogawa N, Sugaya T, Kitamura S, Maeshima Y, Shikata K, Makino H 2011. High glucose increases metallothionein expression in renal proximal tubular epithelial cells. Exp. Diabetes Res. 1–8. doi.org/10.1155/2011/534872.

Osmundsen TC, Amundsen VS, Alexander KA, Asche F, Bailey J, Finstad B, Olsen MS, Hernández K, Salgado H 2020. The operationalization of sustainability: Sustainable aquaculture production as defined by certification schemes. Glob. Environ. Change 60, 102025. doi.org/10.1016/j.gloenvcha.2019.102025.

Predieri G, Elviri L, Tegoni M, Zagnoni I, Cinti E, Biagi G, Ferruzza S, Leonardi G 2005. Metal chelates of 2-hydroxy-4-methylthiobutanoic acid in animal feeding: Part 2: Further characterizations, in vitro and in vivo investigations. J. Inorg.Biochem. 99, 627–636. doi.org/10.1016/j.jinorgbio.2004.11.011.

Rink L, Gabriel P 2000. Zinc and the immune system. Proc. Nutr. Soc. 59, 541–552. doi.org/10.1017/S0029665100000781.

Salgueiro MJ, Zubillaga M, Lysionek A, Cremaschi G, Goldman CG, Caro R, De Paoli T, Hager A, Weill RJ 2000. Zinc status and immune system relationship a review. Biol. Trace Elem. Res. 76, 193–206. doi.org/10.1385/BTER:76:3:193.

Shi B, Xu F, Zhou Q, Regan MK, Betancor MB, Tocher DR, Sun M, Meng F, Jiao L, Jin M 2021. Dietary organic zinc promotes growth, immune response, and antioxidant capacity by modulating zinc signaling in juvenile Pacific white shrimp (Litopenaeus vannamei). Aquac. Rep. 19, 100638. doi.org/10.1016/j.aqrep.2021.100638.

Todd WR, Elvehjem CA, Hart EB 1933. Zinc in the nutrition of the rat. Am. J. Physiol. 107, 146–156.

Xu Y, Wang WX 2002. Exposure and potential food chain transfer factor of Cd, Se, and Zn in marine fish Lutjanus argent imaculatus. Mar. Ecol. Prog. Ser. 238, 173–186. www.jstor.org/stable/24866345.

Yuan Y, Jin M, Xiong J, Zhou QC 2018. Effects of dietary dosage forms of copper supplementation on growth, antioxidant capacity, innate immunity enzyme activities, and gene expressions for juvenile Litopenaeusvannamei. Fish Shellfish Immunol. doi.org/10.1016/j.fsi.2018.10.075.

Yuan Y, Luo J, Zhu T, Jin M, Jiao L, Sun P, Ward TL, Ji F, Xub G, Zhou Q 2020. Alteration of growth performance, meat quality, antioxidant and immune capacity of juvenile Litopenaeus vannamei in response to different dietary dosage forms of zinc: Comparative advantages of zinc amino acid complex. Aquaculture. 2020, 735120. doi.org/10.1016/j.aquaculture.2020.735120.

21 Pharmaceutical and Biomedical Applications of Zinc Oxide Nanoparticles
Current Trends and Future Prospects

Kannabiran Krishnan

21.1 INTRODUCTION

Zinc is the most abundant mineral in human brains, essential for gene expression, enzyme catalysed reactions, protein synthesis, learning process, memory function, and acts as a cofactor for matrix metalloproteinases (MMPs), and carbonic anhydrases (CAs) (Leuci et al., 2020). It is also required for the proliferation and differentiation of cells; apoptosis of cells; as well as immunological, reproductive and neuronal activation functions (Maret and Sandstead, 2006). Animal meat, shellfish, plant legumes, seeds, nuts, mushrooms, spinach, and broccoli are natural sources for Zn (Keerthana and Kumar, 2020). Any alterations in Zn homeostasis due to genetic and dietary deficiency lead to many severe diseases (Anzellotti and Farrell, 2008). Zn is widely used for producing various forms of nanomaterial and particles. Green or biological synthesis of ZnNPs has several advantages over conventional methods with respect to eco-friendliness, nontoxic, and safety aspects (Kalpana and Rajeswari, 2018). ZnO NPs possess several unique properties like different morphology, high surface area to volume ratio, and biocompatibility (Mahamuni-Badiger et al., 2020). ZnO is a safe mineral and possesses UV absorption characteristics, and hence used in personal care products such as cosmetics, food additives, dentistry, textiles, and sunscreens (Ali et al., 2018).It also possesses electrical and magnetic properties. Due to its biocompatibility, less toxicity and minimal cost, ZnO NPs are extensively used for multiple therapeutic applications. The Zn salt is widely used for bioimaging, gene delivery, biosensor applications, plant fertilizer, pesticides, soil improvement, water purification, etc. ZnO NPs possess anticancer (Hu, 2019) antioxidant (Yin, 2019), and antimicrobial (Singha, 2019) anti-inflammatory (Kim, 2014) and antidiabetic (Umrani and Paknikar, 2014) activities. Recently nanocomposites (NCs) of

DOI: 10.1201/9781003412472-21

FIGURE 21.1 Pharmaceutical and biomedical applications of zinc oxide nanoparticles.

ZnO NPs with natural polymers such as cellulose, chitosan, and alginate have been used as tissue scaffolds in wound healing (Alavi and Nokhodchi, 2020).

Chronic exposure to ZnO NPs causes several diseases. The toxicity and beneficial effect of ZnO NPs depend on the concentration and duration of exposure to humans (Keerthana and Kumar, 2020). In this review, the pharmaceutical and biomedical applications of ZnO NPs, current status and future prospects are discussed (Figure 21.1).

21.2   SYNTHESIS ZnO NPs

ZnO NPs are synthesised either by bottom-up or top-down approaches. The physiochemical techniques used for the synthesis of ZnO NPs possess several limitations which restrict its uses (Alamdari et al., 2020). Hence, a green chemistry approach using natural sources (plant and microbial extracts) for bioreduction of Zn is safe, environment friendly, and no toxic by-products are formed. Moreover, these NPs possess expected optical, electronic and chemical properties. ZnO NPs can be prepared with different morphologies such as nanorods, nanoplates, nanospheres, nanoboxes, hexagonal, tripods, tetrapods,

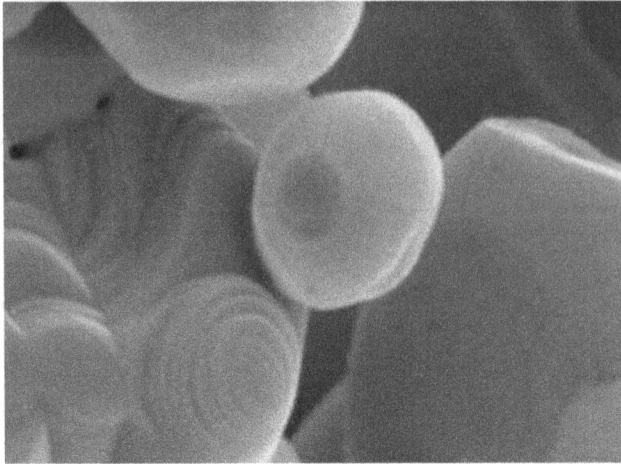

FIGURE 21.2 SEM morphology of ZnO NPs.

nanowires, nanotubes, nanorings, nanocages, and nanoflowers (Siddiqi et al., 2018). The structure of ZnO NPs is shown in Figure 21.2.

21.3 ANTIMICROBIAL PROPERTY OF ZnO NPs

21.3.1 ANTIBACTERIAL ACTIVITY OF ZnO NPs

Several reports are available on broad spectrum antimicrobial activity of ZnO NPs against many pathogens (Figure 21.3).

ZnO NPs produced from *KalanchoeBlossfeldiana* aqueous extract showed activity against *S. aureus*; *Escherichia coli and Pseudomonas aeruginosa* (Aldalbahi et al., 2020). ZnO NPs are used as a therapy against urinary tract infections (UTIs) to combat multidrug-resistant strains (Shalom et al., 2017). The mechanism of antibacterial activity of NPs is due to its small size and electrostatic interactions with the microbial cell membrane. ZnO NPs induce the release of oxygen free radicals by microbes and increased membrane lipid peroxidation which promotes membrane leakage and cellular damage. It also decreases the hydrophobicity of bacterial cell surface and down-regulate the expression of certain genes in bacteria. ZnO NPs prevent hemolysin toxin production and biofilm formation by *Stphylococcus aureus*. The H_2O_2-induced photocatalytic activity and associated bacterial death have been reported (Tiwari et al., 2018). It also affects DNA replication and protein synthesis thereby blocking the bacterial electron transport chain.

Since ZnO NPs possess antimicrobial properties, it is used for food preservation (Azizi-Lalabadi et al., 2019) and food packaging (Sirelkhatim et al., 2015). ZnO is blended into the linings of food cans used for packing meat, fish, corn, and peas for the preservation of food colour and prevention of food spoilage. ZnO NC (Zinc oxide Nanocomposite) along with zeolite (adsorbent) support increases the antibacterial activity against Gram-positive and Gram-negative bacterial pathogens.

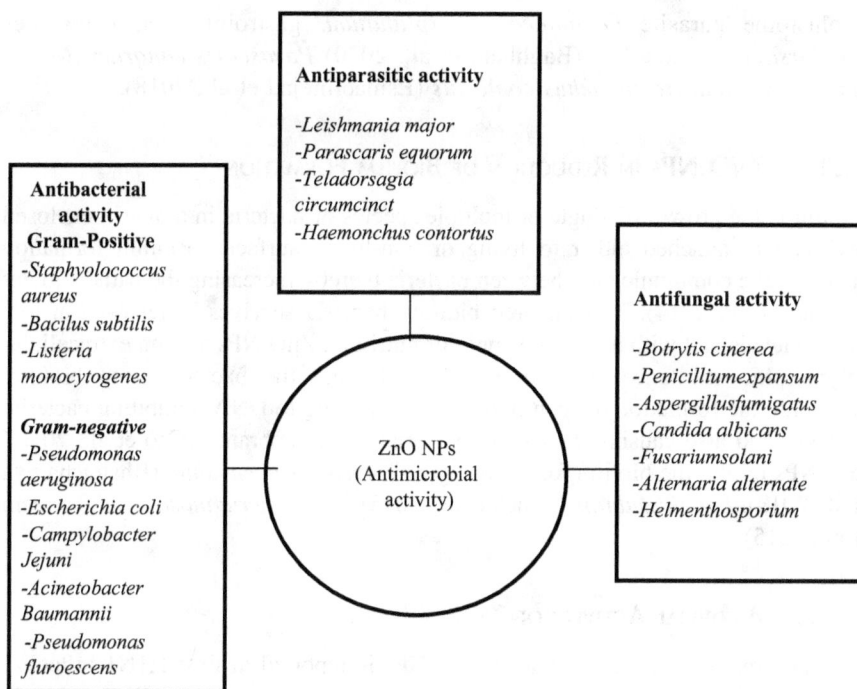

FIGURE 21.3   Antibacterial (Gram-positive and Gram-negative bacteria), antiparasitic and antifungal activity of zinc oxide nanoparticles.

Zeolite\ZnO-CuO NC showed better antibacterial activity on both Gram-positive (*Bacillus subtilis* B29) and Gram-negative bacteria (*Escherichia coli* E266) (Alswat et al., 2017).

21.3.2 ANTIFUNGAL ACTIVITY OF ZNO NPS

Green-synthesised ZnO NPs showed antifungal activity against pathogenic fungi, *Botrytis cinerea* and *Penicillium expansum* (He et al., 2011) and *Aspergillus fumigatus* and *Candida albicans* by inhibiting the cellular functions has been reported (Jasim, 2015). The antifungal activity against *Fusarium solani, Alternaria alternate* and *Helmenthosporium* has been reported by ZnO NPs synthesised using *K. Blossfeldiana* aqueous extract (Sirelkhatim et al., 2015). Fluconazole-resistant *Candida albicans* attachment on the surface of catheters are reduced by treatment with ZnO NPs (Hosseini et al., 2018).

21.3.3 ANTI-PARASITIC FUNCTION OF ZNO NPS

ZnO NPs exposure causes increased membrane permeability, formation of more reactive oxygen free radicals and associated lipid and protein oxidation results in apoptosis in *L. major* (Khan et al., 2015; Nadhman et al., 2016). ZnO NPs inhibit amphistome parasite *Gigantocotyle explanatum,* gastrointestinal nematodes *Teladorsagia circumcinct* (Baghbani et al., 2020) *Parascaris equorum* (Morsy et al., 2019) and *Haemonchus contortus* (Esmaeilnejad et al., 2018).

21.3.4   ZnO NPs in Reduction of Biofilm Formation

Biofilm is the growth of single or multiple species of bacteria in a matrix-enclosed environment attached either to living or non-living surfaces. Biofilm formation increases the communication between bacteria thereby increasing the pathogenicity (Wessel et al., 2014). In a matured biofilm, bacteria survives better even at low oxygen levels and with a limited supply of nutrition. ZnO NPs acts on extracellular polymorphic substances (EPSs), thereby releasing the bacteria from biofilm, increasing the formation oxygen free radicals, binding to DNA, inhibiting bacterial enzymes and also causing damage to bacterial cell membranes (Cao et al., 2020). ZnO NPs inhibit the biofilm formation by *Streptococcuspneumonia* (Bhattacharyya et al., 2018), *Bacillussubtilis* (Hsueh et al., 2015), and *P. aeruginosa* (Abdulkareem et al., 2015).

21.3.5   Antiviral Activity of ZnO NPs

The antiviral activity of PEGylated ZnO NPs is reported against H1N1 influenza virus (Ghaffari et al., 2019); capsid protein of chikungunya virus (Kumar et al., 2018); herpes simplex virus (HSV-1 & HSV-2) (Tavakoli et al., 2018) and RNA viruses (Ishida, 2019).

21.4   BIOMEDICAL APPLICATIONS OF ZnO NPs

21.4.1   ZnO NPs in Wound Healing

Zn plays a major role in wound repair which include fibrin clot formation, increasing cellular inflammation, cell proliferation, epithelization, granulation, angiogenesis, and remodelling of the extracellular matrix (Nethi et al., 2019). The ZnO/cod liver oil paste is found to be very effective in wound healing (Kumar et al., 2012). ZnO NPs composite with a chitosan hydrogel serves as an ideal material for wound dressing (Kietzmann et al., 2006). Alginate/ZnO nanocomposite bandages are used on wounds to control *S. aureus* and *E. coli* infections (Mohandas et al., 2015). ZnO NPs along with collagen and 1% orange essential oil increase wound healing with excellent biocompatibility (Balaure et al., 2019).

21.4.2   ZnO NPs in Drug Delivery

ZnO NPs, due to their smaller size, can easily pass through the fine blood vessels and capillaries and are easily absorbed by the cells, leading to the deposition of drugs at the target sites. Being biodegradable in nature, it allows the drugs to be released at the targeted site for a few days to weeks (Shinde et al., 2012). Drugs delivered through NPs exert a moderate effect on the biological membrane.

Combinatorial drug (cisplatin and docetaxel) loaded, folate and PEG-activated ZnO NPs are used as a targeted drug delivery agent in the treatment of nasopharyngeal carcinoma (Sudhagar et al., 2011). ZnO quantum dots (QD) possess excellent optical properties, smaller in size and are inexpensive for preparation (Martínez-Carmona et al., 2018). It has therapeutic and diagnostic functions due to its properties of producing highly reactive oxygen free radicals, catalytic efficiency, adsorption capability and isoelectric point (Wu et al., 2008). Also used for drug delivery and to treat cancer, bacterial and fungal infections, multidrug-resistant pathogens (MDR) (Preeti et al., 2020), inflammatory diseases, wound healing and diabetes (Guo and Sun, 2020). The ZnO capable of producing reactive oxygen species-mediated oxidative stress is responsible for killing cancer cells (Huarc et al., 2010).

21.4.3 ZnO NPs in Bioimaging

ZnO QDs possess semiconductor and luminescent properties, low-cost, low-toxicity penetration power into the cell nucleus and biocompatibility that's helpful for use in bio-imaging. ZnO QDs used for ZnO NPs conjugated with transferrin is used for imaging cancer cells with very minimal toxicity (Yang et al., 2012). ZnO NPs in colloidal solutions conjugated with cations (Co, Cu, or Ni) are stabilized and used for cellular imaging studies (Liu et al., 2011). ZnO nanorods conjugated with anti-epidermal growth factor receptor antibody are used for imaging of cancer cells (Liu et al., 2011). For cell imaging purposes, when ZnO QDs are placed inside the cell cytoplasm exhibits a stable luminescence under UV light. ZnO NPs are engineered and widely used for multimodal detection using Gd-doped ZnO QDs (6 nm) (Singh, 2011). Similarly Fe_3O_4-ZnO core-shell magnetic QDs are used for imaging of cancer cells and photodynamic therapy. The magnetic property helps in specific targeting and delivery of photosensitizer into the cancer cells.

21.4.4 ZnO NPs in Gene Delivery

ZnO is used as a gene-delivery agent for gene therapy. ZnO (tetrapod structure) is used to deliver pEGFPN1 DNA to human melanoma cells (A375) (Nie et al., 2006a, 2006b). The pDNA (plasmid DNA) binds to ZnO is through electrostatic interactions; the three legs are preferred for internalization and gene delivery within the cells. The pDNA condensed by poly(2-(dimethylamino)ethyl methacrylate) (PDMAEMA) polymers and coated with ZnO QDs are used for gene delivery (Zhang and Liu, 2010). When PDMAEMA is used as a gene vector, ZnO QDs show less cytotoxicity.

21.4.5 ZnO NPs as Anticancer Agents

ZnO NPs showed increased permeability and retention effects on cancer cells (HepG2, PC3 A549, B16F10/A375, HeLa, HNSCC, LoVo/CaCo-2, MCF-7, and T98G) (Jiang and Cai, 2018). Cancer cells are killed by exposure to oxygen free radicals. Photo-stimulated ZnO QDs loaded with paclitaxel and cisplatin are used as therapy against HNSCC cells under UV-A irradiation (Hackenberg et al., 2012).

Doxorubicin and daunorubicin are given to leukemic cells (MCF-7, A549, K562 (sensitive) and K562/A02 (resistant) cells, using ZnO NP-mediated drug delivery systems (Tripathy et al., 2015). Multilamellar liposomes are used to load daunorubicin with ZnO NPs (hexagonal) for pH-sensitive drug release against cancer cells (Sharma et al., 2016). ZnO NPs capped with amino polysiloxane are used to generate synergistic anticancer activity on leukemic cells. ZnO NPs produced from *O. americanum* leaf extracts have been shown to exert anti–proliferative activity on human skin cancer cell lines (A431). ZnO NPs synthesized from *Sargassum muticum* extracts showed cytotoxicity and anti-angiogenesis activity on liver cancer cells (HePG2) (oxygen free radical induced DNA damage) (Chung et al., 2015). Curcumin-loaded ZnO NCs showed anticancer cytotoxic activity against the rhabdomyosarcoma RD cell line and lowest toxicity against normal human embryonic kidney cells (Jin and Jin, 2019). Increased membrane permeability to ZnO NPs leads to DNA fragmentation and apoptosis of cancer cells (Vidhya et al., 2020). ZnONPs are also used in photodynamic therapy (PDT); its exposure to light produces highly reactive oxygen free radicals and its phototoxicity are responsible for killing cancer cells (Yi et al., 2020).

21.4.6 ZnO NPs as Biosensors

Currently, various biosensors containing ZnO NPs find extensive applications in healthcare, food-processing industries and in environmental monitoring gadgets. Biosensors are categorized as electrochemical, photometric, piezoelectric, and calorimetric based on the detection principles (Zhao et al., 2010). Stability, better biocompatibility, higher surface area and electron transfer potential are advantageous to the use of nanomaterials for developing high-performance biosensors for immobilizing antibodies, enzymes, and other proteins (Kumar and Chen, 2008).

Apart from its semiconducting properties, ZnO nanomaterials also exhibit various capabilities such as biosensing, adsorption, isoelectric points (IEPs), and catalytic efficiency. Because antibodies and enzymes possess less IEP, they can be readily adsorbed with ZnO nanomaterials by electrostatic interaction (Wang et al., 2006). Many small molecules, cholesterol, phenol, urea, glucose, H_2O_2, etc., are analysed using ZnO-based biosensors (Figure 21.4) (Pachauri et al., 2010). ZnO NPs attached cysteamine nanosensor (ZnO-Cys) is used for the identification of N-acyl-homoserine lactones (AHLs) produced by *Pseudomonas aeruginosa* in patients having urinary tract infections (UTIs) (Vasudevan et al., 2020).

21.4.7 Biodental Applications

ZnO NPs capped with ZnO-chitosan are used as dentine bonding agents. It reduces *Streptococcus mutans* and *Lactobacillus acidophilus* infections in reinforced dental adhesive discs. It also showed desirable mechanical strength, high resistance to water absorption and solubility, a very high release profile, and variations in shear bond strength values (Javed et al., 2020). Bacteria that cause dental caries in the mouth are controlled by incorporating the ZnO NPs into the composite resin as

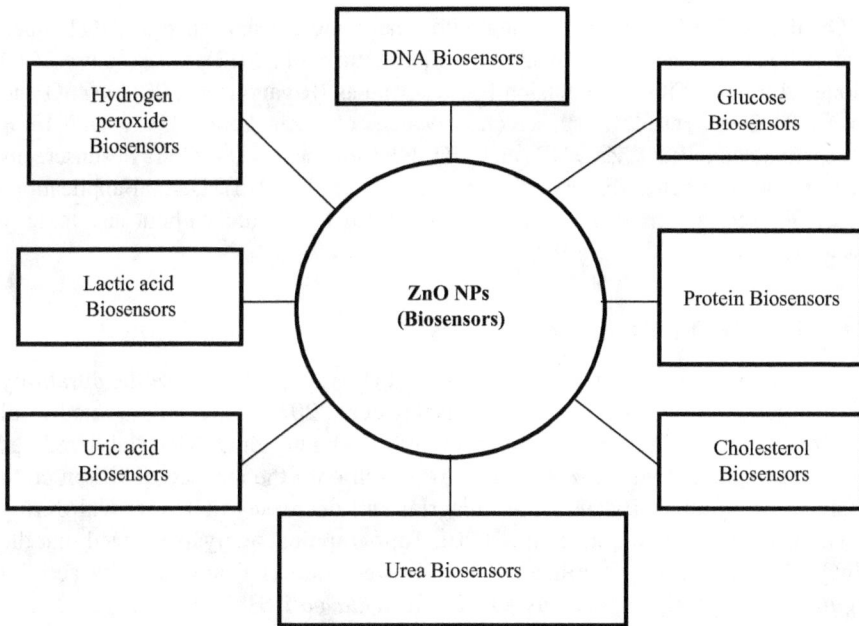

FIGURE 21.4 Zinc oxide nanoparticles in various biosensor applications.

antimicrobial agents (Foong et al., 2020). Antibacterial properties of dental composite resins are improved by adding 1 to 2% of ZnO NPs (Arun et al., 2021).

21.4.8 ANTIDIABETIC ACTIVITY OF ZnO NPs

Chitosan-capped ZnONPs showed comparatively higher antidiabetic activity (69.6%) than ZnONPs (59.4%) and chitosan (46.3%) alone(Arvanag et al., 2019). Chitosan capping of ZnONPs increases chemical bonding and reactivity of surface atoms by increasing the surface area to volume ratio (Javed et al., 2016).

21.4.9 ANTI-INFLAMMATORY FUNCTION OF ZnO NPs

ZnO NPs inhibit inflammatory pathways and pro-inflammatory mediator gene expression; downregulates IL-6, IL-1β, IL10, TNF-α and NF-κB pathways by post-translational modifications (Agarwal and Shanmugam, 2020). ZnO NPs showed antioxidant function (45.47%) by scavenging DPPH and strong anti-inflammatory function by suppressing mRNA and protein expressions of iNOS, COX-2, IL-1β, IL-6 and TNF-α (Nagajyothi et al., 2015).

21.4.10 ZnO IN COSMETIC PRODUCTS

Currently, nano-materials are used in body, hair, skin, oral and sun care products (Fytianos et al., 2020). ZnO is widely used in different cosmetic formulations;

L'Oréal S.A (USA) uses ZnO along with other nano-matrials such as TiO2, silica and carbon black in some of their preparations (Raj et al., 2012). Shiseido uses ZnO along with nano-TiO2 in emulsion-based formulas (Rigano et al., 2016). ZnO and micro-TiO2 are used in sunscreens because of their better UVA and UVB absorption capabilities. ZnO NPs and TiO$_2$ NPs are used as UV filters in sunscreens because of their better dispersion properties (Lu et al., 2015). Dermal applications of ZnO-based sunscreen formulas are considered to be safe without any toxicity issues (Mohammed et al., 2019).

21.4.11 ZnO NPs as UV-absorbers and Antimicrobial Textiles

Cotton and wool exposed to ZnO NPs shield UV rays and increase the durability and air permeability of cotton fabrics (Belay et al., 2020). ZnO NPs-treated wool absorbs the UV light in the range of 300–400 nm. ZnO NPs stabilized (as entrapped species) with 3% soluble corn starch onto the surface of cotton cloth increases the durability of cotton material and decreases their removal during washing cycles (El-Nahhal et al., 2020). Topographical analysis showed that the ZnO NPs treated and stabilized fabric showed increased antibacterial activity *against Staphylococcus aureus* and *Escherichia coli* (El-Nahhal et al., 2020). Antimicrobial textiles are used to increase sanitation, personal protection and to prevent disease transmission. ZnO coating on textile surfaces acts as a light-induced sterilising agent, microbicidal, UV light protection, flame resistant, heat protection, moisture maintenance, hydrophobicity, electrical conductivity, etc. (Verbic et al., 2019).

21.5 CONCLUSIONS

The importance of ZnO and ZnO NPs has been known for years, but its pharmaceutical and biomedical applications were only realized in the recent past. It plays a crucial role as a therapeutic agent, which includes wound healing, anticancer-cytotoxic activity, photodynamic therapy, antimicrobial and antiparasitic activities, antidiabetic, anti-inflammatory properties, drug delivery, gene delivery functions, disease diagnosis and cosmetic preparations. Its biomedical applications include tissue engineering, bioimaging, biosensors, nanosensors, biodental applications and anti-biofilm agents. ZnO NPs coatings in the catheter have the ability to reduce infections in UTI patients. It is also used as a coating on textiles to increase the durability and antimicrobial property of cloths. Apart from this, it is used in varistors to protect the circuits from voltage fluctuations, and also used to produce hybrid solar cells.

ACKNOWLEDGEMENTS

The author would like to thank the management of Vellore Institute of Technology for encouraging me to write this chapter.

REFERENCES

Abdulkareem EH, Memarzadeh K, Allaker RP, Huang J, Prattan J, Spratt D. 2015. Anti-biofilm activity of zinc oxide and hydroxyapatite nanoparticles as dental implant coating materials. J. Dentistry 43, 1462–1469. doi.org/10.1016/j.jdent.2015.10.010.

Agarwal H, Shanmugam V. 2020. A review on anti-inflammatory activity of green synthesized zinc oxide nanoparticle: Mechanism-based approach. Bioorg. Chem. 94, 103423. doi.org/10.1016/j.bioorg.2019.103423.

Alamdari S, Ghamsari MS, Lee C, Han W, Park HH, Tafreshi MJ, Afarideh H, Ara MHM. 2020. Preparation and characterization of zinc oxide nanoparticles using leaf extract of Sambucusebulus. Appl. Sci. 10, 3620. doi.org/10.3390/app10103620.

Alavi M Nokhodchi A. 2020. An overview on antimicrobial and wound healing properties of ZnOnanobiofilms, hydrogels, and bionanocomposites based on cellulose, chitosan, and alginate polymers. Carbohydr. Polym. 277, 115349. doi.org/10.1016/j.carbpol.2019.115349.

Aldalbahi A, Alterary S, Almoghim RAA, Awad MA, Aldosari NS, Alghannam SF, Alabdan, AN, Alharbi S, Alateeq BAM, Mohsen AAA, Alkathiri MA. 2020. Greener synthesis of zinc oxide nanoparticles: Characterization and multifaceted applications. Molecules 25, 4198. doi.org/10.3390/molecules25184198.

Ali A, Phull AR, Zia M. 2018. Elemental zinc to zinc nanoparticles: Is ZnO NPs crucial for life? Synthesis, toxicological, and environmental concerns. Nanotechnol. Rev. 7, 413–441. doi.org/10.1515/ntrev-2018-0067.

Alswat AA, Ahmad MB, Saleh TA. 2017. Preparation and characterization of zeolite\zinc oxide-copper oxide nanocomposite: Antibacterial activities. Colloids Interface Sci. Commun. 16, 19–24. doi.org/10.1016/j.colcom.2016.12.003.

Anzellotti AI, Farrell NP. 2008. Zinc metalloproteins as medicinal targets. Chem. Soc. Rev. 37, 1629–1651. doi.org/10.1039/b617121b.

Arun D, Mudiyanselage DA, Mohamed RG, Liddell M, Hassan NMM, Sharma D. 2021. Does the addition of zinc oxide nanoparticles improve the antibacterial properties of direct dental composite resins? Asystematic review. Materials 14, 40. doi.org/10.3390/ma14010040.

Arvanag FM, Bayrami A, Habibi-Yangjeh A, Pouran SR. 2019. A comprehensive study on antidiabetic and antibacterial activities of ZnO nanoparticles biosynthesized using Silybummarianum L seed extract. Mater. Sci. Eng. C Mater. Biol. Appl. 97, 397–405. doi.org/10.1016/j.msec.2018.12.058.

Azizi-Lalabadi M, Ehsani A, Divband B, Alizadeh-Sani M. 2019. Antimicrobial activity of titanium dioxide and zinc oxide nanoparticles supported in 4A zeolite and evaluation the morphological characteristic. Sci. Rep. 9, 1–10. doi.org/10.1038/s41598-019-54025-0.

Baghbani Z, Esmaeilnejad B, Asri-Rezaei S. 2020. Assessment of oxidative/nitrosative stress biomarkers and DNA damage in Teladorsagiacircumcincta following exposure to zinc oxide nanoparticles. J. Helminthol. 94, 115. doi.org/10.1017/S0022149X19001068.

Balaure PC, Holban AM, Grumezescu AM, Mogoşanu GD, Bălşeanu TA, Stan MS, Dinischiotu A, Volceanov A, Mogoantă L. 2019. In vitro and in vivo studies of novel fabricated bioactive dressings based on collagen and zinc oxide 3D scaffolds'. Int. J. Pharm. 557, 199–207. doi.org/10.1016/j.ijpharm.2018.12.063.

Belay A, Mekuria M Adam G, 2020. Incorporation of zinc oxide nanoparticles in cotton textiles for ultraviolet light protection and antibacterial activities. Nanomater. Nanotechnol. 10, 1847980420970052. doi.org/10.1177/184798042097005.

Bhattacharyya P, Agarwal B, Goswami M, Maiti D, Baruah S, Tribedi P. 2018. Zinc oxide nanoparticle inhibits the biofilm formation of Streptococcus pneumonia. Antonie Van Leeuwenhoek 11, 89–99. doi.org/10.1007/s10482-017-0930-7.

Cao Y, Naseri M, He Y, Xu C, Walsh LJ, Ziora ZM. 2020. Non-antibiotic antimicrobial agents to combat biofilm-forming bacteria. J. Glob. Antimicrob. Resist. 21, 445–451. doi.org/10.1016/j.jgar.2019.11.012.

Chung IM, Rahuman AA, Marimuthu S, Kirthi AV, Anbarasan K, Rajakumar G. 2015. An investigation of the cytotoxicity and caspase-mediated apoptotic effect of green synthesized zinc oxide nanoparticles using Ecliptaprostrata on human liver carcinoma cells. Nanomaterials 5, 1317–1330. doi.org/10.3390/nano5031317.

El-Nahhal IM, Salem J, Anbar R, Kodeh FS, Elmanama A. 2020. Preparation and antimicrobial activity of ZnO-NPs coated cotton/starch and their functionalized ZnO-Ag/cotton and Zn (II) curcumin/cotton materials. Sci. Rep. 10, 1–10. doi.org/10.1038/s41598-020-61306-6.

Esmaeilnejad B, Samiei A, Mirzaei Y, Farhang-Pajuh F. 2018. Assessment of oxidative/nitrosative stress biomarkers and DNA damage in Haemonchuscontortus, following exposure to zinc oxide nanoparticles. Acta Parasitol. 63, 563–571. doi.org/10.1515/ap-2018-0065.

Foong LK, Foroughi MM, Mirhosseini AF, Safaei M, Jahani S, Mostafavi M, Ebrahimpoor N, Sharifi M, Varma RS, Khatami M. 2020. Applications of nano-materials in diverse dentistry regimes. RSC Adv. 10, 15430–15460. doi.org/10.1515/ap-2018-0065.

Fytianos G, Rahdar A, Kyzas GZ. 2020. Nanomaterials in cosmetics: Recent updates. Nanomaterials 10, 979.

Ghaffari H, Tavakoli A, Moradi A et al. 2019. Inhibition of H1N1 influenza virus infection by zinc oxide nanoparticles: Another emerging application of nanomedicine. J. Biomed. Sci. 26, 70. doi.org/10.1186/s12929-019-0563-4.

Guo Y Sun Z. 2020. Investigating folate-conjugated combinatorial drug loaded ZnO nanoparticles for improved efficacy on nasopharyngeal carcinoma cell lines. J. Exp. Nanosci. 15, 390–405. doi.org/10.1080/17458080.2020.1785621.

Hackenberg S, Scherzed A, Harnisch W, Froelich K, Ginzkey, C, Koehler C, Hagen R, Kleinsasser N. 2012. Antitumor activity of photo-stimulated zinc oxide nanoparticles combined with paclitaxel or cisplatin in HNSCC cell lines. J. Photochem. Photobiol. B 114, 87–93. doi.org/10.1016/j.jphotobiol.2012.05.014.

He L, Liu Y, Mustapha A, Lin M. 2011. Antifungal activity of zinc oxide nanoparticles against Botrytis cinerea and Penicilliumexpansum. Microbiol. Res. 166, 207–215. doi.org/10.1016/j.micres.2010.03.003.

Hosseini SS, Ghaemi E, Koohsar F. 2018. Influence of ZnO nanoparticles on Candida albicans isolates biofilm formed on the urinary catheter. Iran. J. Microbiol. 10, 424.

Hsueh YH, Ke WJ, Hsieh CT, Lin KS, Tzou DY, Chiang CL. 2015. ZnO nanoparticles affect Bacillus subtilis cell growth and biofilm formation. PloS One 10, 128457. doi.org/10.1371/journal.pone.0128457.

Hu Y, Zhang HR, Dong L, Xu MR, Zhang L, Ding WP, Zhang JQ, Lin J, Zhang YJ, Qiu BS, Wei PF 2019. Enhancing tumor chemotherapy and overcoming drug resistance through autophagy-mediated intracellular dissolution of zinc oxide nanoparticles. Nanoscale 11, 11789–11807. doi.org/10.1039/C8NR08442D.

Huarc JCB, Tomar MS, Singh SP,Perales-Perez O, Rivera L, Pena S. 2010. Multifunctional Fe3O4/ZnO core shell nanopartilces for photodynamic therapy. NSTI-Nanotech. 3, 405–408.

Ishida T. 2019. Review on the role of Zn2+ Ions in viral pathogenesis and the effect of Zn2+ Ions for host cell-virus growth inhibition. Am. J. Biomed. Sci. Res. 2, 28–37. doi.org/10.34297/AJBSR.2019.02.000566.

Jasim NO. 2015. Antifungal activity of Zinc oxide nanoparticles on Aspergillus fumigatus fungus & Candida albicansyeast. Citeseer 5, 23–28.

Javed R, Usman M, Tabassum S, Zia M. 2016. Effect of capping agents: Structural, optical and biological properties of ZnO nanoparticles. Appl. Surf. Sci. 386, 319–326. doi.org/10.1016/j.apsusc.2016.06.042.

Javed R, Rais F, Fatima H, ulHaq I, Kaleem M, Naz SS, Ao Q. 2020. Chitosan encapsulated ZnO nanocomposites: Fabrication, characterization, and functionalization of bio-dental approaches. Mater. Sci. Eng. C 116, 111–184. doi.org/10.1016/j.msec.2020.111184.

Jiang J, Pi J,Cai, J 2018. The advancing of zinc oxide nanoparticles for biomedical applications. Bioinorg. Chem. Appl. 1062562. doi.org/10.1155/2018/1062562.

Jin SE, Jin HE. 2019. Synthesis, characterization, and three-dimensional structure generation of zinc oxide-based nanomedicine for biomedical applications. Pharmaceutics. 11, 575. doi.org/10.3390/pharmaceutics11110575.

Kalpana VN, Rajeswari VD. 2018. A review on green synthesis, biomedical applications, and toxicity studies of ZnO NPs. Bioinorg. Chem. Appl. 3569758, doi.org/10.1155/2018/3569758.

Keerthana S, Kumar A 2020. Potential risks and benefits of zinc oxide nanoparticles: A systematic review. Crit. Rev. Toxicol. 50, 47–71. doi.org/10.1080/10408444.2020.1726282.

Khan YA, Singh BR, Ullah R, Shoeb M, Naqvi AH, Abidi SM. 2015. Anthelmintic effect of biocompatible zinc oxide nanoparticles (ZnO NPs) on Gigantocotyleexplanatum, a neglected parasite of Indian water buffalo. PloS One. 10, e0133086. doi.org/10.1371/journal.pone.0133086.

Kietzmann M Braun M. 2006. Effects of the zinc oxide and cod liver oil containing ointment Zincojecol in an animal model of wound healing. Dtsch. Tierarztl. Wochenschr. 113, 331–334.

Kim MH, Seo JH, Kim HM, Jeong HJ. 2014. Zinc oxide nanoparticles, a novel candidate for the treatment of allergic inflammatory diseases. Eur. J. Pharmacol. 38, 31–39. doi.org/10.1016/j.ejphar.2014.05.030.

Kumar SA, Chen SM. 2008. Nanostructured zinc oxide particles in chemically modified electrodes for biosensor applications. Anal. Lett. 41, 141–158. doi.org/10.1080/00032710701792612.

Kumar SPT, Lakshmanan VK, Anilkumar TV, Ramya C, Reshmi P, Unnikrishnan AG, Nair SV, Jayakumar R. 2012. Flexible and microporous chitosan hydrogel/nanoZnO composite bandages for wound dressing: In vitro and in vivo evaluation. ACS Appl. Mater. Interfaces 4, 2618–2629. doi.org/10.1021/am300292v.

Kumar R, Sahoo G, Pandey K, Nayak MK, Topno R, Rabidas V, Das P. 2018. Virostatic potential of zinc oxide (ZnO) nanoparticles on capsid protein of cytoplasmic side of chikungunyavirus. Int. J. Inf. Dis. 73S, 3–398. doi.org/10.1016/j.ijid.2018.04.424.

Leuci R, Brunetti L, Laghezza A, Loiodice F, Tortorella P, Piemontese L. 2020. Importance of biometals as targets in medicinal chemistry: An overview about the role of zinc (II) chelating agents. Appl. Sci. 10, 4118. doi.org/10.3390/app10124118.

Liu Y, Ai K, Yuan Q, Lu L. 2011. Fluorescence-enhanced gadolinium-doped zinc oxide quantum dots for magnetic resonance and fluorescence imaging. Biomaterials 32, 1185–1192. doi.org/10.1016/j.biomaterials.2010.10.022.

Lu PJ, Huang SC, Chen YP, Chiueh LC, Shih DYC. 2015. Analysis of titanium dioxide and zinc oxide nanoparticles in cosmetics. J. Food Drug Anal. 23, 587–594. doi.org/10.1016/j.jfda.2015.02.009.

Maret W Sandstead HH. 2006. Zinc requirements and the risks and benefits of zinc supplementation. J. Trace Elem. Med. Biol. 20, 3–18. doi.org/10.1016/j.jtemb.2006.01.006.

Martínez-Carmona M, Gun'Ko Y, Vallet-Regí M. 2018. ZnO nanostructures for drug delivery and theranostic applications. Nanomaterials 8, 268. doi.org/10.3390/nano8040268.

Mahamuni-Badiger PP, Patil PM, Badiger MV, Patel PR, Thorat-Gadgil BS, Pandit A, Bohara RA. 2020. Biofilm formation to inhibition: Role of zinc oxide-based nanoparticles. Mater. Sci. Eng. C Mater. Biol. Appl. 108, 110319. doi.org/10.1016/j.msec.2019.110319.

Mohammed YH, Holmes A, Haridass IN, Sanchez WY, Studier H, Grice JE, Benson HA, Roberts MS. 2019. Support for the safe use of zinc oxide nanoparticle sunscreens: Lack of skin penetration or cellular toxicity after repeated application in volunteers. J. Invest. Dermatol. 139, 308–315. doi.org/10.1016/j.jid.2018.08.024.

Mohandas A, Sudheesh Kumar PT, Raja B, Lakshmanan VK, Jayakumar R 2015. Exploration of alginate hydrogel/nano zinc oxide composite bandages for infected wounds. Int. J. Nanomed. 10, 53–66. doi.org/10.2147/IJN.S79981.

Morsy K, Fahmy S, Mohamed A, Ali, S, El–Garhy M, Shazly M. 2019. Optimizing and evaluating the antihelminthic activity of the biocompatible zinc oxide nanoparticles against the ascaridid nematode, Parascarisequorumin vitro. Acta Parasitol. 64, 873–886. doi.org/10.2478/s11686-019-00111-2.

Nadhman A, Khan MI Nazir S, Khan M, Shahnaz G, Raza A, Shams DF, Yasinzai M 2016. Annihilation of Leishmania by daylight responsive ZnO nanoparticles: A temporal relationship of reactive oxygen species-induced lipid and protein oxidation. Int. J. Nanomed. 11, 2451. doi.org/10.2147/IJN.S105195.

Nagajyothi PC, Cha SJ, Yang IJ, Sreekanth TVM, Kim KJ, Shin HM. 2015. Antioxidant and anti-inflammatory activities of zinc oxide nanoparticles synthesized using Polygala tenuifolia root extract. J. Photochem. Photobiol. B 146, 10–17. doi.org/10.1016/j.jphotobiol.2015.02.008.

Nethi SK, Das. S, Patra CR, Mukherjee S. 2019. Recent advances in inorganic nanomaterials for wound-healing applications. Biomater. Sci. 7, 2652–2674. doi.org/10.1039/C9BM00423H.

Nie L, Gao L, Feng P, Zhang J, Fu X, Liu Y, Yan X, Wang T. 2006a. Three-dimensional functionalized tetrapod-like ZnO nanostructures for plasmid DNA delivery. Small 2, 621–625. doi.org/10.1002/smll.200500193.

Nie L, Gao L, Yan X, Wang T. 2006b. Functionalized tetrapod-like ZnO nanostructures for plasmid DNA purification, polymerase chain reaction and delivery. Nanotechnology 18, 015101. doi.org/10.1088/0957-484/18/1/015101.

Pachauri V, Vlandas A, Kern K, Balasubramanian K. 2010. Site-specific self-assembled liquid-gated ZnO nanowire transistors for sensing applications. Small. 6, 589–594. doi.org/10.1002/smll.200900876.

Preeti Radhakrishnan VS, Mukherjee S, Mukherjee S, Singh SP, Prasad T. 2020. ZnO quantum dots: Broad spectrum microbicidal agent against multidrug resistant pathogens E. coli and C. albicans. Front. Nanotechnol. 2, 576342. doi.org/10.3389/fnano.2020.576342.

Raj, S, Jose S, Sumod US, Sabitha M. 2012. Nanotechnology in cosmetics: Opportunities and challenges. J. Pharm. Bioallied Sci. 4, 186–193. doi.org/10.4103/0975-7406.99016.

Rigano L, Lionetti N. 2016. Nanobiomaterials in galenic formulations and cosmetics. In Nanobiomaterials in Galenic Formulations and Cosmetics. William Andrew Publishing, New York, pp. 121–148.

Shalom Y, Perelshtein I, Perkas N, Gedanken A, Banin E. 2017. Catheters coated with Zn-doped CuO nanoparticles delay the onset of catheter-associated urinary tract infections. Nano Res. 10, 520–533. doi.org/10.1007/s12274-016-1310-8.

Sharma H, Kumar K, Choudhary C, Mishra PK, Vaidya B. 2016. Development and characterization of metal oxide nanoparticles for the delivery of anticancer drug,Artif.Cells. Nanomed. Biotechnol. 44, 672–679. doi.org/10.3109/21691401.2014.978980.

Shinde NC, Keskar NJ, Argade PD. 2012. Nanoparticles: Advances in drug delivery systems. Res. J. Pharma. Biol. Chem. Sci. 3, 922–929.

Siddiqi, KS, Husen, A, Rao, RAK. 2018. A review on biosynthesis of silver nanoparticles and their biocidal properties. J. Nanobiotechnol., 16(1), 14. 10.1186/s12951-018-0334-5.

Singh SP. 2011. Multifunctional magnetic quantum dots for cancer theranostics. J. Biomed. Nanotechnol. 7, 95–97. doi.org/10.1166/jbn.2011.1219.

Singha P, Workman CD, Pant J, Hopkins SP, Handa H. 2019. Zinc-oxide nanoparticles act catalytically and synergistically with nitric oxide donors to enhance antimicrobial efficacy. J. Biomed. Mater. Res. A 107, 1425–1433. doi.org/10.1002/jbm.a.36657.

Sirelkhatim A, Mahmud S, Seeni A, Kaus NHM, Ann LC, Bakhori SKM, Hasan, H, Mohamad D. 2015. Review on zinc oxide nanoparticles: Antibacterial activity and toxicity mechanism. Nano-micro Lett. 7, 219–242. doi.org/10.1007/s40820-015-0040-x.

Sudhagar S, Sathya S, Pandian K, Lakshmi BS. 2011. Targeting and sensing cancer cells with ZnOnanoprobes in vitro. Biotechnol. Lett. 33, 1891–1896. doi.org/10.1007/s10529-011-0641-5.

Tavakoli A, Ataei-Pirkooh A, Mm Sadeghi G, Bokharaei-Salim F, Sahrapour P, Kiani SJ, Moghoofei M Farahmand M, Javanmard D, Monavari SH. 2018. Polyethylene glycol-coated zinc oxide nanoparticle: An efficient nanoweapon to fight against herpes simplex virus type 1. Nanomedicine 13, 2675–2690. doi.org/10.2217/nnm-2018-0089.

Tiwari V, Mishra N, Gadani K, Solanki PS, Shah NA, Tiwari M. 2018. Mechanism of anti-bacterial activity of zinc oxide nanoparticle against carbapenem-resistant Acinetobacterbaumannii. Front. Microbiol. 9, 1218. doi.org/10.3389/fmicb.2018.01218.

Tripathy N, Ahmad R, Ko HA, Khang G, Hahn YB. 2015. Enhanced anticancer potency using an acid-responsive ZnO-incorporated liposomal drug-delivery system. Nanoscale 7, 4088–4096. doi.org/10.3389/fmicb.2018.01218.

Umrani RD, Paknikar KM 2014. Zinc oxide nanoparticles show antidiabetic activity in streptozotocin-induced Type 1 and 2 diabetic rats. Nanomedicine 9, 89–104. doi.org/10.2217/nnm.12.205.

Vasudevan S, Srinivasan P, Rayappan JBB, Solomon AP. 2020. A photoluminescence biosensor for the detection of N-acyl homoserine lactone using cysteamine functionalized ZnO nanoparticles for the early diagnosis of urinary tract infections. J. Mater. Chem. B 8, 4228–4236. doi.org/10.1039/C9TB02243K.

Verbič A, Gorjanc M, Simončič B. 2019. Zinc oxide for functional textile coatings: Recent advances. Coatings 9, 550. doi.org/10.3390/coatings9090550.

Vidhya E, Vijayakumar S, Prathipkumar S, Praseetha PK. 2020. Green way biosynthesis: Characterization, antimicrobial and anticancer activity of ZnO nanoparticles. Gene Rep. 20, 100688. doi.org/10.1016/j.genrep.2020.100688.

Wang JX, Sun XW, Wei A, Lei Y, Cai XP, Li CM, Dong ZL. 2006. Zinc oxide nanocomb biosensor for glucose detection. Appl. Phys. Lett. 88, 233106. doi.org/10.1063/1.2210078.

Wessel AK, Arshad TA, Fitzpatrick M, Connell JL, Bonnecaze RT, Shear JB, Whiteley M. 2014. Oxygen limitation within a bacterial aggregate. mBio 5, e00992. doi.org/10.1128/mBio.00992-14.

Wu YL, Fu S, Tok AIY, Zeng XT, Lim CS, Kwek LC, Boey FCY. 2008. A dual-colored biomarker made of doped ZnO nanocrystals. Nanotechnology 19, 345605. doi.org/10.1088/0957-4484/19/34/345605.

Yang SC, Shen YC, Lu TC, Yang TL, Huang JJ 2012. Tumor detection strategy using ZnO light-emitting nanoprobes. Nanotechnology 23, 055202. doi.org/10.1088/0957-4484/23/5/055202.

Yi C, Yu Z, Ren Q, Liu X, Wang Y, Sun X, Yin S, Pan J, Huang X. 2020. Nanoscale ZnO-based photosensitizers for photodynamic therapy. Photodiagnosis Photodyn. Ther. 30, 101694. doi.org/10.1016/j.pdpdt.2020.101694.

Yin X, Li Q, Wei H, Chen N, Wu S, Yuan Y, Liu B, Chen C, Bi H Guo D. 2019. Zinc oxide nanoparticles ameliorate collagen lattice contraction in human tenon fibroblasts. Arch. Biochem. Biophys. 669, 1–10. doi.org/10.1016/j.abb.2019.05.016.

Zhang, P, Liu, W 2010. ZnO QD@PMAA-co-PDMAEMA nonviral vector for plasmid DNA delivery and bioimaging. Biomaterials, 31(11), 3087–3094.

Zhao Z, Lei W, Zhang X, Wang B, Jiang H. 2010. ZnO-based amperometric enzyme biosensors. Sensors 10, 1216–1231.

22 Molecular Interaction Studies of Some Zinc Compounds with Multiple Drug Targets

Sakshi Bhardwaj and Dhivya Shanmugarajan

22.1 INTRODUCTION

Zinc is an essential element that plays a key role in various biological processes including transcription, lipid metabolism and synthesis of genetic material. Each form of zinc has unique activity such asits ion form in protein structure stabilization, protein-DNA interaction, structural maintenance of cell biomembrane and chromatin. In this study, 18 different complexes of Zn-bound compounds (Cmp1-Cmp18) were utilized to unravel their significant roles against multiple drug targets. Among them, significant molecular interactions were observed with Cmp1 and Cmp13 in all biological targets with variations in the docking score. This chapter explores the role of studied zinc complexes and their therapeutic potential against various targets in diseases such as cancer, oxidative stress, inflammation, tuberculosis, viral, fungal and bacterial infections.

This study aimed to understand the role of zinc-inducing binding activity against various drug targets. To check diverse biological activities, eighteen coded zinc complexes Cmp1–Cmp18 were employed as varied pharmacophores against multiple drug targets.

22.2 BIOLOGICAL SIGNIFICANCE OF MOLECULAR DOCKING

The molecular docking method can be used to simulate the atomic-level interaction between a tiny molecule and a protein, allowing us to characterize how small molecules behave at the binding site of target proteins and to better understand basic biological processes (Meng et al., 2011). Prediction of the ligand structure as well as its placement and orientation within the sites and evaluation of the binding affinity are the two fundamental processes in the docking process. In many instances, the binding site is known before ligands are docked into it. There are numerous useful tools for drug design and analysis which are offered by molecular docking. After thoroughly screening the target, ligand, and docking approach presentations, the likely docking procedure is carried out. Protein-ligand complex

DOI: 10.1201/9781003412472-22

motions can also be studied by using molecular dynamic simulation techniques. The development of scoring functions with high accuracy and minimal computing expense may advance docking applications.

22.3 ZINC FINGER

The term "zinc finger," which was first used to describe a repetitive zinc-binding motif with DNA-binding capabilities in the Xenopus transcription factor IIIA, is now primarily used to refer to any compact domain stabilized by a zinc ion. Zinc, therefore, plays a structural function in the tiny protein domains known as zinc fingers, which help to maintain the domain's stability. Zinc fingers have a variety of structural variations and are found in proteins that carry out a wide range of tasks in different biological processes, including cell replication and repair, transcription and translation, metabolism and signaling, and apoptosis and cell proliferation. Zinc fingers frequently serve as interaction modules and bind to a wide range of substances, including proteins, nucleic acids, and tiny molecules. More information on the locations of these domains may be seen in Figure 22.1

Typically, disulfide bonds or the binding of metal ions are used to maintain the structures of relatively tiny protein domains (most frequently zinc). Although metal binding normally plays no direct role in the function of tiny domains, it does boost their thermal and conformational stability. C2H2 zinc fingers are likely the most well-studied of these domains (Laity et al., 2001).

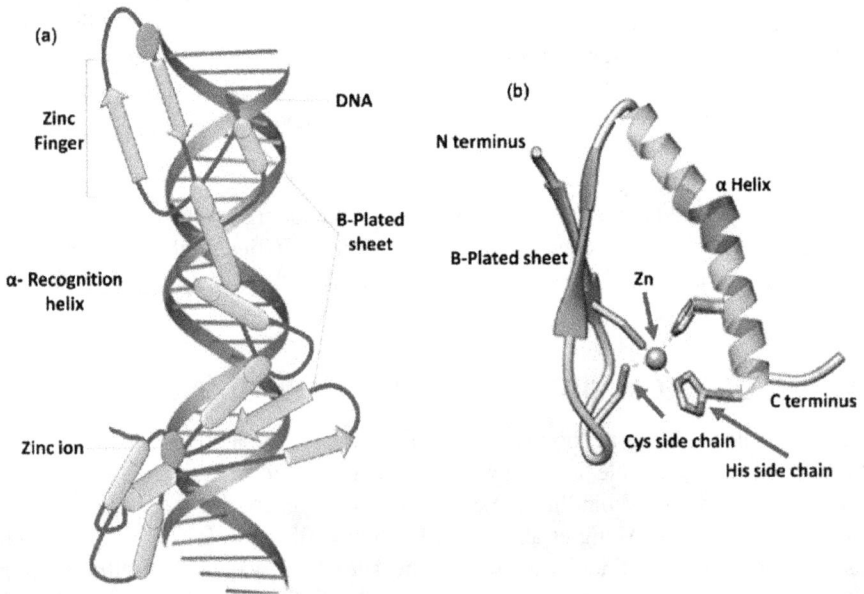

FIGURE 22.1 Explaining Zinc finger: a) DNA binding domain; b) DNA cleavage domain. Source adopted from Genome editing tools: Need of the current era, *American Journal of Molecular Biology* (Aslam et al., 2019).

22.4   GENERAL PROCEDURE FOR DOCKING

The protein and ligand preparation are common steps to be followed before docking with the objective of providing and mimicking *in vivo*-like environment for the biomolecules. The 'Prepare Protein tools' manipulate and interrogate protein structures with the support of tasks like: splitting structures into separate molecules, generating a report summarizing problems and information about the protein, cleaning protein molecules (adding missing atoms, correcting connectivity, correcting names, etc.), inserting missing loops, grafting loops from a template to a target protein, managing conformers (i.e., disordered residues), renumbering sequences, calculate protein ionization and residue pKs, modifying proto-nation of termini and ionizable sidechains, preparing the protein automatically. The 'Prepare Ligands tools' should perform tasks such as removing duplicates, enumerating isomers/tautomers and generating 3D conformations. In the present study, both protein and ligands were prepared using 'BIOVIA Discovery Studio-2022' version (Dassault Systèmes BIOVIA, Discovery Studio Modeling Environment, San Diego). The prepared protein with a defined binding site and prepared ligands were used for receptor-ligand interaction analysis to understand the atomic-level interaction using molecular docking. Through the study molecular dynamicgrid-based docking CDOCKER was implemented (Wu, 2003).

By comparing the target protein to a family of proteins with a related function or proteins co-crystallized with other ligands, one can learn more about the locations. In the absence of information on the binding locations, putative active sites within proteins can be found using cavity detection software or internet servers, such as GRID (Goodford, 1985) POCKET (Levitt and Banaszak, 1992), Sur Net (Glaser et al., 2006), PASS (Brady and Stouten, 2000), and MMC (Mezei, 2003).

22.5   ZINC COMPOUNDS WITH ESTROGEN RECEPTOR-ALPHA

The role that estrogens play in different physiological and pathological pathways is largely influenced by their ability to bind to estrogen receptors (ER) and activate the transcription of estrogen-responsive genes (Gustafsson and Heldring, 2007). When estrogen attaches to ER, the ligand-activated ER moves to the nucleus where it interacts with the responsive region in the target gene promoter to trigger gene transcription (genomic/nuclear signaling). They are exhibited in two subtypes the ERα and Erβ (Kuiper et al., 1996).

It has been discovered that the close intimation of ERα into breast cancer initiation and progression and hence it is also important for breast cancer prevention and treatment. The selected protein target, 2JF9 with 2.10 Å resolution, is an ERα ligand binding domain protein in complex with a tamoxifen-specified peptide antagonist. It is the crystal structure of two peptides that were chosen for their affinity in association with antagonist-bound ERα ligand-binding domains.

In our molecular study of a series of compounds (Cmp1-Cmp18) with selected protein, 2JF9 (Figure 22.2A).

It was identified that Cmp13 had interacted with protein having -CDOCKER_Energy of 8.17517 and -CDOCKER_Interaction Energy of 41.766, through two hydrogen bonds

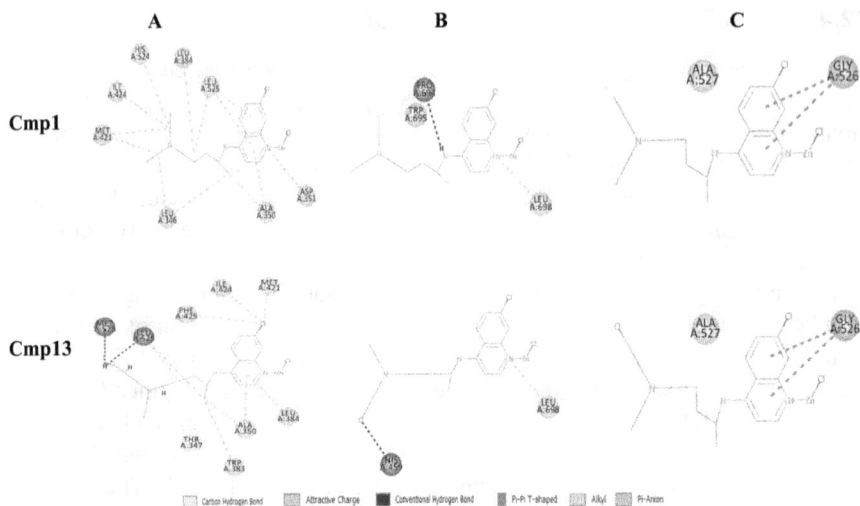

FIGURE 22.2 A) 2D Interacting amino acids diagram of Cmp1 and Cmp13 with A:2JF9). B) 2D Interacting amino acids diagram of Cmp1 and Cmp13 with B: 5OOX. C) 2D Interacting amino acids diagram of Cmp1 and Cmp13 with C:3KK6.

with MET528 and LEU525 amino acid residues. The aromatic ring of Cmp13 linked through pi-pi interaction to ALA350 and LEU384, whereas (Cl)atom had interacted with PHE425, ILE424 and MET421. The compound, Cmp1 having a -CDOCKER_Energy of 52.5728 and -CDOCKER_Interaction Energy of 75.6332, had shown pi-pi interaction with HIS524, ILE424, MET421, LEU346, ALA350, LEU384 and LEU525 amino acid residues, whereas nitrogen of aromatic ring had shown pi-anion interaction with ASP351.

22.6 ANTIOXIDANT ACTIVITY-NADPH OXIDASES WITH Zn COMPOUNDS

The enzyme NOXs-5OOX was utilized for the study. The redox signaling pathways that regulate cell growth and death are impacted by the dysregulation of these polytopic membrane proteins. The docking of Zn compounds at the binding site of NADPH oxidases is favorable for compounds such as Cmp13 that interact with protein amino acid residues through hydrogen bonding and pi-ionic interactions. The oxygen atom substituted for nitrogen at the allylic chain formed a hydrogen bond with HIS459 amino acid residue (Figure 22.2B).

Whereas the nitrogen of the aromatic ring formed pi-ionic interactions with LEU698 amino acid residue. The -CDOCKER_Energy and -CDOCKER_Interaction Energy were found to be 47.5851 and 71.8571 respectively. While in Cmp1 the hydrogen bonding was established to be in between the hydrogen atom of -NH group of the allylic chain with PRO694 amino acid residue and the antibiotic nitrogen on the aromatic ring had contributed to pi-ionic interaction with LEU698. For the compound Cmp1, the -CDOCKER_Energy and -CDOCKER_Interaction Energy were found to be 49.3756 and 71.8328 respectively.

22.7   ANTI-INFLAMMATORY ACTIVITY-CYCLOOXYGENASE-1 WITH Zn COMPOUNDS

Prostaglandins, which are involved in pain-associated inflammation, are produced from arachidonic acid (AA) via cyclooxygenase-2 (COX-2) pathway, whereas thromboxane A(2), which is produced by platelets from AA via cyclooxygenase-1 (COX-1) pathway. The target COX1 (3KK6) was selected in this study to test the anti-inflammatory activity of compounds. Nonselective nonsteroidal anti-inflammatory medications (nsNSAIDs) such as aspirin, have both COX-1 and COX-2 targets; but COX-2 action is preferentially suppressed by COX-2 inhibitors known as coxibs. Figure 22.2C, docking of Zn compounds at the active site of COX-1 is favorable for the Cmp13 and Cmp1 that showed pi-pi interaction with ALA527 and GLY526 with a -CDOCKER_Energy of 21.9687 and -CDOCKER_Interaction Energy of 47.9628 as well as a -CDOCKER_Energy of 21.7341 and -CDOCKER_Interaction Energy of 50.757 respectively.

22.8   ANTITUBERCULOSIS ACTIVITY-ENOYL ACYL CARRIER PROTEIN REDUCTASE WITH Zn COMPOUNDS

It is crucial to find active substances with unique inhibitory effects in the most dangerous pathogens *Mycobacterium TB* and *Plasmodium falciparum* since the spread of multidrug resistance is the major concern. The targeted protein, 1P45, is a crystal structure of InhA complexed with NAD+ and one of the inhibitors. InhA's structure in the presence of the broad spectrum antimicrobial triclosan revealed a special stoichiometry in which the enzyme either held two triclosan molecules bound to the active site or held just one triclosan molecule in a configuration similar to that of other bacterial ENR: triclosan structures. The targeted protein is a dimeric protein with 2.60 Å resolution (Kuo et al., 2003). Figure 22.3A, Cmp13 had interacted with ALA191, PHE149, ILE202, LEU207, ILE215, ALA157, MET105 and MET161 through pi-pi interaction with a -CDOCKER_Energy of 28.0007 and -CDOCKER_Interaction Energy of 52.9804. Additionally, Cmp1 showed pi-stacking with MET103 and pi-pi interaction MET161, MET199, and ILE202 respectively having a -CDOCKER_Energy of 26.8957 and -CDOCKER_Interaction Energy of 48.2334.

22.9   ANTIVIRAL ACTIVITY-INFLUENZA VIRUS NEURAMINIDASE WITH Zn COMPOUNDS

The crystal structure of human 1G04 Fab in complex with the former from A/ Human/02650/2016(H7N9) with 3.45 Å has been taken up for the study. Our studies in this area had given three human monoclonal antibodies that bind with extraordinary breadth to several influenza A and B virus neuraminidases and were purified from a donor infected with H3N2. These antibodies mediate effector activities, are highly protective *in vivo*, neutralize the virus, and block neuraminidase activity by physically attaching to the enzyme's active site. The protein 6Q1Z was utilized for the docking of zinc-based compounds for their antiviral action (Stadlbauer et al., 2019).

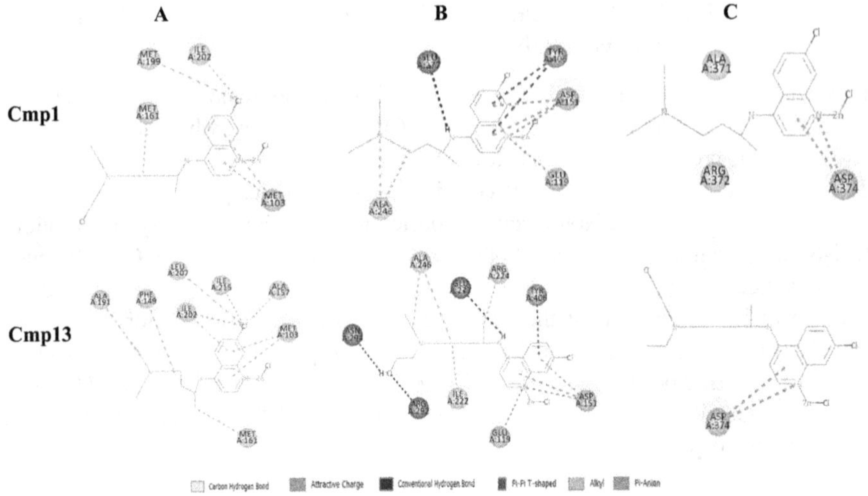

FIGURE 22.3 A) 2D Interacting amino acids diagram of Cmp1 and Cmp13 with A: 1P45. B) 2D Interacting amino acids diagram of Cmp1 and Cmp13 with B: 6Q1Z. C) 2D Interacting amino acids diagram of Cmp1 and Cmp13 with C: 6DEQ.

In our study, the compound, Cmp13 had shown hydrogen bonding with GLU277 amino acid residue through the hydrogen atom of -NH group and with the oxygen atom of -OH group with ARG292 and ASN294 amino acid residues having -CDOCKER_Energy of 59.2146 and -CDOCKER_Interaction Energy of 83.5234. The Cmp1, CDOCKER_Energy of 54.7334 and -CDOCKER_Interaction Energy of 74.9127 and the interactions are stabilized through hydrogen bonding, pi-pi interaction, pi-stacking and pi-anion interactions with the targeted protein. The hydrogen atom of the -NH group-built hydrogen interaction with GLU277 amino acid residue (Figure 22.3B).

22.10 ANTIFUNGAL ACTIVITY -ACETOHYDROXYACID SYNTHASE WITH Zn COMPOUNDS

In this study, acetohydroxy acid synthase, 6DEQ, is the viable new target for the development of antifungal drugs. The low toxicity of its inhibitors to human cells makes them a promising new class of antifungal medication candidates (Garcia et al., 2018). Both the compounds, Cmp13 and Cmp1 (Figure 22.3C) interacted through pi-ionic interaction in between the aromatic ring of the scaffold and ASP374 amino acid residue.

22.11 ANTIBACTERIAL ACTIVITY-ENOLASE WITH Zn COMPOUNDS

The targeted protein is a crystalline structure of *Staphylococcus aureus*-enolase in complex with phosphoenolpyruvate. The crystal structure of enolase (5BOE) with

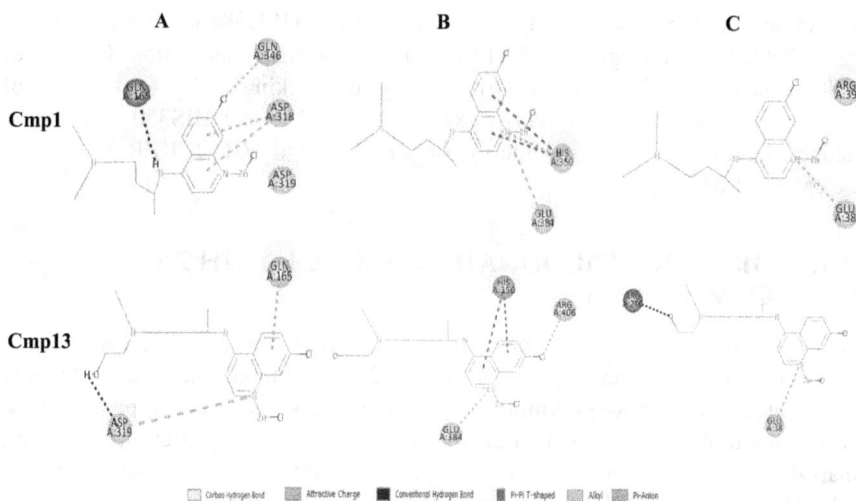

FIGURE 22.4 A) 2D Interacting amino acids diagram of Cmp1 and Cmp13 with A: 5BOE. B) 2D Interacting amino acids diagram of Cmp1 and Cmp13 with B:5WZE. C) 2D Interacting amino acids diagram of Cmp1 and Cmp13 with C: 1DIH.

and without bound phosphoenolpyruvate (PEP) has 1.6 and 2.45 Å resolution, respectively with the octameric arrangement. The octameric form of *S.aureus*- enolase is enzymatically active *in vitro* and probably also *in vivo*, according to biochemical and structural investigations, while the dimeric form is catalytically inert and might be involved in other biological processes (Wu et al., 2015). Figure 22.4A, Cmp13 interacted with protein through one hydrogen bond interaction between the hydrogen atom of the -OH group at the alkyl chain and the ASP319 amino acid residue of the protein. The pi-anion interaction was observed between the nitrogen atom in the aromatic ring and ASP319 amino acid residue of protein, with a -CDOCKER_Energy of 46.1552 and -CDOCKER_Interaction Energy of 74.6917. The hydrogen bonding has been observed between GLN165 amino acid residue and hydrogen atom of -NH group of the alkyl chain of compound Cmp1as also the interaction of halogen atom(-Cl) atom and GLN346 amino acid residue. The compound Cmp1 had -CDOCKER_Energy of 53.1112 and -CDOCKER_Interaction Energy of 89.0982.

22.12 PepP WITH Zn COMPOUNDS

Pseudomonas aeruginosa-PepP is a virulence-associated gene and a prime candidate for anti-*P.aeruginosa* medication development. A tri-metal manganese cluster was observed at the active site, elucidating the mechanism of inhibition by metal ions. The protein, 5WZE, with 1.78 Å resolution has been selected as a drug target (Peng et al., 2017). Figure 22.4B, Cmp13 has shown one halogen interaction between –Cl (which is directly attached to the aromatic ring of the compound) and ARG406 amino acid residue. The pi-pi interactions can be seen between both aromatic rings and HIS 350 amino acid residues of the protein. One pi-ionic interaction between -N atom of the aromatic ring and GLU384 has been observed. The -CDOCKER_Energy of 23.8176 and -CDOCKER_Interaction Energy of 54.8086 have been found for Cmp13. The pi-pi stacking and pi-pi bonding of Cmp1 interact within the protein pocket with GLU384 and HIS350 amino acid residues having -CDOCKER_Energy of 27.4377 and -CDOCKER_Interaction Energy of 51.5177.

22.13   DIHYDRODIPICOLINATE REDUCTASE WITH Zn COMPOUNDS

Escherichia coli dihydrodipicolinate reductase's 3D structure of 1DIH in association with NADPH has been figured out with 2.2 Å resolution. The cofactor binding site is found at the carboxy-terminal edge of the core seven-stranded parallel beta-sheet in the dinucleotide binding domain, which is bordered by four alpha-helices. Analyzing the cofactor binding site of dihydrodipicolinate reductase, which utilizes both NADH and NADPH as cofactors, enables a molecular understanding of the enzyme's dual specificity (Scapin et al., 1995). The compound Cmp13 had shown two hydrogen bond interactions between oxygen atoms of the alkyl chain present in the compound with ARG240 (Figure 22.4C) while GLU38 interacted with pi-ionic interaction with the nitrogen atom on the aromatic ring, with a -CDOCKER_Energy of 37.3711 and -CDOCKER_Interaction Energy of 63.3712. The compound Cmp1 interacted through pi-ionic interaction with nitrogen atoms on the aromatic ring and GLU38 amino acid residue with a -CDOCKER_Energy of 35.6129 and -CDOCKER_Interaction Energy of 61.1476.

22.14   CONCLUSION

In recent years, molecular docking has become a key research tool in drug discovery and its significance is increasing in all domains as this technique helps to screen a large library of compounds in a less painstaking way. This method consists of accurately predicting the structure of a ligand within the constraints of a receptor-binding site and correctly estimating the binding strength through a scoring function. In the present study, molecular dynamics-based docking of eighteen different Zn-bound complexes was screened against multiple drug targets wherein they showed variations in docking scores. It was observed that compound 1 and compound 13 have similar parent molecules, with moiety changes depicts variation in binding. This is mainly due to conformations, structural complementary and the nature of pharmacophore influences the binding of the compounds (Figure 22.5). It has been inferred that zinc is not directly involved, but eventually it sterically and electronically induced other functional group bindings at specific sites of the drug target proteins. The replacement of zinc with other metal docking reduced the binding interaction and docking score thereby supporting the above observation. Hence, it may be confirmed that the role of zinc is crucial for both *insilico* binding as well as *in vitro* or *in vivo* assays.

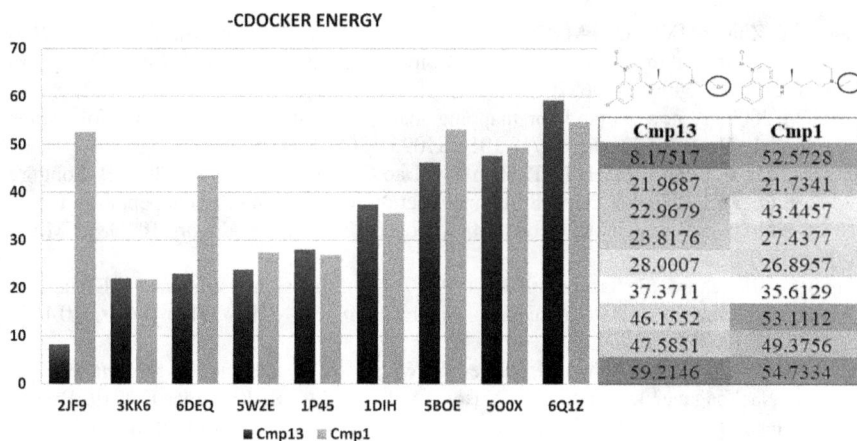

FIGURE 22.5 Docking energy scoring of compounds with various drug targets.

REFERENCES

Aslam S, Khan SH, Ahmed A, Dandekar AM, Aslam S, Khan SH, Ahmed A, Dandekar AM 2019. Genome editing tools: Need of the current era. Am. J. Mol. Biol. 85–109. doi.org/10.4236/AJMB.2019.93008.

Brady GP & Stouten PFW 2000. Fast prediction and visualization of protein binding pockets with PASS. J. Comput. Aided Mol. Des. 14, 383–401. doi.org/10.1023/A:1008124202956.

Garcia MD, Chua SMH, Low YS, Lee YT, Agnew-Francis K, Wang JG, Nouwens A, Lonhienne T, Williams CM, Fraser JA, Guddat LW 2018. Commercial AHAS-inhibiting herbicides are promising drug leads for the treatment of human fungal pathogenic infections. Proc. Natl. Acad. Sci. U.S.A. 115, E9649–E9658. doi.org/10.1073/pnas.1809422115. Epub 2018 Sep 24. PMID: 30249642; PMCID: PMC6187177.

Glaser F, Morris RJ, Najmanovich RJ, Laskowski RA, Thornton JM 2006. A method for localizing ligand binding pockets in protein structures. Proteins 62, 479–488. doi.org/10.1002/PROT.20769.

Goodford PJ 1985. A computational procedure for determining energetically favorable binding sites on biologically important macromolecules. J. Med. Chem. 28, 849–857. doi.org/10.1021/JM00145A002.

Gustafsson JA, Heldring N 2007. Estrogen receptors: How do they signal and what are their targets. Physiol. Rev. 87, 905–931. doi.org/10.1016/J.BBADIS.2011.05.001.

Kuiper GGJM, Enmark E, Pelto-Huikko M, Nilsson S, Gustafsson JÅ 1996. Cloning of a novel receptor expressed in rat prostate and ovary. Proc. Natl. Acad. Sci. U.S.A. 93, 5925. doi.org/10.1073/PNAS.93.12.5925.

Kuo MR, Morbidoni HR, Alland D, Sneddon SF, Gourlie BB, Staveski MM, Leonard M, GregoryJS, Janjigian AD, Yee C, Musser JM, Kreiswirth B, Iwamoto H, Perozzo R, Jacobs WR, Sacchettini JC, Fidock DA 2003. Targeting tuberculosis and malaria through inhibition of Enoyl reductase: Compound activity and structural data. J. Biol. Chem. 278, 20851–20859. doi.org/10.1074/JBC.M211968200.

Laity JH, LeeBM, Wright PE 2001. Zinc finger proteins: New insights into structural and functional diversity. Curr. Opin. Struct. Biol. 11, 39–46. doi.org/10.1016/S0959-440X(00)00167-6.

Levitt DG, Banaszak LJ 1992. POCKET: A computer graphics method for identifying and displaying protein cavities and their surrounding amino acids. J. Mol. Graph. 10, 229–234. doi.org/10.1016/0263-7855(92)80074-N.

Meng XY, Zhang HX, Mezei M, Cui M 2011. Molecular docking: A powerful approach for structure-based drug discovery. Curr. Comput. Aided Drug Des. 7, 146–157. doi.org/10.2174/157340911795677602.

Mezei M 2003. A new method for mapping macromolecular topography. J. Mol. Graph. Model. 21, 463–472. doi.org/10.1016/S1093-3263(02)00203-6.

Peng CT, Liu L, Li CC, He LH Li T, Shen YL, Gao C, Wang NY, Xia Y, Zhu YB, Song YJ, Lei Q, Yu LT, Bao R 2017. Structure-function relationship of aminopeptidase P from Pseudomonas aeruginosa. Front. Microbiol. 8, 2385–2385. doi.org/10.3389/FMICB.2017.02385.

Scapin G, Blanchard JS, Sacchettini JC 1995. Three-dimensional structure of Escherichia coli dihydrodipicolinate reductase. Biochemistry 34, 3502–3512. doi.org/10.1021/BI00011A003.

Stadlbauer D, Zhu X, McMahon M, Turner JS, Wohlbold TJ, Schmitz AJ, Strohmeier S, Yu W, Nachbagauer R, Mudd PA, Wilson IA, Ellebedy AH, Krammer F 2019. Broadly protective human antibodies that target the active site of influenza virus neuraminidase. Science 366, 499–504. doi.org/10.1126/SCIENCE.AAY0678.

Wu Y, Wang C, Lin S, Wu M, Han L, Tian C, Zhang X, Zang J 2015. Octameric structure of Staphylococcus aureus enolase in complex with phosphoenolpyruvate. Acta Crystallogr. D Biol. Crystallogr. 71, 2457–2470. doi.org/10.1107/S1399004715018830/QH5032SUP1.PDF.

Wu G, Robertson DH, Brooks CL, Vieth M 2003. Detailed analysis of grid-based molecular docking: A case study of CDOCKER - A CHARMm-based MD docking algorithm. J. Comput. Chem. 24, 1549. doi.org/10.1002/jcc.10306. PMID: 12925999.

23 Luminescent Nanoscale Materials for the Selective and Sensitive Detection of Zinc Ions

Gouri Tudu, Suvra Sil, and Venkataramanan Mahalingam

23.1 INTRODUCTION

Zinc ions (Zn^{2+}) play a vital role in biological processes. In fact, zinc takes a special place as it is present in all six enzyme classes (Zheng et al., 2020). Many important aspects like apoptosis, immune functions, vision, ion transport, neurotransmission and more rely on the presence of zinc ions. In children, deficiency of zinc may influence their growth and development (Zalewski, 1993; Cuajungco and Lees, 1997). In fact, it is very important to maintain zinc levels for the normal functioning of the various biological processes in humans. In addition, zinc can be found in the drinking water as well. This occurs due to the leaching of zinc from the pipes. The provisional maximum tolerable daily intake (PMTDI) of zinc is 1.0 mg/kg of body weight (Guidelines for drinking water quality, 1996). Thus, detection of zinc ions is crucial as they are involved in various biological processes. Among different analytical techniques used for the detection of zinc at lower concentrations, UV-Vis and fluorescent-based methods are relatively cheaper and faster. Further, currently these spectrometers are available as portable devices thus making it convenient for on-site studies. Although fluorescent-based methods are quite sensitive, care must be taken particularly when there is quenching of fluorescent signals with analyte binding. Appropriate control measurements need to be performed to validate the results.

There are several organic chromophores developed as optical probes for the selective and sensitive detection of Zn^{2+} in the aqueous medium. For example, luminescent Eu^{3+} and Gd^{3+} containing functionalized quinoline complexes are used for selective zinc ion detection (Hanaoka et al., 2004; Meeusen et al., 2011). Although the aforementioned probes are very efficient for Zn^{2+} detection, the scope of organic chromophores is limited as they are less robust. On the other hand, nanomaterials-based probes are interesting as they possess several advantages. They are quite robust, display unique size-dependent optical properties, ease of synthesis with controlled morphologies, easy purification and more. Further, surface functionalization of

DOI: 10.1201/9781003412472-23

nanomaterials with organic molecules (chromophores) is feasible through chemical routes. This allows the attachment of suitable molecules for selective binding of analytes. Such selective binding results in tuning of the optical characteristics in case of noble metal and other luminescent nanomaterials via fluorescent resonance energy transfer (FRET) or aggregation-induced emission (AIE). The following sections discuss reports on different types of nanomaterials and their potential for both *in vitro* and *in vivo* fluorescent detection of Zn^{2+} ions.

23.2   METAL NANOPARTICLES AND NANOCLUSTERS FOR Zn^{2+} DETECTION

One of the widely studied nanoparticle systems is metal nanoparticles (NPs), particularly noble metal NPs. This is largely due to the easy synthesis, and notably their characteristic optical properties (Rosi and Mirkin, 2005). Noble metal NPs particularly Au and Ag display strong absorbance in the visible region of the electromagnetic spectrum which is attributed to the surface plasmon resonance (SPR) (Eustis and El-Sayed, 2006). SPR is quite sensitive to the morphology and to the chemical changes occurring at the surface of the NPs. These characteristic features have been explored for the detection of several analytes with very low detection limits.

Chitosan functionalized Au NPs are used for the selective colorimetric detection of Zn^{2+} ions. Chitosan renders electrostatic stabilization of the NPs against aggregation in addition to providing binding sites for the Zn^{2+} ions. The UV-Vis spectrum of the NPs displayed an intense SPR band at 525 nm which was red-shifted upon Zn^{2+} addition. This shift is associated with a color change caused by the destabilization of the colloidal chitosan-capped Au NPs (Promnimit, 2011). Similarly, Li et al. explored aggregation-induced emission enhancement (AIEE) of glutathione-capped Au nanoclusters for selective Zn^{2+} sensing. The Au nanoclusters showed an emission maxima at 570 nm (λ_{exi} 410 nm). An increase in the emission intensity was noted upon addition of Zn^{2+} in the presence of 2-methylimidazole. The enhanced fluorescence intensity is associated with the formation of Zn-MOF (metal organic framework) preferably caused by the aggregation of the GSH-AuNCs. A linear range in the detection was observed from 12.3 nM to 24.6 nM with a limit of detection (LoD) ~6 nM. The developed nanocluster is very selective towards Zn^{2+} ions (Li et al., 2018b).

Huang and co-workers have shown the use of Au nanoclusters coated paper for the visual detection of Zn^{2+} ions in whole blood and cells. They proposed a headspace solid phase extraction causing quenching of the fluorescence from Au nanoclusters (AuNCs) due to the formation of ZnH_2 (Huang et al., 2018). Diverse metal ions were used to understand the selectivity of the detection and the results clearly confirm the developed Au NCs display change in fluorescence only in the presence of Zn^{2+} ions. There is barely any interference from other metal ions such as Cd^{2+}, Mg^{2+}, Fe^{3+}, Ni^{2+}, As^{2+}, Pb^{2+}, etc., as the co-existence of the aforementioned ions did not influence the fluorescence from the AuNCs. The LoD of Zn^{2+} by nanoclusters is 3 µg L^{-1}.

Bhardwaj et al. prepared salicylaldehyde (SA) conjugated lysozyme-stabilized fluorescent gold NCs (Bhardwaj et al., 2019). The lysozyme stabilized AuNCs possess a red emission close to 650 nm which upon binding to SA shifted to 498 nm due to the imine linkage formation between the aldehyde group of SA and amine group of lysozyme. Interestingly, further blue-shift (~450 nm) in the peak maxima along with enhancement in the emission intensity is observed with the addition of zinc ions. This is attributed to the complexation-induced aggregation. Figure 23.1(a) shows the fluorescence spectra of the modified AuNCs in the presence of different

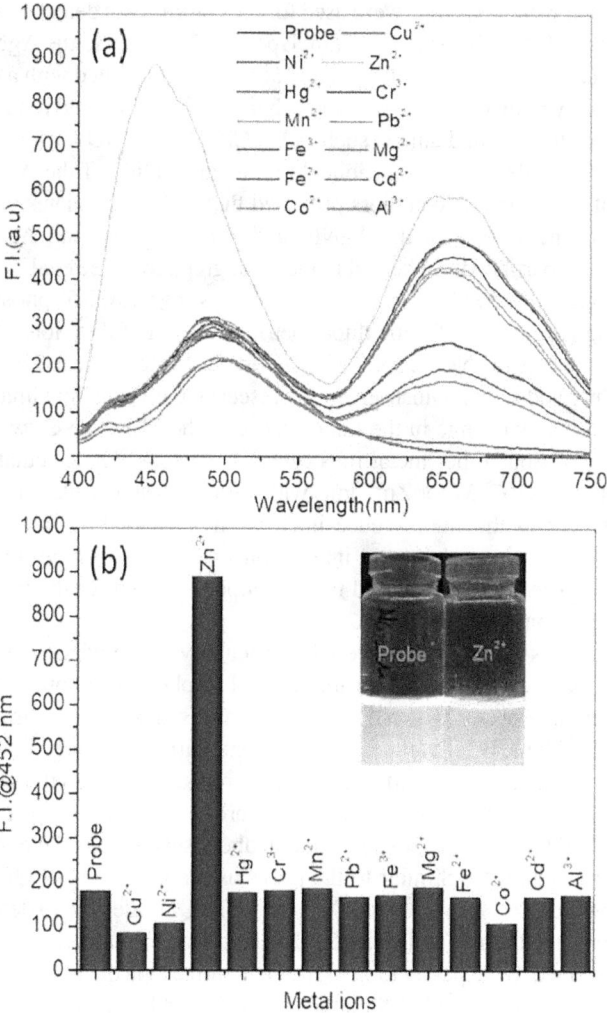

FIGURE 23.1 (a) Fluorescence spectral changes of SA_Lyso-AuNCs in the presence of various metal ions. (b) the bar graph representation of the fluorescence selectivity at 452 nm of SA_Lyso-AuNCs towards various metal ions (inset shows the fluorescent color of probe in the absence and presence of Zn^{2+}). (Adapted with permission from Elsevier, *Microchem. J.* 151 (2019) 104227).

metal ions. The results clearly indicate the enhancement and blue-shift in the peak maxima only with the addition of Zn^{2+} ions. Inset in Figure 23.1(b) illustrates the visual color change in the nanocluster dispersion after the addition of zinc ions. Addition of several other metal ions like Mg^{2+}, Cu^{2+}, Co^{2+}, Ni^{2+}, Mn^{2+}, Cr^{3+}, etc., to the AuNCs displayed hardly any change in the fluorescence intensity suggesting high selectivity of the lysozyme-stabilized gold nanoclusters for Zn^{2+} ions. The estimated limit of detection (LoD) is about 0.137 μM.

Similar to gold, Ag nanoparticles/nanoclusters possess interesting optical properties which are explored for Zn^{2+} detection. For instance, 15 nm silver nanoparticles (AgNPs) capped with BH_4^- ions were used for the colorimetric detection of Zn^{2+} ions. The AgNPs exhibit an SPR peak ~390 nm. Upon Zn^{2+} binding to the AgNPs, the SPR intensity decreases and a new peak appeared at 635 nm associated with a color change of the NPs from yellow to pale green. Such changes were not noted in the presence of several other metal ions and anions such as F^-, Cl^-, Br^-, I^-, NO_2^-, SO_4^{2-}, NO_3^-, and PO_4^{3-}. The shift in the LSPR is attributed to the aggregation of the AgNPs with the attachment of Zn^{2+} ions. The authors proposed that Zn^{2+} coordinates with the BH_4^- ions present at the surface of the AgNPs and pulls neighbouring AgNPs, thereby reducing the interparticle distance and inducing aggregation (Lee et al., 2016). Bothra et al. developed a vitamin B6 cofactor such as pyridoxal-5'-phosphate (PLP) conjugated Ag nanoclusters for the fluorometric detection of Zn^{2+} ions. The colloidal dispersion of the PLP-AgNCs displayed a yellow fluorescence near 520 nm (λ_{Exi} 365 nm). Upon Zn^{2+} addition, a bluish-green fluorescence with peak maxima at 475 nm is observed. Barely any change in the fluorescence of the PLP-AgNCs was noted with the addition of various other metal ions except Cd^{2+} ions. The calculated detection limit is about 50.5×10^{-8} M for Zn^{2+} ions with a good linear correlation ($R^2 = 0.996$). The enhancement in the fluorescence intensity is attributed to the complexation-induced aggregation caused by the interaction of the hydroxyl groups and imine groups of PLP. The authors explored the developed material for the in-vitro cellular imaging of Zn^{2+} ions (Bothra et al., 2017).

Near-infrared (NIR) emitting materials particularly in the window 1.0 to 1.7 μm, are interesting as they possess less scattering and display low autofluorescence. NIR emitting quantum dots(QDs) such as Ag_2S was used as a turn-on fluorescence detection of Zn^{2+} ions. Briefly, the Ag_2S QDs were initially prepared by capping with n-dodecyl mercaptan groups and subsequently replaced by thioglycolic acid. The Ag_2S QDs under UV-Vis irradiation show a broad fluorescence peak centered at 1100 nm. The addition of zinc ions resulted in about seven times the enhancement in the fluorescence intensity. Similar to the previous report, the Ag_2S QDs show high selectivity towards Zn^{2+} as well as to Cd^{2+} ions. The QDs are able to detect Zn^{2+} ions in the aqueous medium to a limit of 0.76 μM. The increase in the fluorescence intensity is ascribed to the passivation of the QDs surface via the formation of Zn-thiol complex. The authors extended the study to real-world samples like lake and tap water. The results concluded a high rate of recovery in the range 90% to 110% (Wu et al., 2017).

In addition to Au and Ag, Cu nanoclusters (CuNCs) have also been reported as probes for Zn^{2+} detection. For instance, Lin et al. have shown that glutathione-capped CuNCs can be used for the selective detection of Zn^{2+} ions (Lin et al, 2017).

An increase in the photoluminescence intensity about 3 times was observed for the CuNCs with the addition of Zn^{2+} ions. The increase in the PL intensity is attributed to the aggregation-induced emission as binding of Zn^{2+} to the glutathione reduced the surface charges causing CuNCs to come together. This is supported by a reduced surface electric potential of the CuNCs from -13.6 mV to -4.97 mV after the addition of Zn^{2+} ions to the CuNCs. The study was extended to cellular imaging of Zn^{2+} ions by using an MTT assay.

23.3 SEMICONDUCTORS AND CARBON QUANTUM DOTS IN Zn ION SENSING

Another interesting class of nanomaterials is semiconductor nanoparticles which display a strong quantum size effect i.e., they display size-dependent optical characteristics (Alivisatos, 1996; Smith and Nie, 2010). In fact, they exhibit strong photoluminescence upon UV or visible light irradiation and the PL peak position can be tuned by varying the size (below Bohr exciton radius). The PL characteristics are explored for the selective sensing of many metal ions including zinc.

Liu et al. developed dual-emission carbon dots (DCDs) as a ratiometric fluorescent probe which showed two emission bands at 470 nm and 655 nm upon 400 nm excitation under different pH environments (Wang et al., 2019). Upon the addition of Fe^{3+} in acidic environments (pH 5.4), a reduction in the fluorescence intensity at 655 nm with a detection limit of 0.8 μmol L^{-1} was noted. Interestingly, upon addition of Zn^{2+} ions in alkaline media (pH 9.4), the fluorescence quenching at 470 nm was noticed with the detection limit of 1.2 μmol L^{-1}. The as-prepared DCDs were found to have excellent chemical and optical stabilities and are suitable to serum samples for practical application.

Core/shell quantum dots like azamacrocycle-activated CdSe/ZnS are reported for Zn^{2+} ion detection (Ruedas-Rama and Hall, 2008). This is the first report on zinc ion sensors where QD nanoparticles in a host-guest and receptor-fluorophore system were utilized. The detection mechanism of Zn^{2+} involves coordination between Zn^{2+} and the lone pair electrons of the nitrogen atom leading to the unavailability of the energy level for the hole transfer. This renders the switching on the QD emission and corresponding increment in the fluorescence intensity. Based on this principle, three zinc ion sensors by using CdSe-ZnS core-shell QD nanoparticles were successfully developed with very good linearity in the range between 5–500 μM, and low detection limits (<2.4 μM). The interference study revealed a good selectivity of the as-prepared QDs towards Zn^{2+} ion in comparison to other cations and transition metals. The Zn^{2+} detection experiments were performed for physiological samples like fetal calf serum, also samples mimicking physiological conditions suggested very good practical applicability. In another report, Yu et al. demonstrated the ratiometric fluorescence method by preparing N,O-co-doped carbon quantum dots (CDs) for detection of Zn^{2+} ions (Bu et al., 2020).

The composite CDs–Zn was prepared by a reaction between the as-prepared CDs and zinc acetate. During the process, a change in fluorescence was observed ratiometrically upon addition of $S_2O_8^{2-}$. The results revealed that, with a change in excitation light, the emission peaks of the CDs and CDs–Zn were unaltered. The materials CDs and CDs–Zn are advantageous due to their good photo- and thermal stability, selectivity, and strong anti-interference ability with low dark toxicity under physiological temperatures. The as-prepared CDs and CDs–Zn were found to have low toxicity, and a facile synthesis strategy in comparison to the other reported ratiometric fluorescent hybrid (organic dyes and inorganic nanomaterials based) nanosensors. A schematic illustration of the work is given below in Figure 23.2.

Jiang et al., developed CdSe/ZnS QDs to detect ziram, a coordination complex of zinc with dimethyl dithiocarbamate, which is popularly used as a fungicide to control fungal diseases in plants (Yang et al., 2018). They described an effective double-decrease effect sensing strategy where upon the interaction between ziram and gold nanoparticles, the zinc ions are released and aggregation of the AuNPs takes place resulting in a decrease in the intensity of the CdSe/ZnS QDs fluorescence. This is due to the overlap between the absorption band of aggregated AuNPs (at 680 nm) and the emission of QDs (at 608 nm). The study reveals relative fluorescence intensity is proportional to the ziram concentration within a wide range of 5 nM to 4 µM in two consecutive linear ranges with a very low limit of detection (~2 nM).

Silicon quantum dots (SiQDs) have also been used for zinc detection. For instance, Wang et al. reported a novel optical nanoprobe based on silicon quantum dots (SiQDs) synthesized from 3-(aminopropyl)-trimethoxysilane (APTMS) and ascorbic acid (AA) (Li et al., 2018a). The prepared SiQDs displayed intense green luminescence in water whereas upon addition of salicylaldehyde (SA) the emission of SiQDs got suppressed considerably resulting in "on-off" change in the luminescence intensity SiQDs. Interestingly, the addition of Zn^{2+} generates emission peaks, and the band at 500 nm gradually blue-shifts to 455 nm. The limit of detection is reported to be 0.17 µM in the linear range between 1 and 100 µM. The study further demonstrated a molecular logic gate (AND) system using SA and Zn^{2+} as two inputs.

Another interesting member of the quantum dots is carbon dots (CDs) display strong photoluminescence in the visible region (Liu et al., 2019). Scientists take advantage of the strong emission for the detection of Zn^{2+} ions by chemical modification of their surfaces. For instance, Banerjee et al. reported carbon quantum dot (CQD) for the first

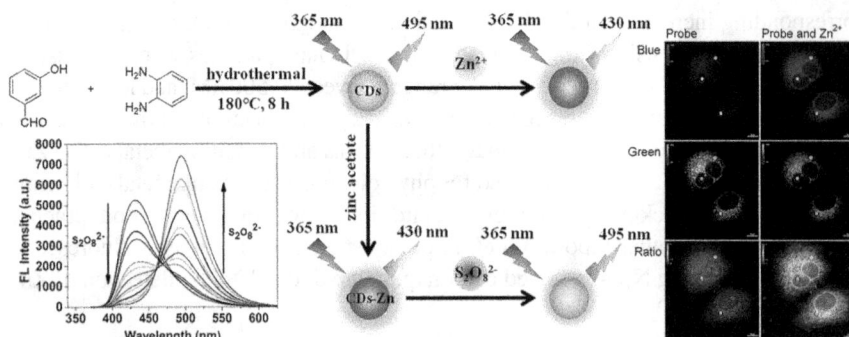

FIGURE 23.2 Schematic illustration for detection of Zn^{2+} and $S_2O_8^{2-}$ using carbon dots (CDs) and CDs–Zn via ratiometric fluorescence methods (Bu et al., 2020). (Adapted with permission from Wiley, Luminescence, 35 (2020) 1319).

time for selective sensing of Zn^{2+} in blood plasma (Dastidar et al., 2021). The CQD was synthesized from onion extract with ~1 nm particle and exhibited enhanced fluorescence intensity upon 325 nm excitation in the presence of Zn^{2+} ions in Tris buffer and blood plasma. However, the addition of other common resulted in decreased fluorescence intensity. Further, the limit of detection and limit of quantitation values for Zn^{2+} ions in plasma were recorded to be 6.4 µM and 21.3 µM, respectively. In another work, Liu and co-workers developed a QD-based FRET assembly for ratiometric sensing and visual detection of Zn^{2+} ion (Liu et al., 2017). The assembly contains QD scaffold with the immediate coating of poly(dA) homopolymer/double-stranded DNA, and meso-tetra(4-sulfonatophenyl)porphine dihydrochloride (TSPP). The biopolymer coating reduces the thickness of the interface (≤2 nm) for QD-TSPP FRET, leading to improved single FRET efficiency up to 60-fold and 4-fold for total FRET efficiency of the QD-biopolymer-TSPP assemblies in comparison with SiO_2-coating-based QD-TSPP assemblies. The dual emission QD-poly(dA)-TSPP assemblies were achieved based on Zn^{2+}-chelation-induced spectral modulation with improved sensitivity and specificity. The sensing could also be visualized as the developed sensor exhibited change in colour from yellow to bright green upon introducing Zn^{2+} ion, both in solution and on a paper substrate. Further, the applicability of the sensor was successfully investigated for exogenous Zn^{2+} ion in living cells.

Yi et al. have developed Zn^{2+}-responsive ratiometric fluorescent core-shell nanoprobe (QPNPs) using an internal reference dye TBAP) encapsulated into a poly (methyl methacrylate) core, and a Zn^{2+}-specific indicator dye (PEIQ) in the shell through graft copolymerization (Chen et al., 2018). The QPNPs exhibited well-resolved dual emissions upon photoexcitation at 380 nm and a green fluorescence peaking at ~500 nm after exposure to Zn^{2+} ions. Further, with the increase in Zn^{2+} ion concentration, the peak intensity increases however, the pink fluorescence of the porphine-derived reference dye at ~650 nm remains unaltered indicating possible detection of Zn^{2+} using both visually and spectroscopically. The reported detection limit for Zn^{2+} ions is 3.1 nM. The nanoprobe could efficiently detect Zn^{2+} in living HeLa cells and in zebrafish.

Gong et al. reported a new nanoprobe based on Cl and N co-doped carbon quantum dots (ClNCQDs) which is able to detect morin, a natural polyphenol and Zn^{2+} ions successfully via fluorometric as well as colorimetric methods (Xiao et al., 2021). The prepared ClNCQDs exhibited a large Stokes shift (177 nm), which is rare for reported CQDs. The effective fluorescence quenching of the ClNCQDs by morin was attributed to the synergistic effect of the inner filter effect (IFE), electrostatic interaction and dynamic quenching process. Further, the quenching can be reversed by adding Zn^{2+} ion due to strong interaction among morin and Zn^{2+} ion. The nanoprobe was found to be very selective and sensitive towards morin and Zn^{2+} ion with detection limits of 0.09 µM and 0.17 µM, respectively. The practical applicability of the nanoprobe was also investigated by taking actual samples (medicinal herb samples, urine samples, and water samples) where high accuracy was achieved for morin and Zn^{2+} ion.

Chen and co-workers have demonstrated the fabrication of sulphur-nitrogen co-doped carbon dots (SNCDs) which act as a nanosensor for "off–on" sensing of hypochlorous acid and Zn(II) with interesting bioimaging properties

(Zhang et al., 2018). The SNCDs displayed favorable optical performances including strong yellow-green fluorescence and excitation wavelength-independent fluorescence behavior with the maximum excitation/emission wavelength at 420/550 nm. Interestingly, the emission of SNCDs was significantly quenched by hypochlorous acid (HOCl/ClO), which was recovered upon the addition of zinc(II) ions. For Zn^{2+} ion detection, the observed quantitative range is from 8.41 to 84.12 µM under "turn on" mode with a detection limit of 0.30 µM. The developed fluorescent nanoprobe was also utilized for assaying HOCl/ClO and Zn^{2+} in water samples and MCF-7 cell imaging with promising applicability. A graphical illustration of the above study is provided in Figure 23.3.

Luo et al. reported red fluorescent (610 nm) CDs, synthesized by solvent thermal treatment of p-phenylenediamine in toluene (Wang et al., 2020). In the presence of quercetin (QCT) the red fluorescence of CDs was slightly altered upon 440 nm excitation, however upon addition of Zn^{2+} ion the quick complexation between QCT and Zn^{2+} ($QCT-Zn^{2+}$) resulted in the bright green fluorescence at 480 nm upon UV light irradiation. A similar phenomenon was observed with Al^{3+} due to the interference of Al^{3+} on Zn^{2+}. To avoid interference by Al^{3+}, fluoride ions were introduced into the system, which resulted in the formation of a more stable complex between F^- and Al^{3+}. The developed $CD-QCT-F^-$ system acts as a ratiometric fluorescent nanoprobe toward Zn^{2+} ion with LoD of 0.14 µM. Furthermore, since adenosine triphosphate (ATP) has a stronger affinity to Zn^{2+} than QCT, the I_{480}/I_{610} value of $CD-QCT-F^--Zn^{2+}$ system gradually decreases with an increase in the ATP concentration. The ratiometric fluorescent nanoprobe toward

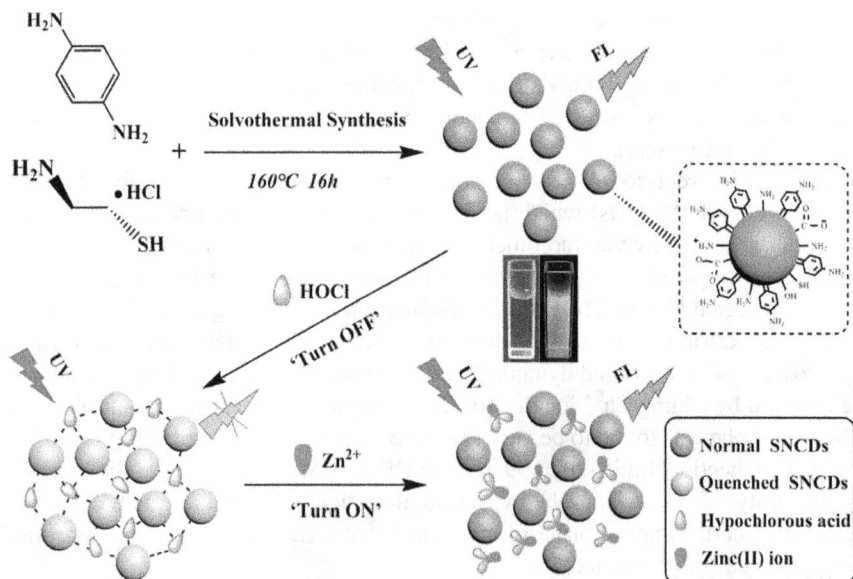

FIGURE 23.3 Schematic of the synthesis of SNCDs and the sensor for hypochlorous acid and Zn(II) ions. (Adapted with permission from RSC, *New J. Chem.*, 2018, 42, 15895).

ATP was established with detection ranges of 0.55–10 μM and 10–35 μM and a LoD of 0.55 μM.

23.4 LANTHANIDE-DOPED NANOMATERIALS FOR Zn^{2+} DETECTION

Lanthanide-doped nanocrystals are an important class of luminescent materials with interesting optical characteristics like sharp transitions with long excited state lifetimes due to the forbidden nature of intra 4 $f{\to}4$ f transitions, display large Stokes shift, wide span of emissions, and less crystals field influence (Bünzil and Eliseeva, 2013). In addition, due to long lifetimes, some of the Ln^{3+} ions possess the ability to undergo downconversion or upconversion through which low-energy photon is converted into high and vice versa (Auzel, 2004; Sarkar et al., 2019). These aforementioned features are explored for metal ion detection.

For instance, in vitro/in vivo detection of Zn^{2+} ions in achieved using an upconverting probe such as poly (acrylic acid) capped Tm^{3+}/Yb^{3+}-doped $NaYF_4$ nanocrystals (NCs) (Peng et al., 2015). In this work, an upconverting core/shell nanocrystal such as Yb/Tm-doped $NaYF_4@NaYF_4$ was prepared. Subsequently, the nanoparticles were modified with polyacrylic acid (PAA) and an organic chromophore (OC), which has a strong affinity towards Zn^{2+} ions. Upon 980 nm excitation, the Yb/Tm-doped $NaYF_4@NaYF_4$ NCs show strong blue emission at 475 nm and a weak red emission at 654 nm originating from intra 4 f→4 f transition of Tm^{3+} ions through Yb^{3+} sensitization. However, upon attachment of the OC, a strong reduction in the blue emission intensity is noted due to the overlap of the strong absorbance by the organic chromophore (~475 nm). Binding of Zn^{2+} ions to the OC, which is attached to the NCs' surface, resulted in the blue shift of the absorbance maxima OC from 475 nm to 360 nm due to intramolecular charge transfer. The blue-shift in the OC absorbance caused a mismatch in the OC absorbance and blue emission from Tm^{3+} ions and led to the recovery of the 475 nm blue emission of the Tm^{3+} ions. A schematic presentation of the above strategy and the results are shown in Figure 23.4.

In another report, cyclodextrin-modified Yb/Tm-doped $NaGdF_4@NaYF_4$ (UCNPs) was used for the selective endogenous detection of Zn^{2+} ions as well as for on-demand photodynamic therapy (Hu et al., 2017). Briefly, a fluorescent organic ligand having an affinity for Zn^{2+} was first prepared. The resulting compound was attached to the surface of the UCNCs. Upon 980 nm radiation of the functionalized UCNCs, the upconversion emission intensity at 475 nm decreases due to radiative energy transfer from upconverting NCs to Zn^{2+} organic complex, as the latter has absorbance centered at 520 nm. The organic ligand attached to the UCNCs is selective to Zn^{2+} ions as addition of other metal ions such as Co^{2+}, Cu^{2+}, Ni^{2+}, Fe^{2+} and more hardly altered the upconverstion intensity.

23.5 IRON-BASED MAGNETIC NANOPARTICLES AND NANOCLUSTERS

Iron oxides (Fe_2O_3) NPs have been used for the detection of Zn^{2+} ions. Use of Fe-based NPs is advantageous as they are very abundant, biocompatible, relatively less

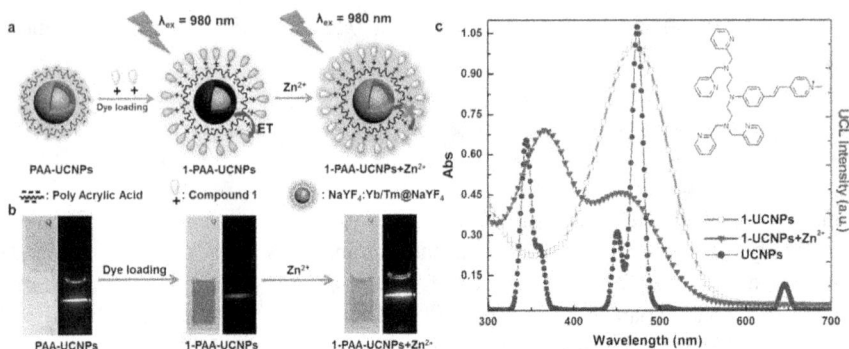

FIGURE 23.4 (a) Schematic illustration showing the synthesis of chromophore-assembled UCNPs and their response to Zn^{2+}. (b) Photographs showing the solution (left) and corresponding upconversion luminescence (right) of PAA-UCNPs (1 mg/mL), 1-PAA-UCNPs (1 mg/mL), and 1-PAAUCNP with Zn^{2+} (120 μM), respectively. (c) UV–visible spectra of compound 1 were measured in the absence (red line, 0.1 mM) and presence (dark yellow line, 120 μM) of Zn^{2+} and UCL spectrum of PAA-UCNPs (blue line). Inserted is the molecular structure of compound 1. (Adopted with permission from ACS, *J. Am. Chem. Soc.* 2015, 137, 2336–2342). Further, the authors have used the developed material for the in vivo detection of Zn^{2+} ions in zebrafish.

toxic and more. For instance, Kang et al. developed a strategy based on photo-induced electron transfer (PET) for the selective detection of Zn^{2+} ions in aqueous media. Briefly, oleic acid (OA) capped Fe_2O_3 NPs with fluorescent BODIPY (578 nm) moiety tagged on the surface were prepared. Thus, developed BODIPY-modified NPs are non-fluorescent in their apo state due to efficient PET-based quenching of the fluorophore caused by the lone pair electrons present on the nitrogen of the benzoyl group. Interestingly, binding of Zn^{2+} ions reduces the PET-based quenching which resulted in enhanced fluorescence (~32 fold). The observed enhanced fluorescence is very selective to the Zn^{2+} ions (Kang et al., 2012).

In another work, Wang and co-workers developed $Fe_3O_4@SiO_2$ NPs which are covalently grafted with 3,5-di-tert-butyl-2-hydroxybenzaldehyde (DTH) by first reacting DTH with aminopropyl trimethoxy silane (APTMS) (Wang et al., 2012). The DTH molecule shows a weak fluorescence peak with maxima close to 450 nm (upon λ_{exi} = 397 nm). However, the intensity of this emission enhanced up to 25 times with the addition of Zn^{2+} ions as well as a red-shift in the emission maxima is noted (470 nm). The study noted hardly any interference from other metal ions like Ag^+, Cd^{2+}, Co^{2+}, etc.

Iron nanoclusters have been used for the detection of zinc ions. Hashemi et al. have developed hemoglobin-incorporated Fe quantum clusters (Hb-FeQCs) as both fluorescent and colorimetric probe for sensing and cellular imaging of Zn^{2+} ions (Hashemi et al., 2018). The developed Hb-FeQCs upon 460 nm excited resulted in an emission peak with maxima at 567 nm. Upon binding of Zn^{2+} ions to the Hb-FeQCs, an enhancement in the fluorescence emission intensity is noted. Further increase in the emission intensity is observed with the increase in the Zn^{2+}

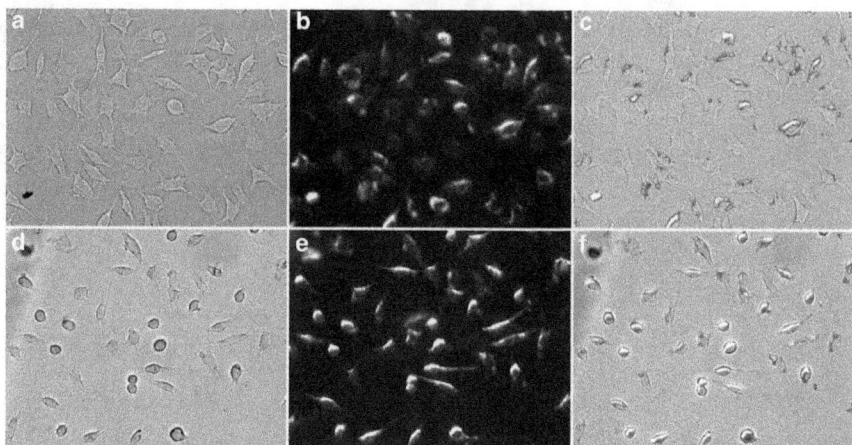

FIGURE 23.5 Live-cell imaging of HFF cells after incubation with Hb-FeQCs (0.5 mg mL^{-1}): (a) Bright-field microscopy image. (b) Greenish-yellow fluorescence image without addition of Zn^{2+} (λ_{ex}:460 nm). (d) Bright-field image of cells treated with 50 µM Zn^{2+} for 30 min. (e) Strong yellow fluorescence image of Hb-FeQCs/Zn^{2+} and (c, f) merged images. (Adapted with permission from Springer). A strong green-yellow fluorescence was observed for the cells exposed to Hb-FeQCs. The intensity of this fluorescence was further enhanced after exposed to Zn^{2+} ions.

concentration. In addition, a blue-shift in the emission maxima is observed from 567 to 520 nm. The high selectivity towards Zn^{2+} is attributed to the strong binding of Zn^{2+} to piperidine moiety. The authors observed a low LoD of 48 nM with a linear response in the 0.04 to 2.2 µM range. Further, in vitro bio-imaging and cellular uptake of Hb-FeQCs and its potential towards detection of Zn^{2+} ions was studied in the cultured HFF cells. The results of the confocal imaging study are shown in Figure 23.5.

23.6 CONCLUSION

With zinc being one of the most essential nutrients for humans, it is quite important to maintain its concentration levels in the body for normal functioning of the various biological processes. Thus, development of probes for the detection of Zn^{2+} at low concentrations is indispensable. This chapter reports the development of various nanomaterials-based optical probes for the selective and sensitive detection of zinc ions *in vitro* and *in vivo*. To name a few, noble metal nanoparticles such as Ag and Au, quantum dots like semiconductor nanomaterials like CdS, CdSe and Ag$_2$S, carbon dots, etc., magnetic materials like Fe clusters and Fe$_2$O$_3$ have been utilized for the detection of zinc ions. In addition, upconverting nanoparticles like NaYF$_4$-Tm/Yb have also been explored. Both Turn-On and Turn-Off of optical signals of these nanoparticles on Zn^{2+} binding has been reported. In most studies, hardly any interference from other toxic metal ions like Ag$^+$, Cd^{2+}, Cr^{3+}, Fe^{3+}, Hg^{2+}, etc., was reported. Despite many reports, the limit of detection with many probes is in the range of 0.1 to 10 μM. Considering the importance of maintaining Zn^{2+} levels in humans for various important biological purposes and enhancing the selectivity in the presence of higher concentrations of other metal ions, it is indeed vital to develop probes to detect zinc ions at lower concentrations (down to nM or even pM). Further, it is important to evaluate the cytotoxicity of the developed probes more systematically and come up with more biocompatible nanomaterial probes for the detection of Zinc ions.

REFERENCES

Alivisatos, AP 1996. Perspectives on the physical chemistry of semiconductor nanocrystals. J. Phys. Chem. 100, 13226–13239. doi.org/10.1021/jp9535506.

Auzel F 2004. Upconversion and anti-stokes processes with f and d ions in solids. Chem. Rev. 104, 139–174. doi.org/10.1021/cr020357g.

Bhardwaj V, Anand T, Choi HJ, Sahoo SK 2019. Sensing of Zn(II) and nitroaromatics using salicyclaldehyde conjugated lysozyme-stabilized fluorescent gold nanoclusters. Microchem. J. 151, 104227. doi.org/10.1016/j.microc.2019.104227.

Bothra, S, Paira, P, Kumar SK, A, Kumar, R, Sahoo, SK 2017. Vitamin B_6 Cofactor-conjugated polyethyleneimine-passivated silver nanoclusters for fluorescent sensing of Zn^{2+} and Cd^{2+} using chemically modified cellulose strips. Chemistry Select, 2, 6023–6029. 10.1002/slct.201701074.

Bu D, Song H, Li Z, Wei L, Zhang H, Yu M 2020. Carbon-dot-based ratiometric fluorescent probe of intracellular zinc ion and persulfate ion with low dark toxicity. Luminescence 35, 1319–1327. doi.org/10.1002/bio.3894.

Bünzli, J G, Eliseeva, S V 2013. Intriguing aspects of lanthanide luminescence. Chem. Sci., 4, 1939–1949. 10.1039/c3sc22126a.

Chen W, Wang Q, Ma J, Li CW, Yang M, Yi C 2018. A ratiometric fluorescent core-shell nanoprobe for sensing and imaging of zinc(II) in living cell and zebrafish. Microchim. Acta 185, 523. doi.org/10.1007/s00604-018-3066-1.

Cuajungco MP, Lees GJ 1997. Zinc metabolism in the brain: Relevance to human neuro-degenerative disorders. Neurobiol. Dis. 4, 137–169. doi.org/10.1006/nbdi.1997.0163.

Dastidar DG, Mukherjee P, Ghosh D, Banerjee D 2021. Carbon quantum dots prepared from onion extract as fluorescence turn-on probes for selective estimation of Zn^{2+} in blood plasma. Colloids Surf. A Physicochem. Eng. Asp. 611, 125781. doi.org/10.1016/j.colsurfa.2020.125781.

Eustis S, El-Sayed MA 2006. Why gold nanoparticles are more precious than pretty gold: Noble metal surface plasmon resonance and its enhancement of the radiative and nonradiative properties of nanocrystals of different shapes. Chem. Soc. Rev. 35, 209–217. doi.org/10.1039/B514191E.

Guidelines for drinking-water quality 1996. Health Criteria and Other Supporting Information. 2nd ed. Vol. 2. World Health Organization, Geneva.

Hanaoka K, Kikuchi K, Kojima H, Urano Y, Nagano TJ 2004. Development of a zinc ion-selective luminescent lanthanide chemosensor for biological applications. J. Am. Chem. Soc. 126, 12470–12476. doi.org/10.1021/ja0469333.

Hashemi N, Vaezi Z, Sedghi M, Manesh HN 2018. Hemoglobin-incorporated iron quantum clusters as a novel fluorometric and colorimetric probe for sensing and cellular imaging of Zn(II) and cysteine. Microchim. Acta 185, Art. No. 60. doi.org/10.1007/s00604-01 7-2600-x.

Hu P, Wang R, Zhou L, Chen L, Wu Q, Han MY, El-Toni AM, Zhao D, Zhang F 2017. Near-infrared-activated upconversion nanoprobes for sensitive endogenous Zn^{2+} detection and selective on-demand photodynamic therapy. Anal. Chem. 89, 3492–3500. doi.org/10.1021/acs.analchem.6b04548.

Huang K, Dai R, Deng W, Guo S, Deng H, Wei Y, Zhou F, Long Y, Li J, Yuan X, Xiong, X 2018. Gold nanoclusters immobilized paper for visual detection of zinc in whole blood and cells by coupling hydride generation with headspace solid phase extraction. Sens. Actuators B Chem. 255, 1631–1639. doi.org/10.1016/j.snb.2017.08.177.

Kang G, Son H, Lim JM, Kweon HS, Lee IS, Kang D, Jung JH 2012. Functionalized Fe_3O_4 nanoparticles for detecting zinc ions in living cells and their cytotoxicity. Chem. Eur. J. 18, 5843–5847. doi.org/10.1002/chem.201200294.

Lee S, Nam YS, Lee HJ, Lee Y, Lee KB 2016. Highly selective colorimetric detection of Zn (II) ions using label-free silver nanoparticles. Sens. Actuators B Chem. 237, 643–651. doi.org/10.1016/j.snb.2016.06.141.

Li Y, Hu X, Zhang X, Cao H, Huang Y 2018a. Unconventional application of gold nanoclusters/Zn-MOF composite for fluorescence turn-on sensitive detection of zinc ion. Anal. Chim. Acta 1024, 145–152. doi.org/10.1021/acs.inorgchem.8b00788.

Li X, Zhou Z, Zhang CC, Zheng Y, Gao J, Wang Q 2018b. Ratiometric fluorescence platform based on modified silicon quantum dots and its logic gate performance. Inorg. Chem. 57, 8866–8873. doi.org/10.1021/acs.inorgchem.8b00788.

Lin L, Hu Y, Zhang L, Huang Y, Zhao S 2017. Photoluminescence light-up detection of zinc ion and imaging in living cells based on the aggregation induced emission enhancement of glutathione-capped copper nanoclusters. Biosens. Bioelectron. 94, 523–529. doi.org/10.1016/j.bios.2017.03.038.

Liu Y, Qu X, Guo Q, Sun Q, Huang X 2017. QD-Biopolymer-TSPP assembly as efficient BiFRET sensor for ratiometric and visual detection of zinc ion. ACS Appl. Mater. Interfaces 9, 4725–4732. doi.org/10.1021/acsami.6b14972.

Liu ML, Chen BB, Li CM, Huang CZ 2019. Carbon dots: Synthesis, formation mechanism, fluorescence origin and sensing applications. Green Chem. 21, 449–471. doi.org/10.1039/C8GC02736F.

Meeusen, JW, Tomasiewicz, H, A Nowakowski, A, Petering, DH 2011. TSQ (6-methoxy-8-p-toluenesulfonamido-quinoline), a common fluorescent sensor for cellular zinc, images zinc proteins. Inorg. Chem. 50, 7563–7573. doi.org/10.1021/ic200478q.

Peng J, Xu W, Teoh CL, Han S, Kim B, Samanta, A, Er JC, Wang L, Yuan, L, Liu X, Chang, YT 2015. High-efficiency in vitro and in vivo detection of Zn^{2+} by dye-assembled upconversion nanoparticles. J. Am. Chem. Soc. 137, 2336–2342. doi.org/10.1021/ja5115248.

Promnimit S, Bera T, Baruah S, Dutta J 2011. Chitosan capped colloidal gold nanoparticles for sensing zinc ions in water. J. Nano Res. 16, 55–61. doi.org/10.4028/www.scientific.net/JNanoR.16.55.

Ruedas-Rama MJ, Hall, EAH 2008. Azamacrocycle activated quantum dot for zinc ion detection. Anal. Chem. 2008 80, 8260–8268. doi.org/10.1021/ac801396y.

Rosi NL, Mirkin CA 2005. Nanostructures in biodiagnostics. Chem. Rev. 105, 1547–1562. doi.org/10.1021/cr030067f.

Sarkar D, Ganguli S, Samanta T, Mahalingam V 2019. Design of lanthanide-doped colloidal nanoparticles: Applications as phosphors, sensors and photocatalysts,. Langmuir 2019, 35, 6211–6230. doi.org/10.1021/acs.langmuir.8b01593.

Smith AM, Nie S 2010. Semiconductor nanocrystals: Structure, properties, and band gap engineering. Acc. Chem. Res. 43, 190–200. doi.org/10.1021/ar9001069.

Wang Y, Peng X, Shi J, Tang X, Jiang J, Liu W 2012. Highly selective fluorescent chemosensor for Zn^{2+} derived from inorganic-organic hybrid magnetic core/shell Fe3O4@SiO2 nanoparticles. Nanoscale Res. Lett. 7, 86.

Wang Y, Lao S, Ding W, Zhang S, Liu, S 2019. A novel ratiometric fluorescent probe for detection of iron ions and zinc ions based on dual-emission carbon dots. Sens. Actuators B Chem. 2019, 284, 186–192. doi.org/10.1016/j.snb.2018.12.139.

Wang B, Liang Z, Tan H, Duan W, Luo, M 2020. Red-emission carbon dots-quercetin systems as ratiometric fluorescent nanoprobes towards Zn^{2+} and adenosine triphosphate. Microchim. Acta 187, 345. doi.org/10.1007/s00604-020-04316-5.

Wu Q, Zhou M, Shi J, Li Q, Yang M, Zhang Z 2017. Synthesis of water-soluble Ag2S quantum dots with fluorescence in the second near-infrared window for turn-on detection of Zn(II) and Cd(II). Anal. Chem. 89, 6616–6623. doi.org/10.1021/acs.analchem.7b00777.

Xiao Y, Dong W, Wang H, Hao Y, Wang Z, Shuang S, Dong C, Gong X 2021. A fluorometric and colorimetric dual-readout nanoprobe based on Cl and N co-doped carbon quantum dots with large stokes shift for sequential detection of morin and zinc ion. Spectrochim. Acta A Mol. Biomol. Spectrosc. 261, 120028. doi.org/10.1016/j.saa.2021.120028.

Yang, L, Zhang X, Wang J, Sun H, Jiang L 2018. Double-decrease of the fluorescence of CdSe/ZnS quantum dots for the detection of zinc(II) dimethyldithiocarbamate (ziram) based on its interaction with gold nanoparticles. Microchim. Acta 185, 472. doi.org/10.1007/s00604-018-2995-z.

Zalewski PDIJ, Forbes IJ, Betts WH 1993. Correlation of apoptosis with change in intracellular labile Zn(II) using zinquin [(2-methyl-8-p-toluenesulphonamido-6-quinolyloxy)acetic acid], a new specific fluorescent probe for Zn(II). Biochem. J. 296, 403–408. doi.org/10.1039/C3SC22126A.

Zhang Z, Pei K, Yang Q, Dong J, Yan Z, Chen J 2018. A nanosensor made of sulfur–nitrogen co-doped carbon dots for "off–on" sensing of hypochlorous acid and Zn(II) and its bioimaging properties. New J. Chem. 42, 15895–15904. doi.org/10.1039/c8nj03159b.

Zheng X, Cheng W, Ji C, Zhang J, Yin M 2020. Detection of metal ions in biological systems: A review. Rev. Anal. Chem. 39, 231–246. doi.org/10.1515/revac-2020-0118.

24 Fluorogenic Sensor for the Determination of Trace Quantities of Zinc in Soil and Water

Sangita Ghosh, Sukanya Paul, and Chittaranjan Sinha

24.1 INTRODUCTION

Zinc is the 24th most abundant element on the earth and fourth in the list of metals next to iron, aluminium and copper. Out of 17 essential elements and eight micronutrients, zinc is unique for the growth of plants and second most abundant micronutrient next to iron. It regulates neural signal transmission, immune function, mammalian reproduction and others. The deficiency of Zn^{2+} ions causes severe harmful effects on eyes and skin, and surplus causes Parkinson's, Alzheimer's and many other diseases.

24.2 ZINC IN SOIL AND PLANTS

It is reported that in almost all varieties of soils like sandy, calcareous and saline, high usage of fertilizers like NPK show a shortage of Zn. Its deficiency in Indian soils is probably upsurge from 49 to 63% by 2025 due to soil erosion, decrease in soil pH, increase in salinity, etc. (Arunachalam, 2013). The dry matter from the plants contains 30–100 mg Zn/kg, and toxicity arises when it is more than 300 mg/kg. Zinc, being a key structural constituent or regulatory cofactor of plants, controls important biochemical and enzyme regulatory pathways *viz.* metabolism of carbohydrates, proteins, fats, auxin photosynthesis, sugars to starch conversion, conception of pollen and preservation of the integrity of biological membranes. Pulses absorb proportionately higher amounts of zinc than rice and wheat. The zinc uptake is 35 to 70 g/ha in case of pulse crops while pigeon pea and green-gram production demand 1200 and 1000 kg/ha.

24.3 ZINC IN INDIAN SOIL

The Indian Institute of Soil Science published a zinc map of Indian soils that truly reflects the soil fertility and type of soil treatment required during the cultivation of

DOI: 10.1201/9781003412472-24

specific crops in particular regions. Zn is the fourth most significant crop-yielding nutrient (next to NPK). About 48.5% of Indian soils and 44% of plants are potentially zinc-deficient. Excess amounts of NPK micronutrients by the farmers cause Zn imbalance in soil, and this is a severe issue in Southern states. Total Zn in Indian soil appears as a sum of different fractions such as water-soluble fraction (WSF), exchangeable portion (EXP), specifically adsorbed amount (SAA), acid-soluble part (ASP), occluded manganese (Mn)-oxide part (Mn-OX), occluded organic matter (OM), bound amorphous iron (Fe)-oxide (AFe-OX), bound crystalline Fe-oxide (CFe-OX) and residual sediments (RES). Soil quality in India varies abruptly and even varies from one side to another of the same land. Total Zn (>90%) generally appears in inactive clay and residual sediments and only a small fraction seemed to be present in other phases (Tapan and Rattan, 2007).

Indian soils are categorized into eight major classes: (i) Alluvial Soil (~40%), (ii) Black Cotton Soil (15%), (iii) Red & Yellow Soil (~18%), (iv) Laterite Soil (4%), (v) Mountainous or Forest Soil, (vi) Arid or Desert Soil (4%), (vii) Saline and Alkaline Soil and (viii) Peaty and Marshy Soil. There are sub-classifications among them viz. Karewa soil, Sub-Montane Soil, Snowfield, Grey/Brown Soil. Zinc shortage in Indian soils is one of the main reasons for poor quality crop yield and hence Zn in Indian people (Chasapis et al., 2012).

Zinc availability in soil is regulated by soil pH, organic matter content, adsorptive surfaces, other physical, chemical and biological features. The solubility of zinc and its salt form, speciation, soil porosity, etc., monitor the mobility of zinc in soil. Hydroxo-Zn (II) is present in dry oxidized soils, while in flooded areas zinc deposits as sulfides and carbonates. In sediments, Zinc is found complexed and sorbed resulting in disruption of its mobility. Zn present in ionic form in acidic sediments and soils, and cation exchange processes influence its fate. Bacteria and fungi also oxidize ZnS to $ZnSO_4$ which makes Zn more available in the soil solution due to higher solubility of its sulphate.

24.4   HOW DOES ZINC ENTER THE SOIL?

Several studies revealed that Zn shortage is the most extensive micronutrient disorder in rice. Use of proper fertilizer improved the food quality and crop yield besides increasing Zn levels. Some of the crops are highly sensitive to Zn deficiency in soil and cannot grow on cooperative support of other minerals (Brian, 1975; Martens and Westermann, 1991). Farmers use different Zn-fertilizers Zn (II) in complex form, which is a better micronutrient than the common compound or mineral form. Humic acid and fulvic acid in soil accelerate the solubility and transportation of Zn in soil. However, stability of these complexes makes them less available to plants (Barrow, 1993; Ganeshamurthy, 2019; Nisab et al., 2019). Zinc shows co-operative micronutrient effects with copper, iron, manganese and boron and influences their concentration in plants. It also shares many metalloenzymes with these elements and regulates the nutrition status of the cells.

24.5   ZINC IN WATER

Zinc in soil is a major source of water pollution. The surface and groundwater (<1% of Global water) also contain Zn and come mainly from municipal wastes, urban runoff, mine drainage, industry, etc. The solution to Zn deficiency in soils, water and plants needs a comprehensive approach to meet the challenges of human health (Jelle and Erik, 2013). World Health Organization (WHO) and Food and Agricultural Organization (FAO) suggested that drinking water must contain Zn < 3 mg/L. Water containing higher amounts produces a fatty film on boiling and has a caustic taste. In irrigation water, Zn would not exceed 2 mg/L. (Pratt, 1972; Lebourg et al., 1998).

The world average of zinc in rivers, oceans and seas is estimated to be 20 to 200 kt/yr (Yoshiaki, 1992; Jerome, 2003). Zinc enters into water from the soil due to natural processes and human activities besides from mines, smelters, municipal drainage, old-galvanized metal pipes, industrial wastes, thermal power plant affluents, coal-fired stations, agricultural run-off, burning of wastes, etc. Water purity may be accounted for by the quantity of zinc. A content of 0.015 mg/L (approx.) stands for fresh and pollution-free water.

Hydrolysis of Zn (II) compounds in water yields different species such as $Zn(H_2O)_6^{2+}$ (dia.,0.8 nm), $Zn(H_2O)_5Cl^+$, $Zn(H_2O)_5OH^+$ (dia.,1 nm). They also form complexes with different humic acid components to give Zn-humate, Zn-cysteinate (dia.,2–4 nm), organic colloids of dia. 100–500 nm and inorganic colloids $Zn^{2+}Fe_2O_3$, $Zn^{2+}SiO_2$ (dia.100–500 nm). This reaction also increases acidity of the water. A small change in pH (0.5 unit) in water might bring or coagulate Zn^{2+} in water. Because of the amphoteric nature of Zn^{2+}, change in pH (in either direction) does not hamper equilibrium too much. Agriculture run-off consisting of phosphates derived from diammonium phosphate (DAP), monoammonium phosphate (MAP), NPKs and SSP may sediment soluble Zn^{2+} in water and also increase its salinity.

24.6   FLUOROGENIC SENSOR FOR THE ANALYSIS OF ZINC

For quantitative estimation of the trace quantity of zinc, classical techniques such as Gravimetry, Volumetry, Electrochemical and Spectrophotometric methods are generally used. Other sophisticated methods include Atomic Absorption Spectroscopy (AAS), Atomic Emission Spectroscopy (AES), Inductively Coupled Plasma-Atomic Absorption Spectrometry (ICP-AAS), Inductively Coupled Plasma-Atomic Emission Spectrometry (ICP-AES), Flame Atomic Absorption Spectroscopy (FAAS), Inductively Coupled Plasma Mass Spectrometry (ICPMS), etc. A spectrophotometric approach based on an optical chemosensor involving UV-Visible and Fluorescence techniques has emerged as the most convenient method for the determination of trace to ultratrace quantity of zinc. Fluorimetric techniques give outstanding sensitivity and fluorescent sensors are 10^5 times more sensitive than absorption spectroscopic techniques besides cost-effective, feasible, operational simplicity, rapid responsive, substantial selectivity and sensitivity, appreciable detection limit and easy analysis (Bernard, 2001; Joseph, 2006).

Use of chemosensors for the detection of zinc is currently a focused area of research. A sensor is a device which measures a physical quantity by converting it into a detectable form as a 'signal' that could be read by an analytical instrument. A Chemical Sensor is 'a portable miniaturized analytical device, which can deliver real-time and on-line information in presence of specific compounds or ions in complex samples.' It consists of a receptor that accepts the analyte selectively and specifically, signalling to a sub-unit that transduces the binding of analyte and a recognizing instrument (Figure 24.1).

The change in the absorbance or fluorescence behaviour dictates its nature of response towards analytes. Chromogenic receptor responses through a colorimetric change in the presence of analytes could be detected by UV-Visible Absorption while fluorogenic receptors respond through a modification in the fluorescence property of the molecule on interaction with analytes could be recognized through Fluoroscence spectrophotometer.

24.7 METHODOLOGY

The first step is the collection of soil samples from uniform areas of ten to 15 different places in a 'V'-shaped cut up to a depth of 15 cm. Plant residues are removed and dried in the shade by spreading them on a clean sheet of paper in the dark. The dried soil sample is pulverized into a fine powder and used for the analysis of micronutrients (iron, zinc, manganese and copper) and other parameters, such as: pH, moisture, electrical conductivity, fertilizers containing K^+, NO_3^-, NH_4^+, different phosphates, organic carbons (humic acid and fulvic acid) as well as bacterial count.

For water samples, the collection site and season of analysis have to be fixed first. Surface water such as flowing river water or the stagnant water of ponds, underground water, agro-field water or fisheries ponds, etc., might be considered for water collection. Commonly, surface water is collected in a DO sampler from 30 cm below the programmed site and Teflon-coated Niskin-X samplers are recommended for this.

24.7.1 USE OF 2,6-BIS-[(6-METHOXY-BENZOTHIAZOL-2-YLIMINO)-METHYL]-4-METHYL-PHENOL (HL) FOR THE DETERMINATION OF TRACE QUANTITIES OF ZN^{2+}

The reaction of 4-methyl-2,6-diformylphenol (PCDF) (1 mmol, 0.164 mg) with 6-methoxy-benzothiazol-2-ylamine (3 mol, 0.360 mg) for 6 h precipitated yellow-colored compound HL (Figure 24.2).

The product was dried *in-vacuo* (Yield: 82%) and characterized by ^{1}HNMR data where it revealed a singlet phenolic OH at δ 12.60 ppm and two singlets of imine-hydrogens (-CH=N-) at δ 9.30 ppm. Other aromatic hydrogens appeared within δ 7.91–7.11 ppm. The six methyl protons of -OCH$_3$ groups were observed at δ 3.92 ppm while the three methyl protons of the di-formyl units appeared at δ 2.41ppm. The corresponding characteristic peaks in ^{1}HNMR spectrum confirmed the formation of the probe.

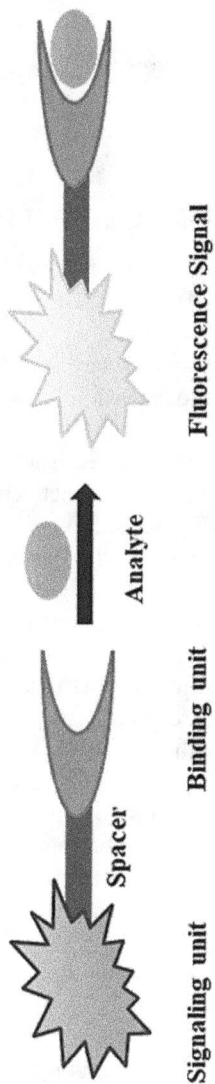

FIGURE 24.1 Schematic representation of chemosensor demonstrating binding of an analyte (guest) through host-guest interaction.

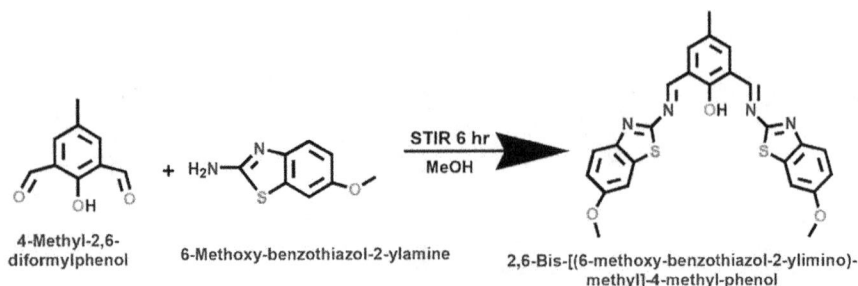

FIGURE 24.2   Synthesis of ligand HL.

24.8   UV–VISIBLE AND FLUORESCENCE SPECTRAL STUDIES

The probe HL (1.83 mg, 0.001 mmol) was dissolved in DMSO (5 ml) the stock solution of metal salts (1×10^{-3} M) prepared in MeOH (50 µL) and diluted to 2 ml (9:1 MeOH-H_2O, v/v) using requisite amount of HEPES buffer (pH 7.5). The spectra were collected at room temperature and the excitation wavelength set at 410 nm (excitation slit = 10.0 and emission slit = 3.0) to collect fluorescence spectra.

UV-Vis absorption spectrum of HL revealed bands at λ 356 and 418 nm which are allotted to π-π* and n-π* transitions, respectively in MeOH-H_2O (9:1, v/v) (HEPES buffer, pH 7.5) medium (Figure 24.3a).

The probe in the same medium showed a weak emission at λ 552 nm on excitation at λ 410 nm. The weak emission might be due to non-radiative decay by Excited State Intramolecular Proton Transfer (ESIPT) between -CH=N and phenolic-OH of the molecule. The fluorescence spectra of the probe (50 µL) in MeOH-H_2O (9:1, v/v) (pH 7.5) medium was recorded in the presence of two equivalent cations (100 µL). Interestingly, the molecule showed 'Turn-On' emission only to Zn^{2+} amongst other cations (Mn^{2+}, Fe^{2+}, Al^{3+}, Co^{2+}, Ni^{2+}, Pd^{2+}, Cd^{2+}, Hg^{2+}, Cr^{3+}, Cu^{2+}, Ba^{2+}, Na^+, K^+, Pb^{2+}) with significant blue-shift and 18-fold enhancement of emission at λ 480 nm (λ_{ex} = 410 nm) while for the remaining cations, the emission of the probe remained almost unaffected. Upon addition of Zn^{2+}, the probe selectively emitted a green colour, which might be due to the rigidity imparted on the molecule because of Chelation Enhanced Fluorescence (CHEF).

The ESIPT mechanism (Figure 24.3b). On gradual addition of Zn^{2+} (2 µL) to the probe solution (50 µL) in MeOH-H_2O (9:1, v/v) (HEPES buffer) medium, the emission intensity gradually increased and varied linearly with the increasing concentration of Zn^{2+} till it attained saturation on 1:1 complexation of Zn^{2+} to the probe. Figure 24.3c. The limit of detection (LOD) was calculated following the 3σ/M method where 'σ' denotes standard deviation and the slope (M) was calculated from the fluorescence titration plot. The LOD calculated from the plot was 35.9 nM and is fairly lower than the recommended limit by WHO (76 µM) (Figure 24.3d).

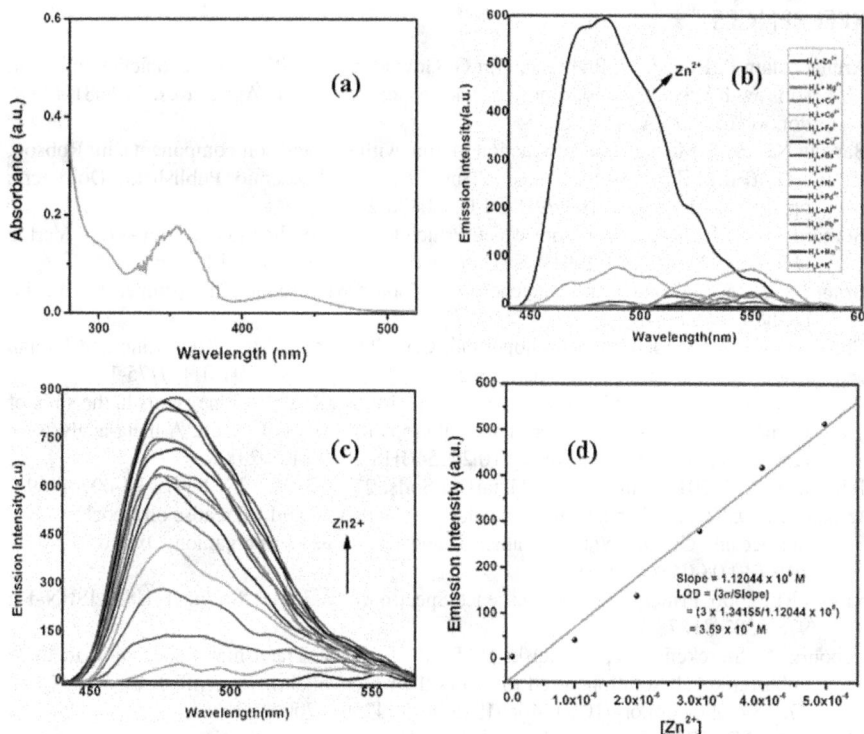

FIGURE 24.3   (a) UV-Vis absorption spectrum of HL in MeOH-H$_2$O (9:1, v/v) (HEPES buffer, pH 7.5) medium. (b) Emission Spectra of HL in the presence of different cations (1:2) in MeOH-H2O (99:1, v/v) (pH, 7.5). (c) Change in the emission Spectra on graduation addition of Zn2+ in MeOH-H2O (9:1, v/v) (pH 7.5). (d) Limit of Detection for Zn2+ Sensing by the probe HL.

24.9   CONCLUSION

The health and environmental importance of zinc is versatile. Plant growth, crop yielding, soil quality for food production, soil erosion, public health and engineering are largely indicated by measuring concentration of Zinc. A highly sensitive fluorogenic sensor has been designed and used for the determination of Zn^{2+} ions as low as 35.9 nM in the solution phase and in the presence of other cations. The method is cheap, sensitive, selective and useful for social well-being and a step toward Sustainable Development Goal.

ACKNOWLEDGEMENTS

The financial and infrastructure facilities from Jadavpur University are gratefully acknowledged. Sincere thanks to Prof. Ethirajan Sukumar for useful discussion and constant interaction.

REFERENCES

Arunachalam P, Kannan P, Prabukumarm G, Govindaraj M. 2013. Zinc deficiency in Indian soils with special focus to enrich zinc in peanut. Afr. J. Agric. Res. 8, 6681–6688. doi.org/10.5897/AJARx12.015.

Barrow NJ. 1993, Mechanisms of reaction of zinc with soil and soil components, in: Robson, AD. (Ed.), Zinc in Soils and Plants. Kluwer Academic Publishers, Dordrecht, pp. 15–32. doi.org/10.1007/978-94-011-0878-2_2.

Bernard V. 2001. Molecular Fluorescence Principles and Applications. Wiley-VCH, Verlag GmbH, ISBNs: 3-527-29919-X (Hardcover); 3-527-60024-8 (Electronic).

Brian JA. 1975. Micronutrient Deficiencies in Global Crop Production. Springer, Dordrecht, Netherlands, pp. 1–39.

Chasapis, CT, Loutsidou, AC, Spiliopoulou, CA, Stefanidou, ME. 2012. Zinc and human health: an update. Arch. Toxic., 86, 521–534. 10.1007/s00204-011-0775-1.

Ganeshamurthy AN, Rajendiran S, Kalaivanan D, Rupa TR. 2019. Zinc status in the soils of Karnataka and response of horticultural crops to zinc application: A meta-analysis. J. Hortic. Sci. 14, 98–108. doi.org/10.24154/JHS.2019.v14i02.003.

Jelle M, Erik S. 2013. Zinc. Heavy Metals in Soils. 22, 465–493, ISBN:978-94-007-4469-1

Jerome G 2003. Trace elements in river waters. in: Drever, JI (Ed.), Treatise on Geochemistry, Surface and Ground Water, Weathering and Soils. Elsevier Pergamon, Oxford, U.K. 10. 1016/B0-08-043751-6/05165-3

Joseph RL. 2006. Principles of Fluorescence Spectroscopy, 3rd ed. Springer, USA, ISBN-13: 978-0387-31278-1.

Lebourg A, Sterckeman T, Ciesielski H, Proix N. 1998. Trace metal speciation in three unbuffered salt solutions used to assess their bioavailability in soil. J. Environ. Qual. 27, 584–590. doi.org/10.2134/jeq1998.00472425002700030016x.

Martens, DC, Westermann, DT. 1991. Fertilizer Applications for Correcting Micronutrient Deficiencies, Micronutrients in Agriculture, 2nd ed. SSSA Book Series, pp. 549–553. ISBN No. 9780891188780 (online) and 9780891187974 (print). 10.2136/sssabookser4.2ed.c15

Nisab CPM, Ghosh GK, Sahu M. 2019. Available zinc status in relation to soil properties in some red and lateritic soils of Birbhum district, West Bengal, India. Int. J. Curr. Microbiol. Appl. Sci. 8, 1764–1770. doi.org/10.20546/ijcmas.2019.805.204.

Pratt PF. 1972. Quality Criteria for Trace Elements in Irrigation Waters, University of California, Division of Agricultural Science, California Agricultural Experiment Station, books.google.co.In/books?id=a1b PzwEACAAJ.

Tapan A, Rattan RK. 2007. Distribution of zinc fractions in some major soils of India and the impact on nutrition of rice. Commun. Soil Sci. Plant Anal. 38, 2779–2798. doi.org/10. 1080/00103620701663032.

Yoshiaki K. 1992. Water chemistry. in: Nierenberg, WA (Ed.), Encyclopedia of Earth System Science, Vol. 4, Academic Press, San Diego, pp. 449–470.

Index

Note: Locators in *italics* represent figures and **bold** indicate tables in the text.

For Product Safety Concerns and Information please contact our EU
representative GPSR@taylorandfrancis.com
Taylor & Francis Verlag GmbH, Kaufingerstraße 24, 80331 München, Germany

www.ingramcontent.com/pod-product-compliance
Lightning Source LLC
Chambersburg PA
CBHW052120230326
41598CB00080B/3893